From Mesmer to Freud

From Mesmer to Freud

Magnetic Sleep and the Roots of Psychological Healing

Adam Crabtree

Yale University Press

New Haven and London

Designed by Sonia L. Scanlon

Set in Times Roman type by The Composing Room of Michigan, Inc., Grand Rapids, Michigan.

Printed in the United States of America by Edwards Brothers, Inc., Ann Arbor, Michigan.

Library of Congress Cataloging-in-Publication Data

Crabtree, Adam.

    From Mesmer to Freud : magnetic sleep and the roots of psychological healing / Adam Crabtree.

       p. cm.

    Includes bibliographical references and index.

    ISBN 0-300-05588-9

    1. Subconsciousness—History. 2. Mesmerism—History. 3. Animal magnetism—History. 4. Altered states of consciousness. 5. Dissociative disorders.

    I. Title.

RC497.C73   1994   93-4017

154—dc20   CIP

A catalogue record for this book is available from the British Library.

10 9 8 7 6 5 4 3 2 1

Contents

Preface

Psychological healing, as it is understood today, had its start with the discovery of magnetic sleep in 1784. Magnetic sleep (artificially induced somnambulism) revealed a realm of mental activity not available to the conscious mind. The development of a consistently successful procedure for inducing magnetic sleep—using the techniques of animal magnetism—made it possible to access and explore that realm systematically. From that point a method of psychological healing based on a psychodynamic model was able to emerge.

The importance of animal magnetism in the development of modern dynamic psychiatry was first compellingly pointed out by Henri Ellenberger, whose *Discovery of the Unconscious* (1970) is a classic in the history of psychology. Ellenberger's pioneering work made it clear that all modern psychological systems that accept the notion of dynamic unconscious mental activity must trace their roots, not to Freud, but to those animal-magnetic practitioners who preceded him by a century.

The historical vision presented by Ellenberger has been given even greater significance by recent developments in psychotherapy. The emergence of multiple personality disorder and other striking dissociative conditions as important psychological pathologies has dramatically increased the interest of psychotherapists in the history of natural and artificial dissociative phenomena. This history necessarily centers on magnetic sleep and the many streams of experimentation that sprang up after its discovery.

Magnetic sleep led directly to the evolution of a new paradigm for understanding the nature of the human psyche and mental disturbance: the alternate-consciousness paradigm. Before 1784 two other paradigms were most commonly called upon to explain mental aberrations: the intrusion paradigm, which took them to be the result of the intervention from without of some spirit, demon, or sorcerer, and the organic paradigm, which ascribed them to physiological dysfunction. The alternate-consciousness paradigm opened up the possibility of an intrapsychic cause of mental disturbance, pointing to the influence of unconscious mental activity as the source of unaccountable thoughts or impulses.

The story of the discovery and evolution of magnetic sleep now needs to be told. With the notable exception of Ellenberger's writings, the importance of animal magnetism and its offshoots has gone largely unacknowl-edged by historians of psychology. References to this rich tradition range from perfunctory recognition to dismissal as a historical dead-end. In modern text-books of psychology and psychiatry, too often the tradition of magnetic sleep is written off as a combination of clever charlatanism and naive credulity. The authorities quoted to substantiate this opinion are writers who have almost no firsthand knowledge of the literature of animal magnetism and the events and circumstances of its evolution. It is one of those strange quirks of academic historiography that a medico-psychological tradition that was investigated and used by practitioners in every country in the western world for one hundred years before Freud came onto the scene, a tradition that found sup-porters among the most brilliant researchers and thinkers during that period and produced thousands of medical treatises describing tens of thousands of cures and ameliorations, a tradition that counted among its offshoots a practi-cable surgical anesthesia and an effective system of psychotherapy, could be dismissed with a few cursory paragraphs. Whatever may be the explanation for this lacuna in the knowledge of our psychological roots, it is time to remedy the situation. That is the purpose of this book.

The story begins with Franz Anton Mesmer and the system of healing he invented, called animal magnetism. In applying that system, Mesmer's pupil the marquis de Puységur fairly stumbled across the phenomenon that would radically alter psychological thinking in the western world. Puységur discov-ered that some people fall into a sleep-like state when being magnetized, and he decided to explore this "magnetic sleep" to see how it could aid in healing. Being more psychologically inclined than Mesmer, by his research Puységur gave birth to the alternate-consciousness paradigm. He eventually developed a method of psychotherapy that contained the principal elements of psycho-logical healing that would come into vogue in the latter decades of the nine-teenth century. The induction of magnetic sleep set up a special relationship (called "magnetic rapport") between magnetizer and subject. This relation-ship played an important part in the healing process but also was fraught with potential for abuse. These events and issues are dealt with in Part I.

Part II takes up animal magnetism as a healing system. Mesmer's theory held that a physical agent (magnetic fluid) produced the cures and ameliora-tions attributed to animal magnetism, but for most practitioners magnetic sleep also held a central role in the healing process. Magnetic somnambulism was employed to diagnose illnesses and prescribe cures. From the earliest times, the magnetic somnambulist was believed to have the ability to read the magnetizer's thoughts and to perceive by clairvoyance hidden objects and see into the body of the ill person. This capacity was accepted by most animal magnetizers but rejected by some. Among the skeptics was the Manchester physician James Braid, who developed an alternative theory to explain the trance and healing phenomena of animal magnetism. He believed that they resulted from a special state of consciousness produced through purely psychological means. He called this state "hypnotism" and used his theory to explain away the paranormal phenomena reported by the magnetists.

Part III reveals that it was not so easy to dismiss these aspects of magnetic sleep. The sheer volume of reports of paranormal phenomena was enough to cause critical investigators to take the matter seriously. Practitioners in Germany and France developed theories of "magnetic magic" and "magnetic spiritism" based on their own experiments, and the rise of spiritualism in the United States in 1848 (made possible partly through the practice of animal magnetism in that country in the 1830s and 1840s) created a new mass of paranormal phenomena that had to be accounted for. Experiments with "table turning," a spiritualistic phenomenon arising in the early 1850s, led investigators in a fresh and unexpected direction. In attempting to explain the seemingly intelligent nature of communications received through the tipping of tables, they propounded some of the very first speculations on unconscious mental activity. At the same time the inquiries into paranormal phenomena arising from magnetic sleep carried out by the Society for Psychical Research in Britain led to important discoveries about the nature of somnambulistic consciousness.

The scene was now set for new ideas about the nature of the human psyche, and Part IV describes the developments that occurred. The subject in the state of magnetic sleep was seen as quite different from the waking subject, and this duality was called "double consciousness." Some of the earliest cases of dual personality or multiple personality were reported just a few years after the discovery of magnetic sleep. This means that the establishment of the alternate-consciousness paradigm through the discovery of magnetic sleep was strikingly illustrated by spontaneous incidents of an alternate-consciousness pathology.

In the 1880s Pierre Janet carried out exhaustive experiments with both spontaneous and artificially produced double consciousness. His theory of dissociation and subconscious acts provided the framework for a truly effective alternate-consciousness-based psychotherapy. Frederic W. H. Myers studied psychological automatisms, particularly automatic writing, and elaborated an even more comprehensive alternate-consciousness view of the psyche. Josef Breuer also built an effective psychotherapy on the principles of the alternate-consciousness paradigm derived from the magnetic tradition. His colleague Sigmund Freud profited from the fruits of that tradition as well, but he set out in his own direction and in the process consigned to oblivion certain valuable insights in the alternate-consciousness tradition.

The Postscript describes a resurgence of interest in issues connected with alternate-consciousness pathologies. Contemporary mental health workers are constantly dealing with dissociative disorders (such as multiple personality) that arise from childhood physical and sexual abuse. For that reason it is most timely to reclaim the discoveries of the alternate-consciousness tradition before Freud.

In tracing the evolution of the alternate-consciousness paradigm I have relied for the most part on primary sources published mainly in France, Germany, Switzerland, Great Britain, and the United States. The relevant literature is vast and varied in its orientation. It has been necessary to resist the temptation to go down byways that are intriguing but not germane to the theme of the book. Magnetic healing, spiritualism, or paranormal phenomena each could easily be the subject of a book in itself. Other areas such as the use of animal magnetism as an anesthetic and medical applications of hypnosis could only be touched on briefly. And though there were many psychological investigators in the last two decades of the nineteenth century who made contributions to understanding the alternate-consciousness paradigm, only a few of the most important could be discussed here.

I would like to express my appreciation for the help and encouragement offered me by family, friends, and colleagues throughout the course of my research. In the forefront I include John Gach, Stephen Braude, Heinz Schott, and the late Eric Carlson. I also want to thank Joel Whitton, Sharon MacIsaac-McKenna, and my other coworkers on the faculty of the Centre for Training in Psychotherapy in Toronto. I want to express my appreciation to Bill Williams, who so generously opened his valuable library of psychical research for my use, and to the workers at the University of Toronto library who patiently aided me in my search for obscure sources. I owe a special debt of gratitude to my wife, Joanne, whose contributions to this work have helped shape it from conception to editing. Finally, for the kind words and encouragement of Dr. Henri Ellenberger—thank you.

Part I

Magnetic Sleep

Chapter 1

Mesmer

The modern era of psychological healing was ushered in by a most improbable figure. Franz Anton Mesmer was an inventive physician and eccentric thinker, but his work centered exclusively around physical healing. Although he exhibited innate psychological savvy, he was no psychologist. It is one of the strange ironies of history that a man who made no attempt to explain the workings of the mind set in motion a series of events that revolutionized the way we view the human psyche.

Mesmer was born on May 23, 1734, in a hamlet on the edge of Lake Constance, near the town of Radolfzell, Germany. He was the third of the nine children born to Franciscus Antonius Mesmer and Maria Ursula Michlin Mesmer. Mesmer's father was gamekeeper to the archbishop of Constance, and this connection was instrumental in his obtaining a superior education for his son.[1]

In 1743 Franz Anton began his schooling, probably at the monastery at Reichenau, concentrating on Latin, music, languages, and literature. In 1750 Mesmer enrolled in the Jesuit seminary at Dillingen in Bavaria to study philosophy and theology. After four years he moved to the Jesuit-run University of Ingolstadt, also in Bavaria, where he continued his focus on theology and probably added mathematics and physics.

It is likely that during this period Mesmer began to educate himself in the more esoteric philosophical and scientific systems, such as the mystical philosophies of the Rosicrucians and Freemasons, which were popular at the time. These societies studied the modern sciences but interpreted them in the light of occult thinking dating from the fourteenth and fifteenth centuries. Although frowned on by the church, such circles of "illuminati"

1. More complete biographical information on Mesmer can be found in Artelt 1965; Benz 1976, 1977; Buranelli 1975; Erman 1925; Figuier 1860, vol. 3; Frankau 1948; Goldsmith 1934; Ince 1920; Jensen and Watkins 1967; Walmsley 1967; Kerner 1856; Kiesewetter 1893; Milt 1953; Rausky 1977; Schneider 1950; Schott 1985; Schroeder 1899; Schürer von Waldheim 1930; Thullier 1988; Treichler 1988; Tischner 1928; Tischner and Bittel 1941; Vinchon 1936; Wyckoff 1975; and Wydenbruck 1947. Gauld (1992) has produced an excellent history of animal magnetism and hypnotism, with useful biographical information on Mesmer.

flourished in these years, and Mesmer would have had ample opportunity to ponder their metaphysical systems.

In 1759 Mesmer began a course in law at the University of Vienna but switched to medicine after a year. The medical faculty at Vienna, under the direction of Gerhard Van Swieten, was one of the best in the world, and Mesmer entered the school at the height of its vitality. Mesmer completed his medical studies in 1766. His thesis, *Dissertatio physico-medica de planetarum influxu,* contains the first glimmerings of a theory of animal magnetism. Although building on the work of Richard Mead (1673–1754), who wrote about the influence of the stars on men, Mesmer was talking not about an occult astrological influence but about a purely physical, scientific one. After a general discussion of the laws of planetary motion, centrifugal force, and gravitation, he expounded his belief that there must be tides in the atmosphere, just as there are in the oceans, and that there are also tides in the human body emanating from the stars.[2] This generalized influence of celestial bodies on the human organism he labeled "animal gravity":

> One must not think that the influence of the stars on us only has to do with diseases. The harmony established between the astral plane and the human plane ought to be admired as much as the ineffable effect of UNIVERSAL GRAVITATION by which our bodies are harmonized, not in a uniform and monotonous manner, but as a musical instrument furnished with several strings, the exact tone resonates which is in unison with a given tone. (Mesmer 1766, trans. Bloch 1980, pp. 14, 19)

Mesmer conceived of this influence as something so subtle in essence that one would hesitate to call it "matter." Nine years later the same concept reappeared, somewhat modified by his experience with magnets, in his theory of animal magnetism. Then he called it "magnetic fluid." Even at this early stage, Mesmer conceived of this influence as the very foundation of life itself, the principle by which organic bodies carry out their vital functions, and stated that the inhibition of this vital force would produce malfunction and disease.

Mesmer's Medical Practice

After completing his medical degree, Mesmer set up his practice in Vienna. On January 10, 1768, he married Maria Anna von Bosch, the widow of an

---

2. Frank Pattie (1956) points out that this idea originated with Mead in his *De imperio solis ac lunae* (1704), and that Mesmer sometimes reproduced Mead's very words, with only slight alterations, without giving him credit.

imperial councillor. Her inherited wealth and social status provided Mesmer with a congenial ambience in which to build his practice and discuss his unusual ideas. Their home, a mansion on the Danube, became a center of social activity and entertainment for the Viennese aristocracy. Among those who frequented the Mesmer household was the family of Leopold Mozart, whose eldest son had already established a reputation throughout Europe as a pianist and composer. When in 1768, at the age of twelve, Wolfgang Amadeus first visited Mesmer's home, it was to perform an opera commissioned by Mesmer, *Bastien und Bastienne*. Over the years of Mesmer's stay in Vienna he maintained a friendship with the Mozarts.

Mesmer had a profound love of music and was a proficient player of the glass harmonica. But Mesmer's interest in music was as much medical as aesthetic. His doctoral dissertation compares the healthy resonance of animal gravity in living things to musical harmony, and the treatment clinics Mesmer later established incorporated music as an essential element of the healing environment. The glass harmonica was Mesmer's instrument of choice as a magnetic aid, and it is not difficult to see why. This one small instrument was capable of producing sounds that could fill a large space, an advantage in an era when amplification was not possible. Well played, the harmonica produced tones of great richness that were both penetrating and soothing.[3] But although Mesmer seems to have been keenly aware of the healing potential of music, it did not become the core technique of his medical system. That position was occupied first by mineral magnets and then by the magnetism of the body of the physician.

Mesmer's interest in mineral magnetism went back at least to June 1774, when he heard about a cure for stomach cramps developed by the Jesuit priest Maximillian Hell, professor of astronomy at the University of Vienna, and involving the use of an iron magnet. For some time Mesmer had been treating Francisca (Franzl) Oesterlin, a young woman who suffered from various hysterical symptoms, including convulsions, vomiting, aches, fainting, hallucinations, paralysis, and trance. Mesmer's more orthodox medical ministrations had produced little relief for Franzl, and he was looking for a new approach when word of Hell's success reached him.

On July 28 Mesmer began using magnets made for him by Hell in Franzl's treatment. He placed three magnets on her body, one on her stomach and one on each leg. She underwent a reaction to the magnets, which she said caused

3. The glass harmonica is made up of a series of bell-shaped glasses played by stroking the rim with moistened fingers. The technique is similar to rubbing a wet finger along the rim of a wineglass to produce a tone. Benjamin Franklin had occasion to hear Mesmer play the glass harmonica in a private meeting in Paris in 1779 (Lopez 1966, p. 170).

severe pain, but the treatment resulted in an improvement in her condition that lasted for several hours. Mesmer continued his experimentation and believed he noted currents of force that moved through Franzl's body when the magnets were in place. He was convinced that he had finally discovered empirically the general force that he had dubbed animal gravity. At the same time he believed that he had found out how to control it for the benefit of the patient's health. Mesmer considered the violent initial reaction to the treatment a "beneficial crisis" that would lead to a balancing of the currents and cure. He continued his experiments, eventually arriving at rules specifying where on the body to apply the magnets for different effects, the optimal duration of the application, and the precautions to be observed. Informed of Mesmer's success with Franzl Oesterlin, Father Hell published a treatise describing the treatment and attributing the beneficial results to his magnets (Hell 1775, in Tischner and Bittel 1941, pp. 34–35).

The story of the magnets' remarkable curative power appeared in various local newspapers throughout the German-speaking world. Unfortunately, in Mesmer's eyes, the chief credit went to Hell and his magnets. On January 5, 1775, Mesmer therefore published a "Letter on Magnetic Treatment" addressed to Dr. Johann Christoph Unzer presenting Mesmer's version of the treatment and cure of Franzl. He recalled statements he had made in his dissertation about the effect of the sun and the moon on the human body, described his further experiments with magnets and the rules he devised for their use, and told how he had communicated those rules to Father Hell and other doctors (Mesmer 1775, quoted in Tischner and Bittel 1941, p. 37).

From here, Mesmer went on to distinguish his ideas from those of Hell and claim the credit for the originality of thought that led to the healing of Franzl: "I have discovered that steel is not the only substance that may be used to receive magnetic power. I have been able to magnetize paper, bread, wool, silk, leather, stone, glass, water, various metals, wood, men, dogs— everything that I touch. And these magnetized objects have produced the same effect on patients as have magnets themselves. I have also filled bottles with magnetic material, just as is done with electrical material" (p. 38).

In saying that he was able to magnetize anything he touched, Mesmer moved the discussion of the effect of magnets to a whole new plane. Minerals are only one kind of magnet that can be used to affect the "tides" of the body. The most important magnet, he insisted, is the human body. He went on to describe how, by using his own magnetic body, he was able to bring about salutary effects in a patient that were superior to those produced by the iron magnet. Other people also may serve as excellent magnets, he said; one in particular was so magnetic that she could not come within ten feet of a patient without causing pain. He added that some people cannot be magnetized, and their presence even impedes the process of magnetization. He described experiments in which he was able to produce magnetic "jolts" in a patient from behind a wall. Finally, he pointed out that sensitivity to magnetism ceases when a state of health has been restored—that the body is affected by magnetic action only when its natural harmony has been disturbed by illness.

In the letter, Mesmer claimed that he had used animal magnetism successfully to treat hemorrhoids, restore regular menstruation, remove paralysis following apoplexy, and cure all kinds of convulsive and hysterical disorders. At the same time he continued to experiment with its application to epilepsy, melancholia, mania, and fever.

Thus, by the end of 1774 Mesmer had finally found the healing technique he had long sought. Clearly distinguished from mineral magnetism, it was supported by a theory that placed the physician and his body at the center of the cure. The physician, like the healers of old, would lay hands on the ill and perform miraculous cures. He could use the magnetism of his own organism to restore the patient's harmony, lost through illness. But now the ancient technique was backed up not by an occult world view but by a physical theory. Mesmer called for broad scientific experimentation to verify his theory and determine the laws of its function.

By May 1775 Mesmer had broadened his technique to include curative baths. Since water could be made a conductor and container of magnetic force, it made sense that baths with magnetized water would be beneficial to patients. Mesmer also made it a rule that those who looked after patients should be tested for magnetic potency. Each was to touch the patient and the effect was to be noted by the physician. Those who were magnetic were to be kept completely out of the patient's vicinity during an attack lest their magnetic vibrations aggravate the paroxysms.

News of Mesmer's innovative approach spread, and he found ample opportunity to try it on patients who had been disappointed with the results of more orthodox techniques. In July 1775 he traveled to Castle Rohow in Hungary at the invitation of Baron Horeczky de Horka, who was troubled with contractions of the throat muscles. Unable to find relief through the ministrations of the best Vienna physicians, as a last resort he turned to Mesmer, who by now had something of a reputation as a "wonder doctor." Mesmer's prescribed technique—repeated touches, iron magnets under the feet, and direct magnetization of the patient—produced periodic crises during which Mesmer would hold the baron's hands. Mesmer also set up a peculiar device that involved a pail of water. He would place one of his feet in the pail while a servant held an iron rod immersed in the same pail. Wearing a silk stocking on his other foot,

Mesmer alternated holding now the hand, now the foot of the baron, who was at the time in a fever.

While Mesmer was staying at the castle, the people of the area heard about the remarkable healer from Vienna and flocked to see him. Mesmer spent long hours magnetizing persons who suffered from various illnesses and obtained many cures. In one case he restored hearing to a man who had been deafened by a thunderstorm. But while Mesmer was successful with the people of the region, he was unsuccessful with the baron. Although Horeczky de Horka believed in the power of animal magnetism, the convulsions, fever, and other unpleasant aspects of the "healing crisis" were too much for him. At last, spurred by his wife and his physician, he asked Mesmer to leave (Tischner and Bittel 1941, pp. 46–47).

Mesmer and Gassner

The late summer of 1775 found Mesmer in Stockach, near Constance, where he treated a sixteen-year-old girl suffering from arthritis and epilepsy. After fourteen days of treatment she enjoyed a "full and lasting" cure. In this case Mesmer seems to have made less use of iron magnets than of the animal magnetism of his own organism: "When he would point his index finger at her, even though from some distance, she would actually fall senseless to the ground—and this would occur even when he was standing behind two closed doors or a wall. The same thing would happen when he would press on her image in a mirror or hold up a mirror to her. A similar effect was produced when he sprinkled a drop of water on her from his hand" (Mesmer 1776, Appendix, p. 36).[4]

The inhabitants of the Lake Constance region could hardly refrain from comparing Mesmer's cures to those of the Catholic priest Johann Joseph Gassner (1727–1779), who had visited the same area the previous year and had demonstrated a striking ability to heal the sick. Gassner developed a theory of disease as possession and cure as exorcism. When presented with an ill person, Gassner would apply a test to determine whether the disease was natural or supernatural. Speaking in Latin and in Jesus' name, he would command that if the disease be of supernatural origin, the pain of it should increase. If nothing happened, he considered the illness to be natural and stated that he could do nothing to help. If the patient's condition worsened,

---

4. Tischner (1928) rightly points out that comments such as these were not made by scientifically oriented observers. There are ways to account for such effects without accepting the existence of an animal-magnetic fluid that acts at a distance through doors and walls. He notes that suggestion and even telepathy or clairvoyance might provide alternative explanations (pp. 49–50).

Gassner interpreted this as a sign that the devil was the cause of the illness. He would then proceed with an exorcism to drive the demon out. Gassner's technique was a great success; many reported that their symptoms disappeared following convulsions produced by the exorcism. The Catholic clergy, however, were themselves divided about the nature of Gassner's cures. Hence when Mesmer came to Meersburg in 1775, Bishop von Rodt suggested that the magnetist observe Gassner at work and make a judgment as to the nature of the resulting cures. Then, at the instigation of Maximilian Joseph III, elector of Bavaria, the Munich Academy of Science invited Mesmer to travel to that city to give his opinion about Gassner.

Mesmer's inclination was to seek natural rather than supernatural or spiritual explanations for human events. Seeing that Gassner seemed to produce reactions in his subjects similar to those he himself had induced through animal magnetism, he concluded that the cures were genuine and were effected through animal magnetism. In Mesmer's view, Gassner was an honest but deluded man who did not recognize the true source of the cures he produced: the healing power of nature itself. Mesmer's encounter with Gassner served to strengthen his own conviction that animal magnetism operates most powerfully and effectively through the organism of the physician.

In 1776 Mesmer had the opportunity to treat the director of the Munich Academy, Peter von Osterwald, who suffered from a lameness in both legs, a severe narrowing of the field of vision, hemorrhoids, and hernia. In his treatment of von Osterwald, Mesmer did not use iron magnets at all. It was reported: "At present, Herr Doctor Mesmer makes most of his cures without any artificial magnets. Rather he carries them out purely by repeatedly touching the ill part of the body, sometimes directly, sometimes indirectly, as circumstances demand" (Mesmer 1776, Appendix, p. 13).

Von Osterwald began the experience a skeptic but ended believing that Mesmer had a true and effective healing technique. He came to accept the existence of a "fluid"—transmitted to him by Mesmer and sometimes augmented by music—which he experienced as a "warm wind." The results of Mesmer's treatment were striking; von Osterwald was able to walk again, and his breadth of vision increased markedly. During his treatment, he spent some time with Mesmer discussing his ideas and observing his treatment of other patients. As a result of his improvement and the cures he had witnessed, von Osterwald spoke in support of Mesmer's ability to treat disease through animal magnetism.

Mesmer was now certain that iron magnets were not essential for producing crises and curative effects; the physician's body was the true magnet for augmenting and controlling animal magnetism. It could be directed at will by the physician. It did not depend on contact—Mesmer's experiments with magnetizing at a distance and even through walls proved that. The essence of the treatment was the induction of a proper balance and harmony of magnetic fluid in the patient's body. Various healing aids could be used to control magnetic fluid: water could store it, mirrors could reflect it, iron rods could direct it. Music could enhance the healing action of the physician and generally promote the desired state of magnetic harmony. This state could eventually be brought about in most patients, although some were resistant—consciously or unconsciously—to the treatment and could not be helped.

In early 1777, with these principles firmly in mind, Mesmer began the treatment of a young woman in Vienna who was afflicted with numerous symptoms, most of them apparently hysterical in origin. This case became a turning point in Mesmer's life, and the outcome forced him to make a decision that had far-reaching effects for the history of animal magnetism.

Maria Theresa Paradis

Maria Theresa Paradis was eighteen when Mesmer began her treatment. Blind from the age of three and a half, the young woman had overcome her handicap to develop a talent for playing harpsichord, organ, and piano. The empress Maria Theresa had given her financial support to develop her musical abilities and had directed Mesmer's friend Dr. Anton von Störck to see what he could do for her vision problem. Maria Theresa's parents seem to have been socially acquainted with Mesmer. In any case, he had known of her through professional channels, having witnessed some of the early medical treatments administered to her. Von Störck and other specialists had tried every technique available—including painful electrical shocks—but eventually announced that they could not improve the young woman's sight. At that point the mother turned to Mesmer. By this time Mesmer was accustomed to being seen as a physician of last resort, and he accepted the challenge of relieving Maria Theresa's suffering, although at first he did not speak of restoring her sight. But, as Mesmer himself would later write (1779, p. 39), he did hope that she would provide him with a striking cure that would convince the world of the truth of animal magnetism.

Maria Theresa's blindness was accompanied by severe spasms of the muscles around the eyes. She was also subject to various hysterical symptoms, including vomiting and "melancholia." In addition, she underwent "fits of delirium and rage" and sometimes believed herself to be mad (Mesmer 1779, p. 40).

Mesmer invited Maria Theresa to live at his house so that he could pursue his treatments without interruption. Mesmer's home had by now become a residential clinic for magnetic treatment. Besides Maria Theresa, a nineteen-year-old woman named Zwelserine and an eighteen year old, Miss Ossine, were staying there to be treated by animal magnetism for hysterical physical symptoms.

By this time Mesmer had entirely abandoned the use of iron magnets. His technique now included touches, pointing with the fingers or an iron rod, and the use of magnetic conductors, music, and mirrors. In Maria Theresa's case it also involved a great deal of personal attention, solicitous care, and patient explanations of the purposes and effects of the various elements of the treatment.

After an initial increase in the intensity of her symptoms, the young woman started to show marked improvement. For the first time in years she began to use the motor muscles of her eyes. She gradually acquired the ability to discern light, then objects. Since she was accustomed to operating as a blind person, the restoration of sight required considerable psychological adjustment. Where she had learned deftness and skill without the benefit of sight, she now experienced clumsiness. Mesmer helped her through the depression of this stage and encouraged her growing ability to see.

The success of the treatment was acknowledged by Maria Theresa's parents and by a number of University of Vienna faculty members who observed the changes, among them Anton von Störck. One of the faculty observers, however, publicly denied Maria Theresa's improvement even though, according to Mesmer, he had admitted privately that she could use her eyes. The Dutch physician Jan Ingenhousze, who had expressed a negative opinion on animal magnetism when he had seen it applied to Franzl Oesterlin some years before, also denied Maria Theresa's improvement.

Mesmer's work received a further blow when trouble developed with Maria Theresa's father. According to Mesmer's account, Ingenhousze persuaded the father to doubt the cure and incited fears that Maria Theresa's pension might be terminated if she were cured. The father asked that his daughter return home, but she refused, at first with her mother's support. Soon the mother reversed her position. Finally, after a great deal of public commotion, Maria Theresa was withdrawn from treatment with Mesmer and lost the improvement she had attained.

By now von Störck had joined the ranks of those protesting Mesmer's medical treatment, and in May 1777 he sent Mesmer a letter demanding that he cease using the methods of animal magnetism. The furor created by the Maria Theresa incident and the opposition of members of the Vienna Medical Faculty made it impossible for Mesmer to continue his work in Vienna. In 1778 he headed for Paris, hoping to find a more hospitable environment for his important discovery.

Chapter 2

Animal Magnetism in France

The forty-three-year-old Mesmer arrived in Paris in February 1778, armed with a letter of introduction from the Austrian minister of external affairs to Florimund Mercy d'Argenteau, the Austrian ambassador to Paris. He took up lodgings at the Place Vendôme and set up an animal-magnetic practice assisted by his footman, Antoine, who was trained to apply animal magnetism.

Soon Mesmer's office was flooded by the ill and the curious. Ailments treated in the first months included vomiting, "obstructions of the spleen and liver," and paralyses of various kinds. While awareness of Mesmer and animal magnetism increased steadily in Paris, acceptance by medical men of stature eluded him. Then a certain Doctor Portal played a trick on Mesmer by presenting himself as a patient with various self-described symptoms. After being treated by Mesmer, he denounced him as incompetent. But as it turned out, this bit of dirty dealing helped Mesmer by arousing sympathy among some physicians and paving the way for a hearing by the Academy of Sciences.

The director, Monsieur Le Roi, invited Mesmer to a meeting of the academy to present a brief paper or *mémoire,* and Mesmer accepted gladly. The meeting, however, was so noisy and disordered that Mesmer declined to discuss his *mémoire* and settled for talking privately with a few interested members. Some of them agreed to attend one of Mesmer's treatment sessions to make a judgment about the reality of animal magnetism.

A friend of Le Roi who suffered from asthma was the subject of the treatment-demonstration. Fixing the man with his gaze and holding his hands, Mesmer magnetized the man. The asthma sufferer described a feeling of tugging in his wrists and, upon being further magnetized, underwent an asthma attack unlike his usual ones. Mesmer then demonstrated the magnetic "poles" of the body and used them to produce a variation in the man's smell and taste perception, creating hallucinatory effects at will in those senses. Although some of the skeptical observers had left before the treatment was finished, Mesmer believed he had convinced those who remained of the reality of the effects of animal magnetism. Another demonstration was arranged for members of the Academy of Sciences. Striking phenomena were produced, but the observers simply attributed them to the patient's imagination rather than animal magnetism.

Discouraged by the futility of these attempts to convince the members of the academy, Mesmer turned his attention to the Société Royale de Médecine de Paris, which had been founded in 1778 in competition with the powerful Faculty of Medicine of Paris. The upstart society, favoring novel therapeutic techniques such as electric shock, was seeking to draw public attention to itself and was ready to listen to Mesmer's proposal for a test of animal magnetism. As subjects Mesmer suggested some patients of a member of the society, Monsieur Mauduit, who treated illnesses with electricity. Treated by animal magnetism, the chosen patients described effects quite different from those produced by electricity. As a result of this trial, the society set up an investigatory committee to examine some patients before Mesmer treated them. This arrangement, however, was short-lived. One trial case, that of an epileptic young woman, Mademoiselle L., led to a dispute between Mesmer and members of the society and Mesmer's eventual second rejection of a proposed committee to investigate his work.

Mesmer then sent certificates of cure signed by patients and physicians to the secretary of the society, Vicq d'Azyr, in the hope of convincing him that animal magnetism should be taken seriously. Vicq d'Azyr returned the certificates unopened, stating that the society could only consider cases that had been previously examined by members.

By this time the large numbers of patients seeking Mesmer's attention had strained the capacity of his first small apartment beyond its limits, and in May 1778 he had moved his residence and professional offices to Créteil, a town just outside Paris. In August he moved back to Paris, this time to the Hôtel Bullion on Rue Coq-Héron, where he had sufficient space for his living quarters, an office, treatment rooms, and even housing for certain patients.

Mesmer by now was making extensive use of the magnetic *baquet,* an oaken tub specially designed to store and transmit magnetic fluid. The tub, some four or five feet in diameter and one foot in depth, had a lid constructed in two pieces. At the bottom of the tub, arranged in concentric circles, were bottles, some empty and pointing toward the center, some containing magnetized water and pointing out to the circumference. There were several layers of such rows. The tub was filled with water to which iron filings and powdered glass were added. Iron rods emerging through holes in the tub's lid were bent at right angles so that the ends of the rods could be placed against the afflicted areas of the patient's body. A number of patients could use the *baquet* at one time. They were encouraged to augment the current of magnetic fluid by holding hands, thus creating a circuit.

The room where the *baquet* was situated was kept dark. Except for the music, carefully chosen by Mesmer, there was silence during the treatment. The patients were seated around the *baquet* on chairs while Mesmer, wearing an ornate robe, moved among them, fixing them with his gaze or touching them with hand or wand. Mesmer considered it essential for healing that some kind of physical reaction or "crisis" result from the establishment of a magnetic current. Sometimes a crisis ignited in one patient induced similar crises in others in the group. A special room was set aside for those who were overtaken by convulsions or other violent forms of crises in an attempt to control this kind of psychic contagion.

Mesmer's magnetic treatment was not always collective. Individual treatments were given to some, particularly the well-to-do. Such attention involved an intricate system of magnetic "passes," measured movements of the hands of the magnetizer over the body of the patient. Typically, the patient was seated with his back to the north. Mesmer would sit facing the patient, knee against knee and foot against foot, to establish "harmony." While fixing the patient with his eyes, Mesmer would place the fingers of one hand on the stomach and make parabolic movements over that area with the fingers of the other hand. While this was happening, the patient would often experience feelings of cold, heat, or pleasure from the passes.

Mesmer considered his hands to be two magnetic poles, one north and the other south. He believed that he caused a current of magnetic fluid to pass out of one hand, through the affected part of the patient, and back in through the other hand. To apply that current to the best advantage, Mesmer would vary his passes according to the problem being treated. For instance, for migraine he would place his hands on the patient's forehead and back of the head; for eye problems, he would touch the patient's temples and move his fingers around the area of the affected eye.

Mesmer also made magnetic passes that traversed the whole length of the patient's body (*magnétisation à grand courant*). Putting his hands together and forming a pyramid with his fingers, he would begin at the top of the patient's head and move his hands down the front to the bottom of the patient's feet. Then he repeated the movement from head to feet in the back. Mesmer believed that these long passes produced a surcharge of magnetic fluid that could lead to a magnetic crisis and healing. Magnetic passes of another kind (*passes longitudinales*) were made at a distance from the patient's body, the magnetic fluid being transmitted and received through the fingers held in the shape of a pyramid or through wands of brass or iron.

By now animal magnetism had become so popular that there was often insufficient room at the magnetic tubs to treat everyone who sought help. Mesmer therefore magnetized a nearby tree with his *magnétisation à grand courant,* tied ropes to its branches, and instructed his patients to hold onto the ropes or touch the trunk of the tree to receive their magnetization. This tree became very popular among the poor, and a fable grew up that it was the last to lose its leaves in the fall and the first to sprout new foliage in the spring (Vinchon 1936, p. 44).[1]

Even in cosmopolitan Paris, Mesmer's medical innovations were bound to raise eyebrows. Rausky (1977, pp. 65ff) has pointed out that an "antimedical" movement was already afoot in the 1770s that was attempting to promote reliance on the healing powers of nature rather than the radical interventions of physicians. It eschewed the strong, sometimes dangerous chemicals often used as medicines, violent purges, and painful electric shocks, which sometimes left the patient in worse shape after medical treatment than before. The antimedical movement attempted to make the relationship between patient and physician more personal, insisting that the ill person be regarded not as a passive receptor of medical action but as an active participant in the healing process.

Mesmer's animal magnetism was perceived as congenial to this antimedical framework. He prescribed few medicines, was little concerned about diet, and used no electrical machines. He spoke of the healing power of animal magnetism as a natural force and of the healing process as an attempt to restore the balance of magnetic fluid originally established by nature. The relationship between magnetic physician and patient was considered central to the treatment. Physical contact, physical manipulation, and eye contact were all part of a personal interaction essential to magnetic healing. The patient had to participate in the procedures, often being required to cooperate with other patients in a communal magnetic seance. All in all, Mesmer's approach fit in well with the antimedical viewpoint, and the orthodox medical men of the day were bound to recognize the kinship and react to it.

Mesmer, however, wanted animal magnetism to gain the approval and support of the orthodox medical establishment of Paris, just as he had sought that of Vienna. His attempts had so far proved unfruitful. The Academy of Sciences had not welcomed his ideas. The Society of Medicine had closed its doors. There was only one organ of the medical establishment—and that the most orthodox of all—left for him to approach: the Faculty of Medicine of

1. See Vinchon 1936, pp. 41ff, for more details of Mesmer's early magnetic procedures in Paris. Caullet de Veumorel (1785, pp. 82ff) has a useful detailed description of magnetic techniques of the early 1780s.

Paris. And in 1779 a possible avenue of entry was presented to him in the person of a young physician who was impressed by Mesmer's demonstrations of animal magnetism.

Charles D'Eslon

Among the medical men who had observed Mesmer's work was Charles Nicolas D'Eslon (1750–1786), physician to the comte d'Artois. He was impressed by Mesmer's treatment of a hypochondriac who regularly made the rounds of the medical treatment centers of Paris, and he had witnessed Mesmer's magnetization of the epileptic Mademoiselle L. (Vinchon 1936, pp. 37, 45). These demonstrations convinced him of the reality of animal magnetism. Constantly in search of ways to benefit his patients, D'Eslon placed himself under Mesmer's tutorship. A disciple of the surgeon J. L. Petit and a highly respected member of the powerful Faculty of Medicine, D'Eslon gave a prima facie legitimacy to the practice of animal magnetism. His high standing with the faculty held out the hope that that august body might be persuaded to put its stamp of approval on Mesmer's discovery.

D'Eslon was struck by the speedy and powerful effects Mesmer could produce in refractory patients. According to him, Mesmer taught that medications were effective only because they happened to be conductors of animal magnetism. That is, they might be helpful or not, depending to a great extent on luck. Mesmer, with his knowledge of animal magnetism, was in a better position than most physicians to tap the healing power of nature rationally and consistently wherever it might be found. Success in this undertaking would be visibly demonstrated by the patient, for, as D'Eslon asserted, in the process of treatment Mesmer unleashed the healing crisis that nature uses to help the body throw off the illness. This crisis could be rather violent, accompanied by massive evacuations or even convulsions (pp. 34–38).

D'Eslon described what he regarded as a particularly impressive treatment. It was the case of a ten-year-old boy who, after a bout of stomach trouble, developed a fever, trembling in his arms and legs, and a severe rash. D'Eslon diagnosed the illness as a *fièvre milliaire*. Perspiration and evacuations ceased and the boy became dangerously weak. As the body wasted, the boy fell into a state of lethargy. Despite all D'Eslon could do for him, after forty-five days of illness he seemed headed for death.

> In this state of despair I engaged Monsieur Mesmer to come and see the patient. We arrived together around midday. He was shocked by the frigidity and wasting of the body, and in secret reproached me for bringing him to be a useless witness of the inevitable misfortune. Nev-

ertheless, he took hold of the boy by the hands, and a few minutes later the stomach and chest were covered with a sticky moisture. Touching the tongue produced an agreeable interior warmth. A half-hour later, the patient urinated. Truly astonished to see animal magnetism produce in such a short time effects that forty-five days of our remedies had perhaps prevented, I pressed Monsieur Mesmer to finish what he had so well begun. But he refused, seeing the boy as beyond hope: he anticipated death. But if his resistance was great, my insistence was stubborn. I implored him, and consequently the boy was placed in a bath. He stayed there for an hour and a quarter, saying gaily that he felt fine. In the evening, warmth returned, the moisture spread to the whole body, and he felt hungry. He ate some crab meat and bread, and drank water mixed with champagne. That night his sleep was calm, the boy only waking to ask for food. Finally, a foul evacuation relieved exhausted nature. The remainder of the cure took three or four weeks. I have not seen much of the boy since, but I have seen him. He was filled out, alert, and had all the signs of good health. (pp. 41–44)

In discussing this striking case, D'Eslon took up the issue of the role of the imagination:

At one time I had rashly decided that Monsieur Mesmer really had no discovery at all, and that if he did extraordinary things, it was by seducing the imagination. I want to observe that this is not the case here. No one knew beforehand of Monsieur Mesmer's arrival. The patient did not know him and had never heard him spoken of. Also, he was too weak to want to concern himself with anything that was going on around him. But besides, if Monsieur Mesmer had no other secret than to be able to cause the imagination to act effectively to produce health, would he not have a marvelous thing? And if the medicine of the imagination is the better one, why are we not using the medicine of the imagination? (pp. 46–47)[2]

D'Eslon was impressed with Mesmer personally as well as professionally. He described Mesmer as a man of great honesty, "of established age, with exquisite judgment, combining facile elocution with an uncommon precision" (p. 18). He knew, however, that Mesmer would need more than these qualities

---

2. Others explicitly spoke about Mesmer's "secret." In his *Traité théorique et pratique du magnétisme animal* (1784) D'Eslon's follower François Amédée Doppet discussed Mesmer's secret, reduced it to the sum total of procedures Mesmer used in the practice of animal magnetism, and purported to reveal it.

to convince the medical establishment, the "savants," of the existence of animal magnetism: "I thought he might more easily make the four great rivers of France run in the same bed than get the savants together to judge in good faith a matter that was outside their fundamental concepts" (p. 21).

Despite his doubts about Mesmer's plan, D'Eslon helped him as much as he could. He began by speaking about animal magnetism informally at meetings of the Faculty of Medicine. After he had created some curiosity about the subject, one of the members suggested that Mesmer read a *mémoire* to a gathering of interested physicians explaining in some detail his theory of animal magnetism. At first Mesmer opposed the idea, since it fell short of the immediate official approbation he was hoping for. But in the end he acquiesced and began work on his presentation.

Mesmer's *Mémoire* of 1779

Mesmer, aided in the organization and presentation of the material by Monsieur de Bachelier, produced a treatise eventually published as *Mémoire sur le découverte du magnétisme animal* (1779). It contained a short description of the history of his practice of animal magnetism and a summary of the theory.

Mesmer first stated that "nature provides a universal means of healing and preserving men" (p. vi). He then described how he came to know this: by reflecting on the interactions that take place between bodies in nature, particularly the effects of heavenly bodies on each other and on the oceans and atmosphere of the earth. From the awareness of this "ebb and flow" in nature he became cognizant of animal magnetism, the property in animal bodies that renders them susceptible to the action of the heavenly bodies and the earth. He viewed disease as a "disharmony" within the organism, and cure of disease as restoration of the body's natural harmony through the application of nature's own agent of harmony, animal magnetism. Mesmer then related how he had gradually evolved a way to control the flow of animal magnetism in human bodies and cure illnesses. He described his initial interest in mineral magnets and the treatments he carried out in Vienna. He recounted the Maria Theresa Paradis story in some detail, presenting his view of the persecution to which he believed he had been subjected.

The rest of the *Mémoire* consists of twenty-seven propositions describing the principal properties of animal magnetism. Among the more significant are the following:

(1) There exists a mutual influence between the celestial bodies, the earth, and animate bodies; (2) The means of this influence is a fluid that is universally distributed and continuous . . . and which, by its nature,

is capable of receiving, propagating, and communicating all impressions of movement; (3) This reciprocal action is governed by mechanical laws as yet unknown . . . ; (8) The animal body experiences the alternative effects of this agent which insinuates them into the nerves and affects them immediately; (9) It particularly manifests itself in the human body by properties analogous to the magnet . . . ; (10) Because the property of the animal body which makes it susceptible to the influence of heavenly bodies and to the reciprocal action of those around it is analogous to that of the magnet, I decided to call it "animal magnetism" . . . ; (23) The facts will show, following the practical rules that I will establish, that this principle can heal disorders of the nerves immediately, and other disorders mediately. (pp. 74–83)

Mesmer was invited to deliver his *mémoire* at a dinner meeting of the faculty, but he had not prepared the way very well. He had sent a letter to the faculty which created the impression that he had little patience for the traditional practice of medicine and that he would present a new theory that would revolutionize medical thinking. The meeting was attended by twelve members. Mesmer's delivery of his *mémoire* was followed by dinner and a lively discussion—without Mesmer—about whether animal magnetism was an important discovery or a delusion. The outcome was a decision to further investigate animal magnetism by delegating three physicians to observe Mesmer at work.

Shortly afterward, the three doctors came to the Hôtel Bullion and witnessed the treatment of two paralytics, a blind girl, a suicidal soldier, and a girl afflicted with scrofula. They had to admit that the condition of each improved, but, not having examined the patients beforehand, they could not determine whether the improvement was the result of treatment with animal magnetism or of the natural progress of the condition. Observations by this committee continued for several months, during which time they saw the effects of the *baquet* and saw Mesmer treat people on an individual basis. They witnessed a number of impressive recoveries. The three physicians, however, could not agree that animal magnetism was the agent of cure. When Mesmer asked for a formal opinion, they withdrew from the project (Vinchon 1936, pp. 51–52).

In the meantime, D'Eslon was getting more deeply involved with Mesmer and his treatments. Working together, the two accomplished what they considered to be remarkable cures and improvements in the health of their clientele. In 1780 D'Eslon published his first work on animal magnetism, *Observations sur le magnétisme animal,* strongly supporting Mesmer and describing cases for his colleagues to ponder. He did not propose an elaborate theory but rather called on the medical world to consider the facts and give animal magnetism a chance to prove itself. He denied that there was any "secret" or mystical force involved in the practice of animal magnetism, as some had claimed, and quoted Mesmer on the matter: "Animal magnetism . . . is not what you call a secret. It is a science, which has principles, consequences, and a doctrine. . . . My object is to obtain from the government a public house to treat patients, where, under the shelter of further discussions, the salutary effects of animal magnetism could be established" (pp. 145–46).

It was becoming clear, however, that the Faculty of Medicine was not going to be Mesmer's supporter and protector, and D'Eslon was not going to be his entrée to official approval. On the contrary, now that D'Eslon had come out publicly in favor of animal magnetism, the faculty turned on him. In general assembly on September 18, 1780, it condemned animal magnetism and censured D'Eslon for supporting a charlatan and insulting the faculty by making light of the teaching of the medical school and the sound practice of medicine. The faculty called for his rejection of the propositions of Mesmer, deprived him of his right to speak in the assembly for one year, and directed him to disavow his *Observations* on pain of loss of membership. D'Eslon nonetheless continued to support Mesmer. In 1782 he received a second condemnation and shortly thereafter a third, which removed his name from the list of *docteurs-régents*.

Mounting Opposition to Animal Magnetism

The publication of Mesmer's *Mémoire* and D'Eslon's *Observations* caused a storm of controversy in the public press. The medical journal *La Gazette de la Santé* published an article by the physician Jean Jacques Paulet (1740–1826), an acerbic writer who eventually produced a number of antimesmeric treatises. While admitting that Mesmer might be able to heal people, Paulet denied that the healing was due to animal magnetism. In the same journal, the physician de Horne attempted to prove that animal magnetism did not exist. In the *Journal de Médecine* Alexandre Bacher stated that he knew the three physicians who had observed Mesmer's work and that if they had witnessed true cures, he was certain they would have said so.

Among Mesmer's defenders was Nicolas Bergasse (1750–1832), a lawyer, philosopher, and political theorist from Lyon. In 1781 Bergasse wrote an anonymously published treatise declaring his belief in the reality of the cures performed by Mesmer and condemning the narrow-mindedness of orthodox medicine. That same year Michel Fournier published a short statement attesting to Mesmer's cure of a young paralyzed woman. The following year

D'Eslon (1782a) published a lengthy work attempting to persuade the Faculty of Medicine to reconsider the potential importance of animal magnetism for medical practice. And Antoine Court de Gébelin (1728–1784) described how he was cured of a serious illness by Mesmer's animal magnetism, appealed for a serious study of that system, and suggested directions such a study might take (Court de Gébelin 1783).

Discouraged by his failure to win official recognition, Mesmer threatened to leave Paris and look elsewhere for the appreciation he sought. This threat caused an uproar among the sick who had come to trust and hope in Mesmer and the curious who would miss the extraordinary diversion afforded by animal magnetism. Word of this situation came to the attention of Marie Antoinette, who appointed the comte de Maurepas to persuade Mesmer to remain in Paris. Accompanied by D'Eslon, Mesmer met with the count on March 28, 1781. Maurepas was prepared to offer Mesmer, in the name of the government, an annual pension and rent money for his treatment building in return for his promise to stay in Paris and accept a number of students to be named by the government. Astonishingly, Mesmer refused the offer, fearing that government-appointed students would sooner or later take control of the operation. He insisted that he would take on only those students he chose and would accept Maurepas's offer only if it included official recognition of animal magnetism. This response led Maurepas to withdraw from the negotiations, and the whole matter was dropped—with, however, one final salvo from Mesmer. He wrote a letter to Marie Antoinette expressing his indignation at the offer and repeating his demands. His arrogant request received no reply.

Meanwhile the forces of opposition were gathering strength. In 1782 Noël de Rochefort Retz, physician ordinaire to the king of France, published a severe criticism of Mesmer and animal magnetism in which he claimed that it was well known throughout France that animal magnetism was a "secret" possessed by Mesmer, and very lucrative for him. Retz considered Mesmer a charlatan along with the peddlers of mystery and mystification. Retz himself had attended mesmeric seances and experimented with the technique. He obtained palpable results from his experiments but attributed the cause to "imagination," not magnetic fluid. In fact, Retz identified Mesmer's "secret" with his ability to manipulate weak imaginations.

Earlier, in 1779, Charles Andry and Michel Thouret had carried out a study of the use of magnets in medicine, favorably noting Mesmer's work. Their findings were published in book form in 1782. But by 1784 Thouret had revised his opinion. In *Recherches et doutes sur le magnétisme animal,* he employed his considerable erudition to show that Mesmer was merely the most recent example in a long tradition of thinkers who posited a hidden power of nature that produces healing effects, citing Paracelsus, Kircher, Maxwell, Fludd, the exorcist Gassner, and Valentine Greatrakes, the "stroking doctor" of Ireland. Thouret's critique was one of the most formidable that Mesmer and his followers had to reckon with.

Jean Paulet's *L'antimagnétisme* (1784) is significant both for its statement of the case against animal magnetism and for its references to earlier works on the subject. Another essay by Paulet published in the same year, *Mesmer justifié,* satirizes Mesmer's technique by pretending to give serious instructions on how to use animal magnetism.

The Rift between Mesmer and D'Eslon

As opposition to animal magnetism was growing, trouble was also developing among Mesmer's supporters. In the early phase of their relationship, D'Eslon considered himself Mesmer's pupil and witness to his prowess, but not a practitioner—an attitude clearly expressed in D'Eslon's *Observations* of 1780. But as time went on and D'Eslon became more experienced, he recognized his own ability as a magnetizer. It pained him to see that Mesmer neglected some patients, being content simply to let them sit at the *baquet* (Rausky 1977, p. 109). In coming to the aid of these patients, D'Eslon began to realize that he too could generate healing magnetic fluid and carry out the operations performed by Mesmer.

Besides, D'Eslon was highly intelligent, and a very good physician. Even though he had been effectively ostracized by the Faculty of Medicine, he was still a man of stature among his contemporaries. Such a person could not continue indefinitely in the position of disciple and student. His personal financial fortunes were also a consideration, for, as a result of the penalties accruing from defending animal magnetism, he had lost a large part of his income and was in need of supplementing his personal medical practice. D'Eslon also was a dedicated physician with a strong drive to help the suffering. He truly believed that animal magnetism offered a revolutionary treatment that would change every aspect of medical practice. Once he discovered that he could effectively practice animal magnetism, he could hardly be expected to refrain from doing so. All of this added up to an inevitable tension between D'Eslon and his old teacher.

In a reflective frame of mind, Mesmer left Paris for Spa in August 1781, planning to ponder his situation, and also probably counting on being sorely missed by the population of Paris. Against his wishes, D'Eslon continued the practice of animal magnetism while Mesmer was gone. This led to a vehement argument on Mesmer's return: Mesmer insisted that no one was capable of replacing him, and D'Eslon insisted that he could not abandon patients who depended on their magnetic treatments. Despite this disagreement, the two men were sufficiently reconciled to keep working together into the summer of 1782.

In July Mesmer accepted an invitation from one of his patients, the marquis de Fleury, to return to Spa and establish a treatment house there. The vacuum left by his departure created anxiety among Parisians who wanted their magnetic treatments to continue. They turned to D'Eslon, urging him to establish his own school of magnetic practice. D'Eslon could hold back no longer. He wrote a strongly worded letter to Monsieur Philip, the new dean of the Faculty of Medicine, demanding proper recognition for animal magnetism. The letter was sent in D'Eslon's name only, not Mesmer's, and the response, on August 20, 1782, came in the form of the faculty's third condemnation of D'Eslon, removing from him the rights and privileges of *docteur-régent*.

Having burned his bridges, D'Eslon began his magnetic practice in earnest. His clientele grew rapidly, probably not least because D'Eslon charged only half Mesmer's fee. When Mesmer got word of D'Eslon's independent overture to the dean of the faculty and his thriving practice, he was furious. A reproachful letter to D'Eslon received no reply. Then, on October 4, 1782, Mesmer wrote to the dean of the Faculty of Medicine protesting D'Eslon's claim to be practicing animal magnetism.

Neither Mesmer nor D'Eslon truly desired the rift that had developed, and in 1783, at the urging of mutual friends, they entered into negotiations aimed at reuniting them in a common magnetic school under Mesmer's direction. But in the end the negotiations failed (Vinchon 1936, pp. 78–79).

The Commissions of 1784

As D'Eslon became independently established, he increased his efforts to gain the official governmental recognition for animal magnetism so long sought by Mesmer. He wrote to the government early in 1784 calling for the formation of an official commission to investigate animal magnetism.

D'Eslon's request was received favorably. No doubt his personal connection with the aristocracy partly accounted for this response, but there was also an attitude developing in governmental circles that this controversial matter should be settled once and for all. The king therefore decided to appoint a commission to determine the scientific status of animal magnetism. The final composition of the commission was five members from the Academy of Sciences—Benjamin Franklin, who was to chair the commission; J. B. Le Roy; G. de Bory; A. L. Lavoisier; and J. S. Bailly, a world-class astronomer who would serve as secretary—and four from the Faculty of Medicine—

Majault, Sallin, J. D'Arcet, and J. L. Guillotin. It was understood that the academy members would carry out the principal investigation, and the faculty members would simply comment on the results. In view of Franklin's advanced age and various infirmities, it was tacitly understood that he would not be able to attend all the investigative sessions. The commission began its meetings on March 12, 1784.

Less than a month later, on April 5, the king appointed a second investigatory commission with a slightly different purpose. This commission, composed of members of the Royal Society of Medicine (P. I. Poissonnier, C. A. Caille, P. J. C. Mauduyt, F. Andry, and A. L. de Jussieu), was to determine the usefulness of animal magnetism in the treatment of illnesses. From the start, it did not have the competence and stature of the Franklin commission (see Rausky 1977, pp. 124–31), and the report it issued was not as significant.

Because Mesmer refused to have anything to do with these commissions, all investigations relied entirely on D'Eslon. This course of action, however, made their eventual reports vulnerable to attack, for Mesmer declared that D'Eslon had neither the true doctrine nor the correct technique of animal magnetism and could not, therefore, be its official representative in any investigations. Mesmer was, after all, the discoverer and principal practitioner of animal magnetism, and his magnetic school, with some three hundred pupils, was larger and more influential than D'Eslon's, which had only sixty.

Nevertheless, the investigations went forward. The Franklin commission concluded that all the apparent effects of animal magnetism were due to imagination and imitation, and that there was no proof of the existence of animal-magnetic fluid. In its report, published in August 1784, the commission defended its sources:

> If someone objects to the commissioners that this is a conclusion reached about magnetism in general instead of being applied only to the magnetism used by Monsieur D'Eslon, the commissioners respond that the intention of the king had been to obtain their opinion on animal magnetism; consequently they have not exceeded the boundaries of their commission. So they respond that Monsieur D'Eslon was prepared to instruct them in what are called the principles of magnetism, and that he was in possession of means of producing the effects and exciting the crises. These principles are the same as those found in the twenty-seven propositions made public by Mesmer in his publication of 1779. If today Monsieur Mesmer would announce a broader theory, the commissioners would not need to know this theory to decide about the existence

and utility of magnetism; they have only to consider the effects. It is by the effects that the existence of a cause is manifested; and it is by the same effects that its utility can be demonstrated. (Bailly 1784a, p. 78n)

Mesmer could, of course, argue that the commissioners' assumption that D'Eslon's effects and his own were identical was not warranted. On the other hand, the commissioners had a job to do, commanded by the king; if Mesmer would not cooperate, they would go forward without him.

According to the report, D'Eslon was engaged by the commission to communicate his knowledge of animal magnetism and prove its usefulness in the cure of illnesses. After he had explained the theory as he understood it, he invited the commissioners to visit his hall of healing. Their report describes the structure of the magnetic *baquet,* the positioning of the patients, the use of the magnetic "chain," and the background music played on the pianoforte. D'Eslon explained in detail the various accoutrements of the magnetic process—the iron rod he employed to direct the magnetic fluid to patients or to magnetize objects, the use of the rope for conducting fluid from patient to patient, and the construction of the *baquet*. The commissioners saw the rows of patients seated around the *baquet* receiving their treatments in a communal magnetic experience; they observed the convulsions the treatment induced in some patients; and they examined the "hall of crises" in which convulsive patients were placed. They noted the great variations in the nature and intensity of these convulsions and the bodily evacuations that often followed. They commented:

Nothing is more astonishing than the spectacle of these convulsions; if one had not seen them, one would have no idea. When one sees them, one is surprised equally by the profound repose of some patients and the agitation that animates others. One is surprised by the various phenomena that are repeated and by the sympathies that are established. One sees the patients seek each other out, engage each other, smile at one another, converse with affection, and mutually soothe the crises in each other. All subjugate themselves to the magnetizer. They may seem satisfied to be in an apparent state of drowsiness, but his voice or a look or sign from him will draw them out. One cannot help but note in these consistent effects a great power that moves the patients and masters them. The result is that the magnetizer seems to be their absolute ruler. (Bailly 1784a, p. 8)

The commissioners set about their next task: to determine the existence of animal magnetism and its usefulness in treating patients. This turned out to be a problematic undertaking, for it was very difficult to prove the nonexistence of a thing:

> It did not take long for the Commissioners to recognize that the [magnetic] fluid is not perceptible to the senses. . . . [D'Eslon] has expressly declared to the Commissioners that he can only demonstrate the existence of magnetism for them through the action of the fluid producing changes in the animated body. Its existence is very difficult to establish by effects that are decisive and have an unequivocal cause; that is, by authentic facts uninfluenced by moral causes, and by proofs that are striking and convincing to the spirit—which alone could satisfy enlightened physicians. (Bailly 1784a, pp. 10–13)

The commissioners attempted to overcome this difficulty by stating that the action of magnetism on the human body could be observed by its effects in the treatment of disease, and by its temporary effects on the organism. They decided that they could not come to any firm conclusions by investigating the effects on disease because the causes of any particular cure are too hard to isolate. So they searched for palpable temporary effects that could be observed and verified as free from all misinterpretation and illusion.

As the first subjects of experiments of this kind the commissioners chose themselves. They subjected themselves to the *baquet* once a week for a number of weeks, but felt no effects. They then experimented with other subjects, some drawn from the general public and others thought to be particularly susceptible and sensitive to animal magnetism. Through a long series of experiments and trials they encountered a variety of effects in the subjects— from no response to an amelioration of symptoms. The commissioners concluded that magnetism did not produce any noticeable effects on those who were skeptical. Using the services of a Doctor Jumelin, who magnetized in a way different from both Mesmer and D'Eslon, the commissioners observed results similar to those already seen and concluded that effects attributed to factors such as the polarity of the body were chimeric (Bailly 1784a, pp. 19–35).

From this point the emphasis shifted to an investigation of the effects of imagination and how the phenomena of animal magnetism might be explained in terms of that faculty:

> The commissioners, especially the physicians, made numerous experiments on different subjects, whom they themselves magnetized or whom they got to believe themselves to be magnetized. They were magnetized in a variety of ways . . . and in every case they obtained the same effects. In all of these experiments no differences were found other

than those due to varying degrees of imagination. They are therefore convinced by the facts that imagination by itself can produce the different sensations and cause the feelings of discomfort or heat—even considerable heat in all parts of the body—and they have concluded that it necessarily enters strongly into the effects attributed to animal magnetism. (Bailly 1784a, pp. 39–41)

To test this conclusion, the commissioners decided to try to produce severe crises through imagination alone. They set up an experiment to be conducted at the estate of Benjamin Franklin at Passy. They asked D'Eslon to choose a subject whom he considered to be particularly sensitive to the magnetic fluid. D'Eslon chose a twelve-year-old boy. While the boy was detained in Franklin's house, D'Eslon magnetized one of the trees in Franklin's garden. The test was to determine if the boy could distinguish the magnetized tree from those that were not.

Then the young man, eyes bandaged, was brought and presented successively to four trees that had not been magnetized, and made to embrace them, each for two minutes, according to the instructions given by Monsieur D'Eslon himself. Monsieur D'Eslon was present and, at a great distance, pointed his walking stick at the tree that was really magnetized. At the first tree the boy, questioned after one minute, declared that he was congested; he coughed, expectorated, and said he felt a small discomfort in his head. He was about twenty-seven feet from the magnetized tree. At the second tree he felt dazed and experienced the same discomfort in the head. He was thirty-six feet from the magnetized tree. At the third tree the dazed feeling redoubled, as did the headache. He said he believed he was approaching the magnetized tree. He was now thirty-eight feet from it. Finally, at the fourth nonmagnetized tree and about twenty-four feet from the one that had been magnetized, he fell into a crisis: he lost consciousness, his limbs became stiff, and he was carried onto a neighboring lawn where Monsieur D'Eslon ministered to him and revived him. If the young man had felt nothing even under the magnetized tree, one could simply say he was not very sensitive—at least on that day. But the young man fell into a crisis under a nonmagnetized tree. Consequently, this is an effect which does not have an exterior, physical cause, but could only have been produced by the imagination. The experience is conclusive: the young man knew that he would be led to a magnetized tree; his imagination was stimulated, successively heightened, and at the fourth tree reached the degree necessary to produce the crisis. (Bailly 1784a, pp. 43–45)

The direction of the report was now becoming clear. Imagination was the adequate explanation for the phenomena. The commissioners also noted that touching, so often a part of the magnetic treatment, can itself have soothing and restorative effects. Those effects had always been known and did not involve a communicated agent such as magnetic fluid: "An effect requires only one cause and . . . since the imagination suffices, the fluid is useless" (Bailly 1784a, pp. 72–73).

The commissioners completed their report with an explanation of the effects of magnetism conducted in a communal setting, noting the importance of imitation. They pointed out that public treatment, such as that held around the *baquet,* created a stimulating atmosphere in which the movements of one participant easily set off similar movements in another, and they compared this contagion to the reinforcement of emotion characteristic of theater audiences. They stated that women are the first to fall into these convulsive crises, and the more sensitive men soon follow.

In summary, the commissioners declared that the true causes of the effects attributed to animal magnetism are touching, imagination, and imitation. Since these are sufficient to explain all of the data, there is no need to resort to a concept of magnetic fluid (Bailly 1784a, pp. 63–70).[3]

In comparison to the well-constructed and thoughtful report produced by the Franklin commission, the report of the Royal Society of Medicine commission (Poissonnier et al. 1784) was anticlimactic. It contained a lengthy examination of the history of ideas considered similar to those of the animal magnetists and a detailed criticism of D'Eslon's assertions and conclusions. The commissioners were quite ready to substitute their own sometimes abstruse theories about the causes of the phenomena produced by animal magnetism, and they showed little of the clarity of thought exhibited by the Franklin commission. Still, the Royal Society commission arrived at similar conclusions: that magnetic fluid did not exist and that the apparent effects were due to irritation, imagination, and imitation. It also pointed out dangers in the practice of animal magnetism that "merit attention."

The commissioners of the Royal Society were not, however, unanimous in their stand. Antoine Laurent de Jussieu drafted a dissenting report in which he distinguished four kinds of effects observed by the commissioners: (1) those general positive effects about whose cause no conclusions could be drawn, (2) those effects that were negative, showing only the nonaction of magnetic

3. Their public report completed, the commissioners also drew up a secret report, intended for the king's eyes only, in which they described the intimacy involved in the magnetic procedure and expressed their concern about possible abuses when a man magnetizes a woman. This report was not published until 1800; it is discussed in more detail in chapter 6.

fluid, (3) effects, either positive or negative, that could be attributed to the action of the imagination, and (4) those positive effects, independent of the imagination, that could be explained only by the action of some unknown agent. The fourth category, in Jussieu's opinion, left the way open for animal magnetism and required further investigation. Jussieu, however, felt the impulse to give his own theory of what animal magnetism might be, developing the notion of an "atmosphere" that surrounds living things and speculating that the "atmosphere" of one individual could act on that of another, even at a distance.

Reactions to the Reports

With the two public reports completed and published, the reaction of partisans on both sides of the issue was swift. From the school of D'Eslon came an impressive reply by Christophe Félix Louis Gallert de Montjoye (1746–1816). This clear-thinking theoretician noted the points of agreement between Mesmer and Bailly and then compared Mesmer's views to those of Descartes and Newton, siding with Newton against Bailly in his view of matter and motion. Citing Mesmer's views on the ebb and flow of magnetic fluid, Gallert de Montjoye (1784) attempted to show that they were in agreement with the best of contemporary physics. Moving to the Franklin report, he discussed at length the action of imagination in magnetic phenomena and the action of the will in directing magnetic fluid, stating that it is principally by the will that the fluid acts at a distance.

D'Eslon himself wrote a reply to the reports of the commissions, dated September 3, 1784, in which he gave his version of some of the experiments carried out for the commissioners and drew very different conclusions from theirs. He also warned that the condemnation of animal magnetism would not hold up against the experience of the many physicians who had been using it with great success.

Mesmer more or less avoided challenging the commissions' findings in detail but objected that their conclusions could not be taken as applying to *his* animal magnetism because he had not been involved in the investigations. As far as he was concerned, the reports applied only to D'Eslon's work, and it was he who would have to bear the responsibility for the conclusions of the commissions (Mesmer 1784b, 1784c, 1784e).

Many of Mesmer's followers, however, felt that the criticisms of animal magnetism in the reports had to be answered. So in 1784 there was a flurry of treatises published in defense of Mesmer and animal magnetism. Late in that year supporters of Mesmer assembled and published attestations of magnetic cures signed by an impressive array of gentlemen, academics, physicians, and surgeons. The work, *Supplément aux rapports de MM. les commissaires de l'Académie, et de la Faculté de Médecine et de la Société royale de médecine,* also contained a well written defense of animal magnetism (Vinchon 1936, p. 93).

Probably the most telling response to the reports was by Nicolas Bergasse (1784a), Mesmer's right-hand man at the time. His objections to the commissions' findings were uncompromising. He particularly criticized their attribution of the effects of animal magnetism to imagination and imitation, protesting that they had not taken all the facts into account.

In 1784 there appeared a work by a more objective observer, one that stood above other critiques for its scholarliness and cogency. It was *Doutes d'un provincial proposés à messieurs le médecins-commissaires chargés par le Roi de l'examen du magnétisme animal,* by Joseph Michel Antoine Servan (1737–1807), a distinguished lawyer and a correspondent of Voltaire and D'Alembert. Himself cured by animal magnetism (or so he believed) when conventional medicine had failed, he felt compelled to defend Mesmer's views. Servan addressed some well-considered questions to the commissioners: Why did they expect quick cures through animal magnetism and dismiss its efficacy on the basis of brief trials? Why did they arbitrarily set up conditions they preferred for the experiments rather than seek to create conditions considered optimal by experienced magnetizers? And why, with their limited experience, did they feel justified in drawing such far-reaching conclusions? Servan's was one of the most significant treatises written in support of animal magnetism in the wake of the reports of the commissions.

A brief but important response to the reports by Antoine Esmonin de Dampierre appeared in September 1784. A theologian, magistrate, and president of the parliament of Bourgogne, Dampierre repositioned the discussion of the adequacy of the commissions' investigations by calling for a reevaluation of the rejection of animal magnetism on the basis of a new phenomenon: "magnetic sleep." The marquis de Puységur had discovered earlier in that year that magnetization could produce an altered state of consciousness—a kind of sleep—in some individuals. He also believed that some magnetic somnambulists could clairvoyantly discern the source of illness in themselves and others and could predict the course of an illness and prescribe effective remedies. In Lyon the chevalier de Barberin (1786), a member of the local lodge of Freemasons, went a step further and taught that magnetization did not require physical contact and in fact was better done without it. Barberin claimed that magnetization could be done at a distance of even many miles.

Experiments of the Lyon group of magnetizers in August 1784, performed under the eye of Mesmer himself, had involved the magnetization of a horse and the determination of the seat of its illness. Dampierre, a lodge member of the Lyon Freemasons, believed that these experiments, the work of Puységur, and the successes of Barberin made the findings of the two commissions obsolete and their notion that touching, imagination, and imitation could explain away the phenomena of animal magnetism unacceptable. First of all, with the technique of Barberin, touch was not involved at all, yet some of the most remarkable phenomena were produced in this way. Next, the contention that imagination and imitation could completely explain magnetic effects failed to account for successes in magnetizing animals. Finally, magnetic somnambulism produced testable clairvoyant phenomena that imagination could not explain (Dampierre 1784, pp. 2–18). Dampierre believed that well-constructed experiments should be conducted with clairvoyant magnetic somnambulists to determine the genuineness of the effects. He proposed a framework for such experiments that would exclude fraud or imagination (pp. 18, 23, 32–43) and suggested that the commissioners reconvene and conduct investigations that would take his new data into account. He detailed ten separate verifiable experiments that would eliminate the influence of touch, imagination, or imitation.

Although Dampierre was a layman, he spoke for a number of practicing physicians at Lyon. Nor should it be concluded that the physicians of France stood solidly behind the findings of the commissions. In an anonymously published work the physician Girardin (1784) described his own attitude with regard to animal magnetism as questioning and undecided. But having been present during some of the investigations of the Franklin commission, he felt he had to call attention to certain defects in its approach. Girardin stated that for animal magnetism to be valid it is not necessary that it be effective in every case. He pointed out that the commissioners had drawn sweeping conclusions from their brief and incomplete experiences with the kind of phenomena animal magnetism had been reported to produce. He also accused the commissioners of having ignored certain consequential phenomena that had occurred in front of their own eyes—phenomena that he himself had witnessed. Finally, Girardin objected that the whole affair was conducted by the Faculty of Medicine of Paris with no consultation with medical colleagues in the provinces. He recommended that a new commission to examine animal magnetism be appointed that would better represent the medical community of the whole country.

By this time physicians all over France were using animal magnetism as a part of their medical armamentarium, and with some success. For instance, the physician Pierre Orelut, two months before the appearance of the Franklin commission report, published a summary of cures performed at Lyon through animal magnetism (Orelut 1784).

Some Paris physicians too were unhappy with the commission reports and with the way they were followed up. François Louis Thomas d'Onglée, a member of the Paris Faculty of Medicine, published a scathing account (1785) of how the faculty had used the findings of the commissions to pressure its own members to renounce animal magnetism. D'Onglée himself had been subjected to what he termed "abuse." He described how the faculty had rushed its members into approving the report and then proceeded to denounce thirty "magnetic physicians," himself included. They were summoned before the faculty to give account:

> They nearly all arrived and were consigned to a hall separated from the assembly. With impatience each awaited the general call and paced back and forth, caught up in their thoughts. Someone told me it was a question of signing some kind of formula. I said that we would see what it said and then either sign or not sign. Finally, the apparitor came and called me as the eldest—my honor in that group. I went in, surprised that none of the others followed me. I was asked to be seated, and the dean began to ask me if I had paid money to be taught animal magnetism. Even more surprised by this question, I answered with respect that Monsieur D'Eslon did not take money, that he only received physicians to observe and help; that no one could be more honest, modest, and obliging, and that furthermore the faculty was aware of that fact. I will not tire the reader with further details of other questions. I was interrogated like a criminal and then, I believe, moved to the Chamber of the Tower.

The faculty then presented d'Onglée with a formula to sign: "No Doctor may declare himself a partisan of animal magnetism, through writings or through practice, under penalty of being removed from the role of *docteurs-régents*." D'Onglée responded in his account with righteous indignation: "Can anyone find a more absolute despotism of opinion or fanaticism of imagination? Tremble!—you doctors and physicians who seek to improve your minds!" (pp. 4–8).

Clearly, while the government had hoped to settle the issue of animal magnetism once and for all, the physicians and intellectuals of the day continued to be strongly divided. It took more than the reports of official commissions to silence the controversy.

The Societies of Harmony

Over the years Mesmer had been ambivalent about instructing others in the technique he had discovered. He wanted animal magnetism to receive full acceptance and universal use by the medical profession, yet he was reluctant to teach people how to use it, apparently for fear that his discovery would be exploited by others. He could not tolerate the possibility that others might take credit for something he felt originated with him. From his conflict with Father Hell to his rift with D'Eslon, one can see Mesmer's attempt to exercise sole proprietorship over animal magnetism.

D'Eslon always insisted that there was no "secret" in the practice of animal magnetism, and more than once Mesmer himself made statements to that effect. Yet when it came to the point of allowing animal magnetism to be practiced by anyone else, Mesmer acted as though there were some subtle point of magnetic practice that he had not yet passed on, thus giving the impression that anyone who might attempt to work independently did not know the whole story and so could not competently practice the art.

It appeared that Mesmer hoped to make money from the "special something" that only he could provide to students of animal magnetism. Financial security was always a worry for Mesmer. Separated from his wife and her monetary resources, he feared that he would not be able to gain sufficient funds by himself to be truly comfortable. So he sought to make his precious discovery the source of the security he longed for.

Among the patients who accompanied Mesmer on his return to Spa in July 1782 were two men to whom he confided his financial worries. Nicolas Bergasse, a Lyon lawyer, was being treated for a melancholic mental condition. He had for some time been a firm supporter of Mesmer and a believer in the efficacy of animal magnetism and was delighted to bring his business acumen to Mesmer's aid. (Bergasse's first work in support of animal magnetism, the anonymously published *Lettre d'un médecin de la Faculté de Paris à un médecin du College de Londres,* was written in 1781.) Guillaume Kornmann was a banker whose young son Mesmer had successfully treated for an eye ailment. Somewhat later Kornmann was himself in need of treatment for emotional stress, and when the opportunity arose he was more than happy to aid his healer.

When D'Eslon threatened to set up his own independent—and potentially competitive—magnetic practice, Bergasse and Kornmann leapt to action. They drew up a plan that would ensure Mesmer's financial security and provide a vehicle for teaching and propagating animal magnetism. The structure devised involved three elements: a clinic, a teaching establishment, and a society to promote the doctrine of animal magnetism. Upon returning to Paris, Mesmer set about putting the plan into action.

Bergasse and Kornmann worked out the details of the contract of the proposed society, first referred to as the Loge de l'Harmonie and eventually the Société de l'Harmonie Universelle. The plan was to enlist one hundred charter members, each of whom would pay Mesmer a one-time membership fee of one hundred *louis d'or*. Others would be able to join for a fee, in return for which Mesmer would agree to teach them how to apply animal magnetism. At the end of a series of sessions, the students would receive a diploma to practice animal magnetism, and this would alone legitimize practitioners; only they would be allowed to set up magnetic clinics. The form of the contract was finalized on March 10, 1783.

Bergasse worked assiduously to promote the society, and the one hundred charter members were soon found. The society's first membership lists included some illustrious names, such as the marquis de Lafayette, the marquis de Chastellux, the comte de Puységur, and the comte de Choiseur. Also on the lists was Père Charles Hervier of the Augustine convent in Paris, who was himself cured of a nervous disorder by animal magnetism and went on to promote the system and encourage others to take treatment. The initial members of the society came from a variety of backgrounds, but by and large they were men of intellect and means.

When the Society of Harmony began its meetings, Nicolas Bergasse was the main speaker. Mesmer, not very confident in his powers of elocution, was glad to leave that task to Bergasse.[4] In the process of carrying out his duties, Bergasse wrote a philosophical guidebook to the theory of animal magnetism for the society's adepts entitled *Théorie du monde et des êtres organisés suivant les principes de M . . .* (1784c). In its structure and in the conduct of its meetings, the Society of Harmony resembled the many secret mystical societies of the day, particularly masonic groups (see Darnton 1968, pp. 75–76; Amadou 1971, pp. 361–99). Bergasse's treatise, composed in a coded format reminiscent of alchemical works, could be understood only by initiates of the society, who had been given the key to the code. Bergasse elaborated on Mesmer's basic principles but developed political and social implications far beyond anything Mesmer ever taught.

After the first Society of Harmony was formed in Paris, others were founded in the provinces, at Lyon, Strasbourg, Bayonne, Montpellier, Dijon, Nantes, Marseille, Bordeaux, and Lausanne. Outside of France, there was a

---

4. For excerpts from Bergasse's lectures to the Paris Society of Harmony, see Darnton 1968, pp. 183–85.

Society of Harmony in Turin, a city that some decades later would become a thriving magnetic center, and one in the French-speaking part of the island of Hispaniola (modern Haiti). By early 1784 societies outside Paris numbered some two hundred members (Vinchon 1936, p. 86). The harmony of this system of societies was not to last long, however. In August 1784 Mesmer went to Lyon to demonstrate animal magnetism to the members of the local Society of Harmony. In his absence, Bergasse decided to open up some of his lectures at the Paris society to nonmembers. Discovering this on his return, Mesmer was infuriated; he felt that Bergasse's action was a direct violation of their agreement about Mesmer's control of the dissemination of the doctrine of animal magnetism. Bergasse interpreted their contract differently and was genuinely surprised by Mesmer's reaction. He considered that the original obligations for Mesmer's financial welfare had been properly discharged by the payment of the fees of the first one hundred subscribers directly to Mesmer. That done, Bergasse believed that the society could move on from being the holder of a secret under Mesmer's control to becoming a teacher of the world. Bergasse accused Mesmer of greed and argued that he had no further right to control the propagation of his doctrine. In this position Bergasse was supported by Kornmann, Chastellux, and other influential members of the Paris society.

Mesmer was unmovable. As far as he was concerned, Bergasse was a traitor. With the D'Eslon rift still on his mind, and having only recently received the condemning report of the Franklin commission, Mesmer was attempting to protect what little security he believed he had left—control over the teaching of animal magnetism. As he had done with D'Eslon, he gave the impression that he had not yet revealed everything there was to know about animal magnetism and that without him instruction would necessarily be incomplete.

Was Mesmer acting in good faith? Was there really something that only he could provide in the dissemination of animal magnetism? Writing in early 1785, Mesmer continued to publicly deny a "secret," but he insisted that a special teaching was needed: "There are always complaints about a secret that I am keeping regarding my discovery. This seems to me to be mistaken. All the societies that are today the depositories [of the discovery] have never refused to communicate with those who can make good use of it. If I have ever published anything on this point it is that I am very convinced that it is only through teaching and experience that every kind of prejudice that opposes the progress of my views can be overcome. I do not simply have speculative truths to make known; I also have a very delicate practical application to develop, one that demands from those who would become involved in it a new kind of education" (Letter to the editors of the *Journal de Paris,* January 16, 1785, in Amadou 1971, p. 261).

The implication was, of course, that Mesmer's unique skill was essential to any full teaching of animal magnetism. Whether this was a defensible position for someone who claimed to be teaching an art rather than a science, or whether it was based on an unreasonable desire to maintain complete personal control over the propagation of animal magnetism, remains debatable. In any case, Bergasse and his supporters no longer believed they needed Mesmer. They felt they had the full doctrine and knew all that needed to be known about the technique. They could not be forced to comply by threats. With both sides entrenched in their positions, the Society of Harmony of Paris split in two. Bergasse himself announced the rift and presented his side of the controversy to the public in *Observations de M. Bergasse sur un écrit du docteur Mesmer* (1785b; see also Eprémesnil 1785; Mesmer 1785). The new group, centered around Bergasse and Kornmann, began to meet at Kornmann's Paris home, while the Mesmer group continued at their meeting place, the Hôtel Coigny.

To make matters worse for Mesmer, in 1785 a former student of D'Eslon and a member of the Paris Society of Harmony, the physician Caullet de Veumorel, published a compilation of "class notes" taken from Mesmer's lectures. Titled *Aphorismes de M. Mesmer,* the work was a veritable gold mine of information about the theory and practice of animal magnetism. It became very popular and appeared in many editions. Mesmer, needless to say, denounced the *Aphorismes* as "full of grave faults. It contains distorted passages susceptible of false interpretations. . . . My ideas are turned around, my words are transposed, their meaning is altered. Therefore, Sirs, I disavow it" (Letter to the editors of the *Journal de Paris,* January 4, 1785, in Amadou 1971, p. 259).

By the middle of 1785 Mesmer, greatly discouraged, once again thought of leaving Paris. This time there was no letter from the queen to dissuade him, no outcry from throngs of patients to deter him. So Mesmer departed Paris in 1785. In 1786 he toured the Societies of Harmony in the provinces, and 1787 found him in Germany and Switzerland. He eventually traveled to Vienna in 1793 and returned for a stay in Paris in 1798. He made his final home in Meersburg on Lake Constance. Mesmer would write two more works on animal magnetism (1799, 1812) and with Karl Christian Wolfart produce a final summary of his theory (Mesmer 1814), but from 1785 to his death in 1815 he was no longer the central figure in animal magnetism.

Animal magnetism itself was far from vanquished by the events of 1784 and 1785. It still had many devoted practitioners and supporters in France—whether followers of Mesmer, D'Eslon, or Bergasse—and their successes in treating illness were a stronger factor than the theoretical considerations of the reports. But there was another factor, beyond the loyalty and devotion of the magnetizers, that now entered the scene. It was in the year 1784 that the marquis de Puységur made a discovery about animal magnetism that revolutionized magnetic practice and ensured its continued acceptance.

Chapter 3

Puységur and the Discovery of Magnetic Sleep

On a spring evening in 1784 Armand Marie Jacques de Chastenet, marquis de Puységur, entered the dwelling of Victor Race, one of the peasants of his estate, who was suffering from congestion in his lungs and a fever. Puységur began to magnetize the young man and after seven or eight minutes, to his great surprise, Victor fell peacefully asleep in his arms. Puységur soon discovered, however, that Victor had not fallen into a normal sleep but had slipped into an unusual state of consciousness; he was awake while asleep. In *Mémoires pour servir à l'histoire et à l'établissement du magnétisme animal* (1784) Puységur described his discovery:

> He spoke, occupying himself out loud with his affairs. When I realized that his ideas might affect him disagreeably, I stopped them and tried to inspire more pleasant ones. He then became calm— imagining himself shooting a prize, dancing at a party, etc. . . . I nourished these ideas in him and in this way I made him move around a lot in his chair, as if dancing to a tune; while mentally singing it, I made him repeat it out loud. In this way I caused the sick man from that day on to sweat profusely. After one hour of crisis I calmed him and left the room. He was given something to drink, and having had bread and bouillon brought to him, I made him eat some soup that very same evening—something he had not been able to do for five days. He slept all that night through. The next day, no longer remembering my visit of the evening before, he told me how much better he felt. (pp. 28–29)

Puységur immediately realized that he had stumbled across something important—but at the time he could hardly imagine just how important. His background had not prepared him for the adventure upon which he was about to embark. Puységur (1751–1825) was an artillery officer, a colonel of the regiment of Strasbourg, and a member of an old and distinguished family.[1] He had inherited a large property at Buzancy near Soissons, where

---

1. For biographical information about the marquis de Puységur, see Ellenberger (1970, pp. 70–76) and Lapassade and Pédelahore (1986, pp. i-xlv [preface], i-xxiii [epilogue]).

he spent most of his time looking after his land and occasionally carrying out experiments with electricity. Having heard about animal magnetism from his brother Antoine Hyacinthe,[2] Puységur went to Paris to be trained by Mesmer at his Society of Harmony.

After learning the basics, he took his freshly acquired skills to the provinces and began a series of animal-magnetic experiments that would change the course of the history of psychiatry and psychology. His first experiment involved the daughter of his estate manager, who was suffering from a toothache. After ten minutes of magnetizing, her pain disappeared. Puységur next treated the wife of his watchman, who was also suffering from a toothache. She too was quickly cured. Buoyed by his success, Puységur was looking to address a more serious illness when the opportunity arose to tend to the twenty-three-year-old Victor.

Puységur set forth the basic characteristics of the hitherto undefined condition he had observed in Victor, which he called "magnetic somnambulism" or "magnetic sleep": a sleep-waking kind of consciousness, a "rapport" or special connection with the magnetizer, suggestibility, and amnesia in the waking state for events in the magnetized state.[3] A little later in his account Puységur mentioned a fifth characteristic of magnetic sleep—a notable alteration in personality: "When [Victor] is in a magnetized state, he is no longer a naive peasant who can barely speak a sentence. He is someone whom I do not know how to name" (p. 35).

These five characteristics of magnetic sleep, along with paranormal phenomena (mental communication and clairvoyance), turn up again and again in the literature of the next hundred years. Their importance warrants a more detailed examination.[4]

2. Antoine Hyacinthe, a naval officer, had the distinction of introducing animal magnetism to the native population of Santo Domingo (modern Dominican Republic), combining it with native voodoo techniques (Lapassade and Pédelahore 1986, p. ii). This appears to have been the first use of animal magnetism in the Western Hemisphere.

3. Puységur gave this newly discovered state various names: a "crisis" or "magnetic crisis" (1784, pp. 28, 36), "peaceful sleep" (p. 30), the "magnetic state" (p. 35), "magnetic sleep" (p. 180), the "state of somnambulism" (p. 25), and "magnetic somnambulism" (pp. 193, 230). He used the verb "to touch" as the equivalent of "to magnetize" (p. 196), and to designate the person in the state of magnetic sleep he used "somnambulist" (p. 187), "somnambule" (p. 186), and "magnetic being" ("être magnétique," p. 187).

4. Over a period of three decades Puységur developed a detailed theory of magnetic sleep and its characteristics, which he outlined in several works (Puységur 1784, 1785, 1811, 1812, 1813, 1820 [1807]). In his first works, Puységur did not give a close description of his magnetic technique. For that we must go to his Recherches, expériences, et observations (1811, pp. 14 ff) and especially his Du magnétisme animal (1820, pp. 161–169), where the technique is presented in question-and-answer format:

Characteristics of Magnetic Sleep
Sleep-Walking Consciousness

Puységur could not help but note the striking similarity between Victor's magnetically induced condition and the naturally occurring state of "sleep-walking," or "somnambulism." In fact, he came to believe that the induction of magnetic sleep was simply a way of mobilizing and controlling natural somnambulism. In one of his works Puységur compared the two states, concluding that they were the same in their essential nature but different in the way they were produced and in the matter of "rapport" (1811, pp. 73ff).

For Puységur, somnambulism involves a special state of consciousness that is neither sleep nor waking. The subject is awake while sleeping and capable of carrying out ordinary human activities. Both magnetic and natural somnambulism involve a "sleep of the exterior senses" (1811, p. 76). Somnambulists act intelligently but manifest an apparent disregard of what is going on around them. They may speak, drink, eat, and move around; they may read, write, distinguish colors, and carry out various other mental activities. Indeed, both natural and magnetic somnambulists seem to be capable of per-

---

Think of yourself as a magnet, with your arms and especially your hands as the two poles. Touch the patient by placing one hand on his back and the other, in opposition, on his stomach. Then imagine magnetic fluid circulating from one hand to the other, passing through the body of the patient.

Question: Should one vary this position?

Answer: Yes, you can place one hand on the head without moving the other, always continuing to maintain the same attention and having the same will to do good. . . .

Question: What is an indication that a patient is susceptible to somnambulism?

Answer: When, during magnetization of a patient, one notices that he experiences a numbness or light spasms accompanied by nervous shaking. Then if the eyes close, you should lightly rub them and the two eyebrows with your thumbs to prevent blinking. Sometimes it is not necessary to rub the eyes. A little action performed at a small distance is all that is needed. . . .

Question: Are there different degrees of somnambulism?

Answer: Yes. Sometimes you only produce simple drowsiness in the patient. Sometimes the effect of magnetism is to cause the eyes to close so that the patient cannot open them; if he is aware of everything around him, he is not completely in the magnetic state. This state of demi-crisis is very common. . . .

Question: How does one bring a patient out of the magnetic state?

Answer: When you magnetized him, your goal was to put him to sleep, and you accomplished that solely by the act of your will. So now it is by an act of your will that you awaken him.

Question: You mean you only need to will him to open his eyes in order to awaken him?

Answer: This is the principal operation. After that, in order to connect your idea to its object, you might lightly rub his eyes, while willing that he open them; and the effect never fails to occur.

forming intellectual tasks beyond the sleeper's usual abilities. This is apparently due in part to an extraordinary concentration of attention that also makes them largely unresponsive to stimuli from their environment.

Natural somnambulism, said Puységur, occurs spontaneously, arising from a state of normal sleep. Magnetic somnambulism, on the other hand, is artificially induced. The induction of magnetic somnambulism by an operator results in a "rapport," a state of special connection between magnetized and magnetizer. Whereas natural somnambulists do not respond to any attempts to communicate with them, magnetic somnambulists have a profound and immediate communication with the magnetizer. For this reason, Puységur called the former "independent somnambulists" and the latter "subordinated somnambulists" (1811, p. 83).

Rapport and Suggestibility

One of the things Puységur noted first about the magnetic situation was that he and Victor seemed to have a direct connection between their nervous systems (1785, pp. 2–3). "In this state, the ill person enters into a very intimate rapport with the magnetizer, one could almost say becomes a part of the magnetizer" (p. 17). Puységur conceived this becoming a part of the magnetizer quite literally, explaining that the magnetizer can cause the magnetized to perform specific actions by a simple act of will. The magnetized person, he believed, is functionally inseparable from the magnetizer. Just as our body executes our will (as when I will my hand to pick up a book), when the magnetizer wills something, the magnetized person executes that command (1785, pp. 17–20).

In Puységur's view, rapport causes the somnambulist to respond to and obey only the magnetizer and find the approach of anyone else very uncomfortable (1784, pp. 205–6). In fact, if anyone else touches the somnambulist, he or she will awaken (p. 52). The magnetizer, however, has the power to pass this rapport on to others, so that the magnetized person will also experience a special connection with them (p. 206).

Rapport was considered an essential feature of magnetic sleep. In fact, Puységur believed that the presence of its principal characteristics (hearing and obeying only the magnetizer, reacting with revulsion to anyone else) was the surest indication that a person was in the somnambulistic state (1784, p. 206). Puységur found that his somnambulists experienced the strongest dependence and rapport when they were most ill. As they progressed toward good health, the strength of rapport gradually decreased. He believed that this variability was so consistent that the degree of rapport could be used as an index of the degree of illness (1784, p. 206; see also 1820, pp. 2–3). In fact, susceptibility to being magnetized in general usually decreased as the individual regained health. Victor Race himself predicted that, with his cure, he would no longer be magnetizable (1784, p. 226), a prediction that proved to be true (p. 228).

The control of magnetizer over magnetized was considerable. The magnetizer could make the magnetized person move about, sing, carry things from one place to another, and the like (1784, pp. 181–82). Puységur discovered this suggestibility the first time he came across magnetic sleep, by persuading Victor to imagine that he was engaged in a number of fantasy situations suggested by Puységur and to conduct himself as though those situations were real (p. 28).

Lack of Memory and Divided Consciousness

Following his first magnetic treatment, Victor Race could remember nothing of what had taken place during the session. Puységur discovered that this inability to recall typically follows magnetic sleep, and that it persists despite assiduous attempts "to tie their ideas together in their passing from one state to the other." This brought him to conclude that "the demarcation is so great that one must regard these two states as two different existences" (p. 90).

Puységur noted that there was a continuity of memory within the individual in the state of magnetic sleep. Whereas the waking person can remember nothing of the magnetic state, the somnambulist remembers both the waking state and all that has occurred in previous magnetic states. The somnambulistic memory chain is quite separate from the memory chain of the waking person, which is limited to things that happen in normal consciousness.

It was a peculiarity of the magnetic condition that the subject, while somnambulistic, spoke of himself in the waking state with a certain detachment, as though speaking of another person. It is therefore not surprising that Puységur called these two states "two different existences." With this comment, he presented the germ of an idea that would be developed into the notion of "divided consciousness" or "double consciousness."

The "two different existences" were also characterized by a striking contrast in personality traits.[5] As mentioned, Puységur said of Victor that in the

---

5. The contrast in an individual's personality characteristics between the somnambulistic state and the waking state had been previously noted in the spontaneously occurring somnambulism of the hysteric. Petetin (1787) refers to a case reported by Sauvage in 1742: a girl of very timid disposition who spoke during attacks of cataleptic hysteria "with a vivacity and spirit one did not see in her outside this state" (pp. 72, 74). Ellenberger (1970) explains the radical alteration in personality of Victor (and others) in the somnambulistic state by referring to the "peculiar relationship" between nobleman and peasant at the time. Wishing, says Ellenberger,

magnetized state he was no longer a naive peasant who could barely speak a sentence. The transformation of personality was so great and Victor's wisdom so augmented that Puységur found himself relying on the magnetized Victor for advice in his healing work:

> I continue to make use of the wonderful power that I owe to M. Mesmer, and I bless him for it every day, because it makes me useful and I am able to help and benefit many ill people in the neighborhood. . . . I have only one regret—not being able to touch [magnetize] everyone. But my man [Victor Race], or perhaps I should say my intelligence, calms me. He is teaching me the conduct I must follow. According to him, it is not necessary that I touch everyone. One look, one gesture, one feeling of good will is enough. And it is a peasant, the narrowest and most limited in this locality, that teaches me this. When he is in the crisis, I know no one as profound, prudent, or clear-sighted. (1784, pp. 32–33)

The double memory chain and alteration of personality in magnetic sleep created the impression of a striking discontinuity between the waking state and the state of magnetic sleep. It was as if the magnetic subject lived in one world when somnambulistic and another when awake. From another point of view, it was as if the individual possessed two selves—the magnetic self and the waking self—which were difficult to merge.

Paranormal Phenomena

From his very first experience with magnetic sleep, Puységur believed he had witnessed certain phenomena that would today be called parapsychological: mental communication and clairvoyance. Because of the lack of sufficient scientific control of the magnetic situation, it is impossible to know whether his evaluation of the facts was valid. The discussion here must be limited to examining Puységur's experiences as he himself described them, with no attempt to judge whether genuine paranormal phenomena occurred or not.

I have already noted Puységur's description of the mental commands he gave to Victor in his first magnetic session. Later he says of Victor: "I do not need to speak to him. I think in his presence, and he hears me and answers me.

---

to be like his master (identify with him), the personality that emerged in magnetic sleep imitated him, displaying more intellectual brilliance and less inhibition (pp. 189–191). Although this explanation may throw light on certain aspects of the phenomenon (such as the use of a more correct grammar), it does not fully account for the dramatic change in mental acuity. As we shall see, radical changes in personality in somnambulism are found throughout the history of animal magnetism and early hypnotism, even when there is no significant social difference between magnetizer and magnetized, and when the somnambulist is of the highest social class.

When someone comes into the room, he sees them if I want him to; he speaks to them, saying things that I want him to say, not always what I dictate to him, but whatever truth demands. When he wants to say more than I believe prudent for the listener, I stop his ideas, his sentences in the middle of a word and totally change his thought" (1784, pp. 35–36; see also pp. 163–65).

But, according to Puységur, somnambulists are capable of much more than just picking up the thoughts and mental commands of their magnetists; they are able to perceive objects and conditions not available to the senses. Puységur held that somnambulists possess this ability in the form of a special sensation, a "sixth sense," which is activated during magnetic sleep (p. 90).[6] This sixth sense enables the somnambulist to carry out four important activities: to diagnose the illness from which he or she suffers; to diagnose the illnesses of others;[7] to prescribe treatment for one's own illnesses and those of others; and to predict the course of the illness and time of cure, both in oneself and in others (pp. 34, 73–74, 87–90, 99–100, 109, 111–13, 137–38, 147, 160, 166, 168, 180). Puységur compared the experience of the sixth sense in a somnambulist to giving sight to a man born blind (pp. 90–91). In both cases the experience produces an entirely new and surprising view of things. The only difference is that the blind person retains his new-found ability whereas the somnambulist loses all memory of it when awake.

Puységur was inclined to attribute the puzzling phenomenon of amnesia on awakening to the operation of the sixth sense: "I have pointed out . . . that in the magnetic state [somnambulists] have both the idea and memory of everything that happens to them in the natural state, whereas in the latter state they do not remember anything that happens to them while in the magnetic state. This confirms . . . the existence of an additional sensation in the latter state. With six senses (if one can put it that way) they can recall the sensations gained through the five senses, but with five senses, they cannot remember ideas formed with six" (p. 90).

6. The view that animal magnetism involves a sixth sense was first put forward by Mesmer in 1781. He described it, however, as an additional faculty of sensation available to the magnetizer as well as the magnetized: "Animal magnetism should be considered in my hands as a sixth artificial sense. The senses do not define or describe themselves; they sense themselves. One would try in vain to explain the theory of colors to a blind man. A person must be able to see them—that is, sense them—to understand. It is the same with animal magnetism. It should first of all be conveyed by sensing. Sensing alone makes the theory intelligible. For example, one of my patients, accustomed to experiencing the effects that I produce, has an aptitude for understanding me greater than anyone else" (pp. 24–25).

7. Puységur (1784) noted that some magnetizers could sense the seat of the disease in their patients, although he himself could not (p. 181). He described this ability to "sense and pre-sense" the nature and course of the illness and lamented that if he had not developed this ability, it was not for lack of study (1785, pp. 114–15).

Puységur believed that the degree of access to the sixth sense, like the ability to be magnetized, is related to the degree of illness. That is, only those who were themselves ill were capable of using that faculty to diagnose and prescribe for the sick. Puységur equated the sixth sense with "clairvoyance" (p. 33) or "clear seeing," a capacity most often employed to discover the seat of disease in the somnambulist or other afflicted persons. Because he believed the somnambulist's clairvoyant diagnosis to be accurate, Puységur was in the habit of using somnambulists to aid him in determining the illness and remedy for those who came to him for cure (pp. 136, 138, 147–48). Although he was enthusiastic about the potential for good in this sixth sense, he believed that it was a part of nature and limited in its range, existing only to aid nature's healing power (pp. 186–87).

Another seemingly paranormal aspect of Puységur's practice was his acceptance of the reality of magnetization at a distance. He agreed with Mesmer that the magnetic fluid was not impeded by physical obstacles, so that the practitioner could magnetize through walls and from one dwelling to another. Yet Puységur was not ready to accept the testimony of others on the matter: "This is the kind of thing that it is impossible to prove by rational arguments and for which experience alone provides certainty. For that reason, it is to men who are aware of this small part of their power that I now direct some recommendations about the best way to use it" (1785, pp. 112–13). In this connection, Puységur discussed a subject that would be seriously investigated only many decades later: the induction of artificial somnambulism at a distance.[8] After taking up the possible detrimental effects of this activity, he continued: "Apart from this inconvenience, there is another one very much to be feared—the risk that some extraneous factor will interfere with the effect produced at a distance. If, for example, the effect one produces is somnambulism, one must know very well how susceptible this peaceful state is to being disturbed by the least extraneous circumstance, which can then cause truly miserable confusion" (1785, p. 113).[9]

8. Dupotet (1826) attempted to produce somnambulism at a distance in more controlled conditions in 1820, and in the mid-1880s Janet (1886a, 1886b), Richet (1886), Héricourt (1886), and others described their experiments with long-distance somnambulistic induction.

9. Early in his practice of magnetic healing, Puységur noted a curious phenomenon: sometimes his patients would fall into magnetic sleep before arriving at his house. This happened so often with patients he was treating that he hardly gave it special attention. On one occasion, however, it happened to a man who had never been magnetized. This experience gave Puységur enough pause to ponder the cause of the phenomenon. He explained it to himself in this way: "From the moment Thuillier [the patient] decided to be on his way to get magnetized, he was already entering into the beginning of the action which would naturally terminate in somnambulism" (1785, p. 83).

Psychotherapy

While treating Victor Race, Puységur made the very first attempt to use magnetic sleep as an adjunct to psychotherapy:

> This man has an internal trouble. This trouble is caused by his sister, with whom he lives and who is contesting an endowment left him by his mother. This sister is the most spiteful woman of the district, and she enrages him night and day. I learned all of the details from him without his having the slightest memory that he had given me this knowledge. I have tried to get him to absorb the comforting idea of lightening the burden of looking after his affairs in order to clear up his problems. This morning a woman came to his house while I was magnetizing him. I wanted him to know that this woman was there and that she felt friendly towards him. He said hello, and after that he said to her, "Angelique, dare I ask you to do something that would greatly please me?" "Gladly," [she said] (I told this woman to answer him just as she would in his normal state). [Victor:] "Monsieur has good feelings for me. He comes to see me and takes care of my health; he knows well that I am very troubled." [Angelique:] "Yes, he knows it and will try to ease it." [Victor:] "Ah! what goodness! It's my sister who is causing it, you know, Angelique." [Angelique:] "Have patience, it will all turn out well." [Victor:] "Angelique?" [Angelique:] "Yes?" [Victor:] "I would greatly like to put something in the hands of Monsieur. Would you bring it to him, because I would never dare to take this liberty myself." [Angelique:] "What is it?" [Victor:] "You'll find it in my wardrobe, in a drawer under a large paper folded like this (he motioned to her). It's a deed of gift of this house that my mother gave me *inter vivos* in order to reward me for the care I gave her in her old age." Angelique looked in the armoire, found a parchment just as he had said, and showed it to him, asking him if that was what he wanted to give me (you will note that he always had his eyes closed, something I regularly try to maintain during a crisis, in order not to tire the eyesight). He replied in the affirmative, giving over to her the secret regarding his sister, who would surely have burned the paper if she knew that he had it in his possession. He urged Angelique to bring it to me, etc. . . . . I took the deed of gift from her hands, and barely had it in my pocket when I saw this man's face take on an expression of serenity, an air of jubilation. I left a few minutes later with the usual precautions, and I have not told him since then what he had done. (1784, pp. 36–38)

Puységur came across Victor the next day in a state of depression because he could not find the deed of gift he kept in his wardrobe. Amnesia had completely wiped the incident of the previous day from his memory. Puységur told Victor what had happened and restored him to a good frame of mind.

Puységur's account contains certain important features that will turn up much later, when psychotherapy and artificial somnambulism are effectively combined to treat forms of hysteria. First, the work is based on the intimate rapport established between magnetizer and magnetized. This rapport brings with it a trust of the magnetizer that is childlike in its intensity and extent. Second, the psychotherapy is carried out through the revelation of information or emotional attitudes unavailable to the normal waking consciousness of the patient. Third, this is accomplished through contacting a hidden part of the individual, a consciousness different from that person's ordinary consciousness. The psychotherapeutic work done in the state of magnetic sleep is not remembered in the waking state.

Magnetic Sleep and Healing

Like Mesmer before him, Puységur used animal magnetism to heal. The magnetic treatment of his first somnambulist was aimed at healing the young man of his physical complaint, and his discovery of and experimentation with magnetic sleep did not distract him from that primary concern. Again and again he said that magnetic sleep was to be used for healing and for no other purpose.

Healing Effects of the Trance

Puységur differed strongly from Mesmer in his view of the magnetic healing crisis. Mesmer held that a violent or convulsive crisis during treatment was a sign that the healing was working, even insisting that it was necessary for cure (1799, pp. 34–36). Puységur, on the other hand, felt that such crises were not necessary for cure and could even be harmful (1784, p. 22). Victor's healing took place through the calming effect of magnetic sleep and without convulsive crisis. From this very first experience Puységur formed the opinion that the gentle magnetic crisis was the true healing crisis (p. 42).

Despite his great admiration for Mesmer, Puységur stood firm in his opposition to the induction of violent or convulsive crises in the ill. A man named Lehogais, whom Puységur taught to magnetize, had the frightening experience of seeing his magnetized patient fall into convulsions. He told Puységur about the incident, wondering what had happened. In answer, Puységur pointed out that the followers of Mesmer regularly tried to induce such convulsions, but rarely with any beneficial effect, and sometimes with dangerous ones (1784, pp. 47–48, 49).

Puységur was particularly bothered by the practice of letting patients go into convulsions and then leaving them on their own. Many who were trained by Mesmer would set up, in imitation of their master, a crisis room for patients in convulsion. Puységur believed these rooms to be contrary to both decency and health: "The crisis room, which ought rather to be called a 'hell of convulsions,' ought never to have existed. Mesmer never had them. It was only when there were so many sick people coming to him in his new quarters, and he had to divide his attentions so much, that he thought it best to have a place where, although he had to abandon his patients, they would at least not be touched by anybody, which he knew would be very harmful. It is a pity that such an unfortunate practice has resulted from what was originally dictated by humane concern." In Puységur's view, convulsive reactions, if they occur at all, should be merely passing experiences that take place while the patient is in the hands of the magnetizer. The true crisis is the "calm and tranquil state which, to the onlooker, reveals only a picture of well-being and the peaceful work of nature effecting a return to health" (pp. 97–99).

This does not mean, however, that patients do not suffer in the healing process. Some suffer a great deal, but not in the degrading manner of violent convulsions. Though their bodies suffer, their souls are at peace. In the magnetic state they perceive their suffering as necessary and the precursor of cure, exhibiting the remarkable objectivity so characteristic of the state of magnetic sleep (1784, pp. 98–99).

Suffering should not be created by induced convulsions. On the contrary, everything possible should be done to keep the patient calm. This means that the magnetizer should be in a position to provide continual care for his patient for as long as the cure takes. Puységur gently reproached Mesmer for not being able to do this because of the hectic nature of his practice in Paris, and he stated that Mesmer would not have been able to effect some of Puységur's cures precisely because of this discontinuity (p. 96). Again and again in his writings Puységur revealed that he sustained this demanding commitment to his patients, even to the extent of taking them with him on his travels rather than leaving them at critical points in their cure. Puységur believed that once the cure had started the availability of the magnetizer throughout its course was crucial; without it the patient could fall into a dangerous "disorganization" harmful to his or her health, a disorganization that required the continual intervention of the magnetizer (1784, p. 96).

Healing Aids

In treating the ill with animal magnetism, Mesmer had employed a number of instruments and aids to enhance the magnetic effects, such as the *baquet* and the magnetized tree. One of the first things Puységur did when he returned from his training in Paris was to magnetize a tree on his estate "in order to give these poor people continuing help, and in order to save my energy. . . . After attaching a rope to the tree, I tried its effectiveness on some sick people. The first patient came and as soon as he had put the rope around himself, he looked at the tree and said, with an air of surprise which cannot be duplicated, 'What do I see there?' Next his head dropped and he was in a perfect state of somnambulism" (1784, pp. 30–31; see also Cloquet 1784, pp. 7–8). The production of magnetic sleep through contact with a magnetized tree from that point on became commonplace on Puységur's estate.

Another aid devised by Mesmer to conserve the strength of the magnetizer was the "magnetic chain." This was formed by having magnetized persons hold hands and make a chain through which, it was believed, the magnetic fluid could circulate to the benefit of all. Puységur used this device frequently (see, e.g., 1784, pp. 31, 75). Most often the chain was formed around a "magnetic reservoir," a version of Mesmer's *baquet*. Puységur described in detail how to construct such a reservoir (p. 15) and made extensive use of it in his healing (pp. 75, 136–137, 146, 169).

Puységur also used "magnetized water" as a healing aid, produced by applying magnetic passes to a container of water, which was then considered to be charged with magnetic fluid. Although Puységur himself could not tell magnetized water from ordinary water and relied on people in the state of magnetic sleep to distinguish the two, he believed in its reality because of the effects he observed it having on those suffering from a variety of illnesses.

Healing and the Will

Puységur's view of the role of the will of the magnetizer was an original contribution. Like Mesmer, he believed in a universal fluid that pervades all space and vivifies all nature. This fluid is in constant motion, and modifications of that motion produce various phenomena. He considered the phenomenon of electricity to be a specific modification of the fluid in motion which gives the most palpable and direct experience of the fluid. Mineral magnetism also, in his view, presents an impressive demonstration of its reality (1784, pp. 9ff, 24–25).

To Puységur, human beings are electric animal machines. In fact, they are the most perfect electric machines possible because of their capacity for thought, which controls all the actions of the machine and alone can traverse space. Through thought and the power of the will we determine where our animal electricity should operate in our bodies. This power of thought is also the means of acting in a beneficial way on our fellow human beings: "Our electric organization is so perfect that with the help of the will alone we can produce phenomena which, while being very physical, have the air of the miraculous" (p. 13).

He described how this power of will is conveyed from magnetizer to subject: "The ill person in this state [of magnetic somnambulism] enters into a very intimate rapport with his magnetizer, so that one could say he becomes a part of him. So when [the magnetizer] wants to move a magnetic being [a somnambulist] by a simple act of the will, nothing more astonishing takes place than what happens in our ordinary actions. I will to pick up a piece of paper on the table; I order my arm and my hand to take hold of it. Since the rapport between my principal driving force—my will—and my hand is very intimate, the effect of my will is manifested so quickly that I have no need to reflect on the operation" (1785, p. 17). In the same way, the intimate rapport of animal magnetism establishes a connection so close and so immediate that the will of the magnetizer is instantly carried out by the magnetized. Since magnetic rapport makes the somnambulist a part of the magnetizer, much as his hand is a part of him, there should be no surprise if by a mere act of will the magnetizer can direct the actions of the somnambulist as he chooses (pp. 17–20, 229–30).

Puységur came to believe that all the effects of animal magnetism are directed by the will of the magnetizer. He eventually formulated his view in his treatise *Du magnétisme animal* (1820):

> If one touches an ill person without intention or attention, one effects neither good nor bad. . . . There is only one way always to magnetize usefully: that is strongly and constantly to will the good and the benefit of the ill person, and never to change or vary the will. . . . The magnetic action is directed and sustained by a firm will to relieve the sufferings of the ill person. . . . The compassion which an ill person inspires in me produces a desire or thought to be useful to him. And from the moment I make up my mind to try to help him, his vital principle receives the impression of the action of my will. (pp. 153, 155, 159)

Puységur believed that one could not explain the action of animal magnetism through a purely physicalist theory. Any true theory must take into account that the will exists and that it can direct our "vital principle." And since will is beyond matter, there must be a nonmaterial principle operating at the heart of human action and therefore at the heart of magnetic healing. In this way, Puységur believed, animal magnetic healing refuted materialism:

> The direct communication of the will to the vital principle is . . . no longer in doubt for us. . . . So if a man (in perfect health, as we have said) possesses within him a most fecund source of movement and the best possible conductor to carry his beneficent electricity to his fellow men, it is to him alone that we should look to find the greatest help for illnesses. It is through his nervous electricity that he operates successfully on others. And the science of putting this electricity into action is, properly speaking, that designated by the name of animal magnetism. (1785, pp. 62–63)

Puységur's belief in the power of the magnetizer's will over the somnambulist and the great trust the patient places in the magnetizer led him to speak repeatedly of the need for the magnetizer to have "good will" in doing his healing work. Speaking of his work with Victor, Puységur wrote: "Here a man is forced to give me a document, his most precious possession—and this took place because I had well and strongly desired to obtain the means for making him happy. . . . I do not know if one can will evil as powerfully as good. If this were so, should not one fear the effects of animal magnetism in the hands of dishonest men?" (1784, p. 39).

It is clear that Puységur set high standards for the practice of animal magnetism. In the conclusion of his *Suite des mémoires* (1785) he set out the qualities he considered essential for magnetic healers: perseverance, sensibility, and good will. "By believing, every man acquires the faculty of healing his fellow men just as he acquires the faculty to reproduce. These two faculties are the result of pity and love, two sentiments that are very compelling and certainly common to all men. . . . Without love, there is no reproduction; without friendship, there is no consolation in our miseries; and without sensibility, there is no sure healing of our ills. These three attributes of man are the sole source of our existence and every beneficent effect flows from them" (pp. 214–16).

Puységur and the Splitting of the Magnetic Traditions

Mesmer's approach to healing and his healing theory were physically oriented. His explanation of the phenomena of animal magnetism was consistently formulated in terms of matter and motion, and he believed that every aspect of animal magnetism could sooner or later be verified through physical experimentation and research. With his emphasis on magnetic sleep (the "gentle crisis"), magnetic rapport, and the place of the will in magnetic healing, Puységur turned animal magnetism in a new and clearly psychological direction. He explicitly opposed a materialist philosophy of nature and believed that the phenomena of animal magnetism provided strong evidence against it.

Mesmer emphasized the importance of understanding the nature of the "magnetic fluid" in order to grasp the essence of healing through animal magnetism. Puységur, while not denying the existence of the fluid, thought knowledge of it irrelevant to magnetic healing: "I do not know any longer if there is a magnetic fluid, an electric fluid, a luminous fluid, etc. I am only sure and certain that to magnetize well it is absolutely useless to know whether a single fluid exists or not" (1820, pp. 155–56).

To say that Mesmer's orientation was physicalist is not to deny him his due on the level of psychology. From certain points of view one might see Mesmer as a master psychologist. His instincts told him he had to pay attention to the psychology of his patients and enhance his treatment with drama where possible. He certainly knew how to create a mood of mystery and expectation of cure in his healing salons, how to orchestrate the patient's environment to produce a state of great suggestibility, and how to employ paraphernalia and costume to focus the patient's attention on himself as master healer.

But Mesmer was not the one to develop the potential of his system to explore the human psyche. First of all, Mesmer's psychological canniness was more instinctual than thought-out. He exercised it on the level of his magnetic practice but never attempted a theoretical discussion of this aspect of his technique. Second, Mesmer's psychological canniness had a very narrow focus: to create an atmosphere of confidence and heightened expectation of cure. He did not look more deeply into the minds of his patients to see why this was important, what it might indicate about the nature of human psychological functioning, and what further potentials for psychological healing his system might reveal. Third, Mesmer did not recognize the significance of magnetic somnambulism. This was a crucial oversight, undoubtedly produced by a flaw in Mesmer's own psychology: since he was not the first to point out the importance of magnetic somnambulism, he could not bring himself to give it its due.

With the publication of the *Mémoires* in 1784, Puységur split the magnetic tradition in two. Many continued to follow Mesmer's approach to the letter, ignoring or downplaying Puységur's findings on magnetic sleep. Others received Puységur's discoveries enthusiastically and began to carry out research of their own in the same psychological direction. These followers of Puységur also tended to maintain his emphasis on good will and the intention to benefit the client as crucial to the proper practice of animal magnetism.

Puységur himself never expressed the feeling that he was at odds with his teacher. He constantly praised Mesmer for his discovery and said that his own work was merely a continuation of Mesmer's, augmenting and developing it to make it even more effective as a treatment for illness. Mesmer, from his side, never publicly acknowledged the importance of the discovery of magnetic sleep. He had to respond to the keen interest in artificial somnambulism provoked by Puységur, but his response was reluctant and unenthusiastic. Although after 1784 Puységur and his findings were well known all over France and beyond, Mesmer made no mention of Puységur in any of his writings up to his death in 1815.

The literature on animal magnetism from 1784 to 1815 shows quite clearly the split into two orientations: the physicalist direction of Mesmer and the psychological one of Puységur. But at this early stage the two orientations embodied a difference in emphasis more than an incompatibility of theory.[10]

10. Forms of the fluidic theory persisted side by side with the psychological theory throughout the nineteenth century (Ellenberger 1970, p. 148). The views of Charcot on the nature of hypnotism are a revised form of the physicalist-fluidic theory (McGuire 1984, pp. 12ff).

Chapter 4

Early Developments in France

Puységur's discovery of magnetic sleep had produced reverberations throughout France and beyond. Several people immediately recognized that something of great significance had been unearthed and began experimentation of their own. Mesmer himself also theorized about magnetic sleep, but with reluctance; he was worried that the extraordinary phenomena associated with magnetic somnambulism (such as thought reading and clairvoyance) would paint animal magnetism with the brush of occultism. His fears were not without substance.

Fournel

Jean François Fournel (1745–1820) attempted, in *Essai sur les probabilités du somnambulisme magnétique* (1785), to determine whether the phenomenon of artificial somnambulism was worthy of investigation. He concluded that it was.

Fournel defined magnetic sleep as a state midway between waking and sleeping that participates in both and produces a number of phenomena found in neither. Even though the phenomenon had only recently been brought to light, Fournel estimated that there were already some six thousand somnambulists in Paris and the provinces and that approximately five hundred persons had witnessed Puységur's work at Buzancy in the space of a few short months. In any case, by 1785 magnetic sleep had become something of a craze. Fournel feared that by producing a superficial imitation of the symptoms, false somnambulists were beginning to discredit the genuine phenomenon.

Fournel noted certain characteristics consistently found in magnetic sleep. First was an "intimate rapport" between the magnetizer and the ill person, so powerful that the somnambulist could read the magnetizer's thoughts.[1] This rapport could be communicated from one somnambulist to another by simple contact. Another characteristic was a change in the sen-

---

1. The term is identical to Puységur's and was most likely borrowed from his *Suite de mémoires* (1785).

sitivity of the sense organs. In some cases, Fournel said, the senses were dulled or even deadened, while in others they were heightened. In this way hearing or sight could be blocked, or the sense of touch could attain an unusual subtlety or be totally inoperative. The somnambulist also showed signs of possessing a "sixth sense" that surpassed the bounds of normal sensation, enabling him or her, for example, to see through blindfolded eyes (pp. 3, 45).

Fournel mentioned other characteristics noted previously by Puységur: somnambulists would experience hallucinations suggested by the magnetizer (pp. 48–49); in the magnetic state, somnambulists would reveal intimate secrets that they would not otherwise speak about (p. 63); and on their return to their normal waking state, somnambulists were subject to forgetfulness or amnesia (p. 69). Fournel asked whether these characteristics were genuine or might be produced by fraud on the part of the somnambulist (pp. 3ff). Given the consistency of the phenomenon and its widespread occurrence, the theory of fraud could be supported only if there were a broadly based conspiracy to deceive. But since the phenomenon occurred in every class and in people of every level of education, that hypothesis was unacceptable (pp. 13–14, 17). "It requires less effort for me to conceive [of magnetic somnambulism as] a natural phenomenon which, after all, is susceptible of explanation, than to conceive of a conspiracy to deceive so lacking in purpose or motive, so complicated in its scope, and so impractical in its execution" (p. 9).

Fournel based his contention that magnetic somnambulism was "susceptible of explanation" largely on the ground that the phenomenon, with all its characteristics, already exists in nature in the form of sleep. Somnambulism, he wrote, "is nothing other than a modification of sleep, and there is no somnambulism without sleep" (p. 32). There are two kinds of sleep: perfect and imperfect. In perfect sleep all sensations are reduced to the point that the only signs of life we perceive are respiration and pulse. In imperfect sleep there remains some access to the use of the external organs of the body. Even in the most profound sleep, the sleeper retains a degree of wakefulness and can execute some movements, such as changing position or pulling up the bedclothes. This state would have to be called somnambulism, since there is sleep accompanied by purposeful movement. It had long been known that many people regularly go beyond these simple movements in their somnambulism, to the point of being able to talk, work, and even carry out complicated intellectual tasks while asleep. Fournel emphasized that if this kind of somnambulism is accepted as a natural phenomenon, the somnambulism of magnetic sleep should not strain credulity (pp. 32–34). He concluded that there is no sleep without somnambulism and that all sleep is a kind of somnambulism.

It is a very small step from there to the belief that somnambulism can be induced through artificial means, just as sleep can be encouraged through reading or hot drinks (pp. 26–30, 35). That, said Fournel, is precisely what happens when one uses animal magnetism to produce magnetic sleep. Somnambulism induced through magnetic passes is no different from natural somnambulism; and, Fournel asserted, close study of the medical literature concerning natural somnambulism would reveal all the remarkable phenomena currently causing such a stir when observed in magnetic somnambulists (pp. 35–36). Not one of the phenomena of magnetic somnambulism, he claimed, is omitted from the descriptions of natural somnambulism (pp. 37–70). From this he concluded that objections to accepting the reality of the phenomena of magnetic somnambulism do not hold up:

> To listen to the violent public denunciations against magnetic somnambulism, presented as a miserable deceit unworthy of credence, we are supposed to believe that the phenomena are without precedent and that they are found only in magnetic somnambulism. This opinion is established in the public, even in the sane part composed of people who are respected for their knowledge and their virtue, but who, not being familiar with these physiological phenomena, are compelled to think this way. These people would perhaps change their minds if they knew that there exists in nature a state identical with the state they regard as simulated. . . . It is necessary to realize that there exists a state of natural somnambulism, recognized and accepted by physicians, during which sleepers carry out acts that would be impossible for a waking person . . . [and] which are perfectly analogous to those observed in magnetic somnambulists. (pp. 37–39)

Tardy de Montravel

In 1785 another treatise on somnambulism, destined to become more widely known than Fournel's, was published by A. A. Tardy de Montravel. His *Essai sur la théorie du somnambulisme magnétique* was an attempt to construct a comprehensive theory of magnetic sleep.

Not much is known about Tardy de Montravel. His dates of birth and death are unavailable and only the initials of his first names have come down to us, but his treatise shows him to be a man of intelligence and a good observer. He knew and appreciated Fournel's work (p. 9) and had observed many somnambulists. Tardy based his theory of magnetic somnambulism primarily on his experiences with Mlle N., the first person he seriously attempted to magnetize (p. 5). This twenty-one-year-old woman brought a variety of symptoms to her treatment, including a low fever, violent coughing, frequent hemorrhaging through the nose, and extreme emaciation. When Tardy met her, her physicians had given up, not expecting her to live much longer. She already had some experience with animal magnetism, having spent about two hours each evening at the magnetic *baquet* for the previous six months (pp. 11–12).

Tardy de Montravel began to magnetize Mlle N. at the end of March 1785. In the first session the young woman fell into a somnambulistic state after forty-five minutes. This peaceful crisis replaced the convulsions she was accustomed to from her regular magnetic treatment. From then on, she was placed in the state of magnetic somnambulism each day; gradually the symptoms of her illness disappeared until she was completely cured (pp. 13, 21).

Tardy de Montravel confesses that at first he did not know what to make of magnetic somnambulism, apparently having only the descriptions of the somnambulists of Buzancy to guide him.[2] With so little to go on, he decided to gather as much information as possible from his own experiences, and particularly from the subjective descriptions provided by Mlle N. He posed questions to her in four principal areas: the cause of her illness, her interior state, the remedies she suggested, and the course the illness would take. Her answers, delivered in the state of magnetic somnambulism, were that her problem was due to the suppression of her menstrual period, which produced an interior state of agitation with all the menstrual symptoms but no flow; all she needed for the cure was to be magnetized regularly and to drink magnetized water; her cure would come on May 15 at 8:30 in the evening, when her period would resume its regularity. Everything occurred exactly as predicted (p. 15). (Later Mlle N. stated that the root cause of her irregularity was the presence of worms in her stomach, a condition that was eventually cleared up through medication.)

In the process of working with his somnambulist, Tardy de Montravel discovered that she apparently possessed the ability to diagnose the illnesses of others, prescribe remedies, and predict the course of their illnesses. She claimed to be able to see into the body of the sufferer and perceive the conditions causing illness. She put herself in rapport with ill persons by touching them and feeling a sense of harmony with their physical state. Sometimes the experience would prove to be most arduous for her, since she would frequently suffer the disturbance of the person she was trying to help (pp. 23–24).

2. Tardy de Montravel calls the material he consulted the "Journaux de Buzancy." It is unclear whether this refers to Cloquet's *Detail des cures opérées à Buzancy* or to the *Mémoires* of Puységur, but it is likely to be the latter.

When in the somnambulistic state, Mlle N. had a peculiar ability to "see" magnetic fluid emanating from Tardy de Montravel's hair in a brilliant gold. Although she described this vision as beautiful and most pleasant, she felt that if his head approached her too closely, she would become overcharged and fatigued. When Tardy de Montravel used an iron rod to magnetize her, she saw fluid radiating from it in a golden column and throwing off sparks. She saw the same emanations from his thumb and the other fingers, but in reduced quantity, with the middle finger producing no emanation at all (pp. 26ff).

Tardy de Montravel believed that Mlle N.'s perceptions indicated that the magnetic fluid was more powerful in the magnetizer than in the magnetized, and that as the cure progressed they became equal:

> I place my right thumb opposite the left thumb of my somnambulist, creating some distance between the two horizontally. She perceives the fluid emanating from both thumbs, but notes a difference: hers is less brilliant and has less vitality than mine, so that there is a difference in the portion of the column of fluid that proceeds from her thumb in comparison to mine. In the early days, the portion coming from my thumb made up about three-quarters of the distance between the two thumbs. I noticed through repeated tries that as my patient advanced towards health, her portion of the fluid moved more towards the midpoint between our thumbs, and her fluid became livelier and more brilliant (pp. 27–28; see also Tardy de Montravel 1786a, 1786b).

Tardy de Montravel also believed from observing his somnambulist that the stomach is the seat of the senses. In his view, everyone possesses an "interior sense" that is differentiated into the five exterior senses. This interior sense resides in the stomach area, specifically in the solar plexus. When any of the exterior senses receives an impression, it is conveyed to the interior sense, where perception takes place. Somnambulists, however, short-circuit this process and perceive directly with their interior sense, so that they see, hear, and smell from the stomach area. The somnambulist may believe she sees with her eyes, for example, but that is only because the visual perception happening at the stomach is communicated back to the eye through "prolongment." Tardy de Montravel believed his experiments showed that the stomach is the true organ of perception and that somnambulists "see" objects placed before the stomach area and hear sounds made in that area (1785, pp. 54–55, 90–91).[3]

3. In this matter Tardy de Montravel anticipated the work of Petetin (1787), who is usually cited as the first to claim that somnambulists hear and see in the stomach area. Petetin also

Harmonic Rapport

From his experiences with Mlle N. and other somnambulists, Tardy de Montravel developed a rather sophisticated theory of the nature of animal magnetism in general and somnambulism in particular. Like Mesmer, he began with an examination of the action of the "magnetic fluid." According to Tardy de Montravel, when one individual magnetizes another, a special connection is brought about through a harmonization of the two nervous systems involved:

> For the fluid to circulate freely from one body to another it is necessary that the organs of the two bodies be so alike and similarly disposed that they modify the fluid in a similar and analogous way. The two bodies in this case are said to be in harmony. Two human beings, having organs of the same nature, can be placed in harmony by forcing, over a certain period of time, the fluid which circulates in each of them to circulate indifferently from one to the other. . . . The nerves are the conductors of the universal fluid in the human body. . . . The nerves of the two human beings can, in this instance, be compared to chords of two musical instruments placed in the greatest possible harmony and union. When the chord is played on one instrument, a corresponding chord is created by resonance in the other instrument. (pp. 34–35)

Tardy de Montravel called this state of mutual resonance "rapport harmonique" and used this musical analogy to clarify the mode of action of animal magnetism (p. 37).

Following Mesmer, Tardy de Montravel believed that the magnetic fluid is found everywhere in nature, the principle and cause of movement and life. All healthy living things have the ability to appropriate what they need of the fluid to sustain themselves, but the state of illness prevents the free circulation of the fluid from the environment to the organism and through the nerves, causing interior stagnation. The cure for the illness must involve the restoration of that exchange. That is precisely what the application of animal magnetic techniques accomplishes. By placing the patient in harmony with the

---

believed, like Tardy de Montravel, that his somnambulists were able to perceive the insides of their own bodies. When he began working with persons suffering from "essential hysteria," he came across the same phenomena that the magnetists were discovering in their patients. His patients sometimes fell into a state of catalepsy, lying motionless and without sensation, but with the apparent ability to perceive with the "epigastrium." He believed the state of "essential hysteria" was caused by a superabundance of "electric fluid" in the head, a condition identical to that produced by the application of animal-magnetic passes (1787, Part I, pp. 56–62). Petetin's experiments with this extraordinary perceptive ability were, like those of other contemporary magnetists, lacking in the controls that are essential for scientific reliability.

magnetizer, the magnetizer forces the fluid to circulate in the patient's nervous system in the same manner that it travels through the magnetizer's, breaking down the obstacles created by the illness. When this is done repeatedly, the body can heal itself and maintain full and free circulation of the fluid on its own (pp. 34–39).

It is clear that the magnetizer must be a vital person, capable of regenerating his own magnetic fluid: "One man cannot magnetize another without prejudice to himself. Two men once placed in harmony can be compared to two branches of a siphon in which the fluid seeks its own level. A strong and healthy man, having no superabundance [of fluid], cannot furnish a weak man that portion of fluid he needs without altering his own equilibrium, his own source of health. For one man to magnetize another fruitfully, he must have some means of augmenting in himself not only the intensity of the universal fluid, but also the vitality and flow of that fluid" (pp. 39–40).

Tardy de Montravel stated that Mesmer had discovered a way for magnetizers to accelerate the current of fluid from head to extremities, a sort of self-magnetization. Using this technique (which he never defined precisely), a person could keep himself well charged with the fluid and use it to heal those around him (pp. 40–41).

The successful use of animal magnetism in healing required a knowledge of the way nature works in the body. For if the magnetizer is going to place himself in harmony with the patient and direct the fluid to the seat of the illness, he must know something about the body's structure or, lacking that, be able to discern the natural flow of fluid through the body and where that flow encounters obstacles due to illness. He should then force the fluid through those areas, thereby breaking down the obstacles (pp. 41–42). This seemed to suggest that magnetizers should have a sensitivity to magnetic fluid that Puységur claimed he himself lacked.

Magnetic Somnambulism

Following the teachings of Mesmer, Tardy de Montravel held that when animal magnetic techniques were applied to overcome the obstacles produced by illness, the organism would react with a crisis. Resulting as they do from restoration of the natural circulation of the fluid, crises are to be considered as essential to the healing process, not detrimental to the patient. The healing crisis may take the form of convulsive movements, immoderate laughter, natural sleep—or magnetic somnambulism, "the most astonishing and interesting of these crises." Tardy de Montravel called magnetic somnambulism a beneficial "type of catalepsy" provided by nature for healing. Somnambulism is an "illness," but in some cases nature itself produces it to bring about a healing of grave illness (p. 45).

Magnetic somnambulism, according to Tardy de Montravel, has its seat in the brain. When this nerve center is charged with an excess of magnetic fluid, somnambulism occurs. Since in a healthy person there is a free circulation of fluid throughout the body, no such buildup can occur. For that reason, only the ill are subject to magnetic somnambulism (pp. 45–46). When magnetic treatment is successful and patients become healthier, they are less susceptible to magnetic somnambulism.

Tardy de Montravel believed that one should find more female than male somnambulists since somnambulism is basically a condition of the nervous system, the nervous systems of women are "more irritable" than those of men, and "most of their illnesses have a rapport with the womb, which has an intimate sympathetic connection with the brain" (p. 46).

Sixth Sense

Like Puységur, Fournel, and Mesmer, Tardy de Montravel spoke of a sixth sense in the magnetic somnambulist. This "exquisite" sense operates with the other five senses "and they seem to operate through it. So that in the waking state, when a person is in his normal condition, the five senses, which it usually makes use of, suffocate the sixth sense" (pp. 46–47).

Tardy de Montravel gave the sixth sense a rather important role in the metaphysical constitution of human beings. He saw it as one of three constitutive parts: the first was the nonmaterial soul, the "intellectual man"; the second was the material body, the "machine which operates by means of the five senses"; and the third was the sixth sense, the intermediary between the other two, which determines our physical actions. The sixth sense has a dual role: it serves as the source of the physical instincts, looking after the physical functioning of the body, and as "conscience," striving for good (pp. 47–48). In this schema, perception occurs when objects make their impressions on the five senses, and these in turn act on the interior (sixth) sense, which conveys them to the soul. But it is possible for the interior sense to receive impressions directly from objects, without the intervention of the five senses, and convey them to the soul. At the same time the interior sense may affect the five senses with these impressions so that they seem to perceive the object directly when in fact the perception is taking place in the interior sense, which activates the other senses secondarily. In this way the person may "see" with eyes blindfolded and "hear" sounds produced out of earshot (pp. 48–49, 54–55).

Individuals in the state of magnetic somnambulism perceive the world through this sixth or interior sense, exhibiting the capacity to see or hear distinctly without the benefit of the five senses. According to Tardy de Montravel, the interior sense operates freely in the somnambulist because of the state of the nervous system in somnambulism. When the nerves of the ill person are saturated with magnetic fluid through magnetization, they become very irritable. This causes extreme sensitivity in the interior sense, making it far more active than normal. The interior sense can, however, operate freely only if the five senses, which are mere extensions of the interior sense, cease to be engaged with exterior objects. This takes place in the somnambulistic state, and as long as that state endures, all perception of the external world happens directly through the internal sense. Although the sixth sense is distributed throughout the body, its principal seat is in the stomach.

Tardy de Montravel compared the sixth sense to the sense of touch, which is also distributed throughout the body and is the surest and most trustworthy of the five senses. In fact, he calls the sixth sense a faculty of "interior touching" (p. 50). Permeating the whole interior of the body, the sixth sense perceives that interior distinctly. Although the somnambulist will speak of "seeing" the interior of his or her body, Tardy de Montravel believed that it is really an interior touching that occurs, which then gives the impression of vision. An ill person in a state of magnetic somnambulism can concentrate the sixth sense on the diseased part of the body and judge its condition. The somnambulist can also use a knowledge of causes and effects available through that sense to see how nature would treat this condition and determine the remedies most appropriate. He or she can use the same faculty to foresee how the illness will progress and when a cure will take place (pp. 50–53).

For Tardy de Montravel, the extraordinary abilities demonstrated by the magnetic somnambulist were not a kind of magic but merely the natural application of a faculty or sense not ordinarily available: "If one accepts what I am saying about the causes of magnetic somnambulism, it is not hard to conceive of the effects, as astonishing as they may seem at first sight. One would then no longer regard the somnambulist—through a ridiculous abuse of words—as a sorcerer. . . . Those who would ridicule magnetism would have us find miracles in somnambulism. We on the contrary see in the somnambulist merely an admirable, but purely mechanical, instinct for truth" (p. 64).

Somnambulism and the Will

Like Puységur and Fournel, Tardy de Montravel wrote at length of the "rapport" between somnambulist and magnetizer. But he also used another term employed by Puységur, "perfect analogy" (p. 69). He explains this "analogy" as stemming from the "sympathy" believed to be present in nature between two individuals of the same species, based on the ability of such beings to mutually modify their magnetic fluids (pp. 69–70). Once magnetizer and somnambulist are joined in magnetic rapport, the somnambulist is obedient to the will of the magnetizer, executing those actions willed by the magnetizer so long as they are not contrary to the somnambulist's morals. The magnetizer need not speak his commands, for they are communicated directly to the nervous system of the somnambulist through the rapport of their magnetic fluids. Thus, it is not a matter of the somnambulist "divining" the thoughts of the magnetizer; rather, the somnambulist has the immediate experience of the magnetizer's will (pp. 76–77).

This view of the magnetic state explains how the somnambulist diagnoses the illnesses of others. The intimate connection between the physical organisms of magnetizer and somnambulist is reproduced in the rapport between a somnambulist and one whom the somnambulist touches for purposes of diagnosing disease. As soon as the somnambulist makes physical contact with an ill person, she can see the interior of that person's body in the same way that she can see her own, and she is able to perceive the nature of the illness and its extent (pp. 78–79).

Because the will of the magnetizer was believed to have far-reaching effects on the somnambulist, Tardy de Montravel emphasized the importance of the magnetizer's "will to do good": "It is this disposition in magnetizers that gives their nerves the greatest elasticity and energy and gives the fluid a uniform and constant motion" (p. 75).

Mesmer and Magnetic Sleep

Mesmer's closest adherents insisted that he had known about magnetic somnambulism all along. Würtz (1787), for instance, wrote that Mesmer produced the first somnambulists but believed it best not to give the matter any publicity, foreseeing the harmful enthusiasm to which it could give rise and the superstitious errors it could promote (pp. 50–51).

Unquestionably, Mesmer was aware of the existence of magnetic somnambulism before 1784, since even the Franklin commission's report noted somnambulistic phenomena in connection with the practice of animal magnetism. The experiment in Benjamin Franklin's garden at Passy described the induction of what appears to be magnetic sleep. It is hard to believe that the somnambulistic state could have escaped Mesmer's acute powers of observation. The important question is whether he recognized its significance before he was forced by the sheer popularity of the phenomenon to make a statement on the matter.[4]

Girardin (1784) indicated that somnambulism resulting from animal magnetism was known to both D'Eslon and Mesmer. He described how the ten-year-old patient of D'Eslon mentioned earlier (chapter 2) regularly fell into somnambulistic states after brief exposures to the *baquet*. He would first have light convulsions and then become somnambulistic, eyes open and fixed, lips tightly closed and unable to speak a word. Then he would act out the role of magnetizer, doing such a good job of it that some people preferred to be magnetized by him instead of D'Eslon. The boy would also speak about what he thought to be the cause of the illnesses of other persons, as did Puységur's somnambulists, and would have no memory of what had happened when returning to his normal state (pp. 15–16). Similar effects were observed in patients of Mesmer (p. 20).

The marquis de Dampierre (1784) also took up the issue of somnambulism in Mesmer's practice. Lamenting that the commissioners had passed up the chance to question magnetic somnambulists regarding their ability to diagnose diseases, Dampierre noted that experiments with magnetic somnambulism, which could be decisive in eliminating imagination as an explanation, had been carried out in his presence by Mesmer himself and repeated many times at Lyon (pp. 6–7).

Another witness who claimed that Mesmer knew of magnetically induced somnambulism before Puységur was Dr. Aubry, a Paris physician who had assisted Mesmer when he divided his healing salon into two sections: one for the rich, who could pay, and one for the poor, who could not. Mesmer was in charge of the center for the poor and Aubry of the other. As late as 1840 Aubry testified that cases of magnetic somnambulism were found in the Paris clinic (Gauthier 1842, 2:246ff).

Reports reaching Switzerland in 1784 also indicated that magnetic somnambulism was a frequent occurrence during Mesmer's treatments. Without using the term "somnambulism" or "magnetic sleep," a Zurich paper described how subjects in the "artificial crisis" would describe details of their illnesses and afterwards have total amnesia for what they said or did during the crisis (see Milt 1953, pp. 28–29).

---

4. Fournel (1785) wrote that Mesmer seemed to feel that somnambulism was not an essential part of animal magnetism but a mere accessory that may or may not be present (p. 15). Tardy de Montravel (1785) asserted that Mesmer must have known of the existence of magnetic somnambulism but did not recognize its importance (p. 4n). Lutzelbourg (1786, p. 27) wrote that Mesmer either did not know about magnetic somnambulism or, if he did know, neglected this very important phenomenon.

However much Mesmer may have ignored somnambulism before, in 1784 Puységur himself informed him of the phenomenon, taking his somnambulist Victor Race to Paris to see Mesmer on two occasions. Although Mesmer's reaction is not recorded, it is reasonable to suppose that Puységur made Mesmer aware that he considered his findings to be very significant (Puységur 1785, p. 211).

At some point Mesmer realized that he had to respond to the great popularity of magnetic somnambulism and state his position on the matter. He finally put his ideas into print with his *Mémoire* of 1799 and further developed his thoughts in *Allgemeine Erläuterungen über den Magnetismus und den Somnambulismus* (General commentary on magnetism and somnambulism) of 1812. In the *Mémoire* Mesmer used the term "critical sleep" to designate somnambulism, both natural and magnetic. He stated that somnambulism is simply a dangerous development of certain illnesses and that all mental disorders, such as madness, epilepsy, and convulsions, are in some fashion connected to this condition (pp. vii–ix).

The famous "crisis," he said, is an effort of the living body to throw off an illness, and in the process of healing, the ill person will experience "critical symptoms" that are signs of this effort. These are to be distinguished from "symptomatic symptoms," which are the signs of the sickness itself and signify resistance to healing (pp. 34–35). "The crisis is the general action and effort of Nature to restore the disturbed harmony between fluid and solid parts" (1812, p. 36).

Mesmer wrote that he had noted in chronic illnesses the phenomenon of critical sleep or somnambulism, a sleep state in which a person's faculties are not suspended by sleep but actually enhanced. Persons in such a state are able to walk, talk, and carry on their affairs with skill, and even exercise abilities beyond the normal. They can, he affirmed, foresee the future and penetrate the past; their senses can reach out in all directions and to any distance, unencumbered by obstacles that block the ordinary senses; often they can see the interior of their own bodies and those of others, perceiving the nature of illnesses present there and recommending remedies; they receive impressions of will by other than conventional means and generally are in touch with Nature itself (Mesmer 1799, pp. 59–61).

Mesmer pointed out that these abilities have been acknowledged throughout history but have always been seen as supernatural or occult powers. He now intended to give them a completely natural, even mechanical, explanation (pp. 59–68). The heart of his explanation was that we are all endowed with an "internal sense," a faculty that perceives the inner structure of things. As the five senses are stimulated by movements in the fluids we call water, air, and ether, so the internal sense is stimulated by the fluid that animates the nerves of all living beings. This nervous fluid is affected both by the fluids that stimulate the five senses and by thought and will. It can be transmitted at any distance and is not hindered by obstacles. It is through this fluid that the inner sense is placed in direct rapport with objects, near or far, and receives information about them beyond the reach of the five senses (Mesmer 1799, pp. 80–87; 1812, p. 20).

Mesmer believed that this faculty was present in animals as instinct. In human beings it is not often used because we are preoccupied with reason, which depends completely on the five senses. When we do make use of this internal sense, extraordinary things result. It is because of this inner faculty, this human species of instinct, that a somnambulist can perceive the nature of a disease and distinguish substances that can cure it. This internal sense accounts for the direct communication of will or rapport connecting one person to another, because the subtle fluid to which it responds can extend to any distance and allows the inner sense of one person to affect the inner sense of any other person. Since through this faculty the somnambulist is in touch with all of nature, he is in a privileged position to perceive all connections of cause and effect and so be aware of happenings of the past and the potentialities of the future (pp. 85–88).

But why, Mesmer asks, is the somnambulist or a person in a state of critical sleep more in touch with this internal sense? He finds the reason in the nature of sleep itself. With natural sleep in a healthy person, the functions of the senses are suspended, and this leads to the cessation of all related activity, such as imagination, memory, and movement. But in disturbed sleep (somnambulism) and magnetically induced critical sleep these activities do not cease, even though perception through the five senses is suspended. The result is that the internal sense becomes the sole organ of sensation. Now that subtle faculty, which is usually eclipsed by the activity of the five senses, is able to make its impressions strongly felt, and its remarkable powers, normally not available for use, begin to be exercised (pp. 90–92).

Critical sleep is an intermediary state between wakefulness and perfect sleep. It exists in varying degrees depending on the end of the spectrum to which it is most proximate. The closer to perfect sleep, the stronger the functioning of the internal sense. The closer to wakefulness, the more the five senses operate and the more obscured the internal sense will be (pp. 94–96).

The various kinds of mental illness are simply gradations of imperfect sleep that produce disintegration and confusion. With mental disturbances the internal sense is obstructed and there is a disharmony of the "common sensorium." Animal magnetism, says Mesmer, can cure these conditions by removing the obstacle and bringing about conditions necessary for perfect sleep (pp. 100–101).[5] How is it that animal magnetism sometimes produces somnambulism when applied to the ill? With the application of animal magnetism, says Mesmer, the whole body is vivified and the functions of the "machine" (body) are restored. Then a crisis or bodily action contrary to the disease occurs and obstacles to the proper functioning of the internal sense are removed. This heightening of the internal sense causes sleep in the five senses and induces a state of somnambulism (Mesmer 1799, pp. 100ff).

With the concept of an internal sense that operates according to the ebb and flow of nervous fluid, Mesmer hoped to remove once and for all any notion that animal magnetism was connected with the occult. He believed that his explanation of critical sleep and the inner sense provided a purely natural explanation for phenomena often observed in magnetic somnambulism, of which he lists these four: perceiving the nature of an illness and foreseeing its course of development; without medical knowledge, prescribing effective medications for a disease; seeing and sensing distant objects; and receiving the impressions of the will of another person (pp. 69–70; 1812, p. 58). In ages past, Mesmer pointed out, these phenomena were described as the products of ecstasies and visions and explained in terms of ghosts, gods, demons, and genies (1812, pp. 9, 11).

Thus, while not denying the reality of the extraordinary phenomena of magnetic somnambulism, Mesmer sought to remove occult overtones through a purely physicalist, mechanical theory of the inner sense. He also did away with any reference to the spiritual or nonmaterial aspect of human experience. In sharp contrast to these views, Puységur (1785, pp. 60–63) and Tardy de Montravel (1785, pp. 65–66) believed that the phenomena were striking proof of the spiritual nature of the soul. In the matter of magnetic somnambulism, their views were far more influential than Mesmer's.

Magnetic Sleep and the Occult Traditions

Mesmer's efforts to disassociate animal magnetism from occult traditions were doomed from the start, for animal magnetism stood in clear continuity with the various scientific-occult systems dating back to Paracelsus and van Helmont. Some of Mesmer's contemporaries who were involved with occult, spiritual, and religious traditions were quick to exploit the potential of animal magnetism to confirm their philosophical tenets. So once again Mesmer, who

5. This connection between mental illness and somnambulism was not elaborated further by Mesmer. Puységur developed it into a theory of mental disturbance as "disordered somnambulism" (see chapter 5).

considered himself a true son of the Enlightenment, found his beloved system being used to bolster what he considered to be irrational and unprovable superstitions. The French mystic and philosopher Louis Claude de Saint-Martin made this telling comment: "It is Mesmer—that unbeliever Mesmer, that man who is only matter and is not even a materialist—it is that man, I say, who opened the door to sensible demonstrations of spirit. . . . Such has been the effect of magnetism" (cited in Viatte 1928, 1:223).[6]

By 1784 there were many magnetizers at Lyon (Orelut 1784). Mesmer himself went there in August 1784 to take part in dramatic magnetic experiments. Prince Henry of Prussia had been invited to the demonstration, which it was hoped would convince him of the reality of animal magnetism. The first experiment involved the magnetization of a horse and diagnosis of its disease—a diagnosis confirmed by autopsy. This was followed by the magnetization of a company of soldiers by their commander, Comte Tissart du Rouvre. Henry found this demonstration particularly unconvincing. Finally Mesmer himself attempted to magnetize Henry, but the prince felt no effect. Even touching a magnetized tree left him unmoved.

Lyon's magnetizers and Mesmer himself were discouraged by the lack of success (Figuier 1860, 3:228–29). The effects of this negative experience did not, however, last in Lyon. The magnetizers of the city had developed an easy alliance with the masonic groups of Lyon and found in the masonic leader Willermoz a strong defender against the opposition that developed among certain local physicians. The magnetizers formed a harmonic society, La Concorde, which had many members in common with a masonic group directed by Willermoz, Les Frères de la Bienfaisance. This group, whose members were principally of the aristocracy, had been formed to make available to the poor the benefits deriving from its research into healing and other philanthropic activities.

Among the magnetizers of La Bienfaisance was the chevalier de Barberin, an important theoretician and practitioner of a brand of magnetic healing

6. Saint-Martin had himself been trained in the practice of animal magnetism by Mesmer in 1784 and had read the works of Puységur. He agreed with Puységur that the phenomena of magnetic somnambulism were tantamount to proof of the spirituality of the soul—the final destruction of materialism. As an officer of the Regiment de Foix, Saint-Martin developed a deep interest in the doctrine of the Society of Freemasons and was drawn to settle in Lyon, a masonic stronghold centered around J. B. Willermoz. Signing his written works "the Unknown Philosopher," Saint-Martin developed a visionary philosophy, strongly influenced by Jakob Boehme and Emanuel Swedenborg, which was explicitly opposed to the materialist and rationalist spirit of the Enlightenment. He emphasized the importance of introspection, intuition, and religious experience for arriving at a true philosophy of life.

strongly oriented toward mystic and spiritualistic concerns. At Lyon the *baquet* was soon abandoned and the somnambulistic crisis of Puységur replaced the convulsive crisis of Mesmer. Barberin emphasized the spiritual and psychic dimensions of the magnetic experience, minimizing the notion of a physical magnetic fluid and of physical contact in the magnetic process. Following Mesmer's assertion that magnetization could take place at a distance, Barberin effected magnetization without touch (see Barberin 1786). In fact, he and his followers regularly magnetized subjects situated in distant locales. Taking the lead of Puységur, he emphasized the importance of the will in effecting and directing the magnetic action. But Barberin went beyond Puységur by expecting the magnetizer himself to get in tune with the patient so as to diagnose sympathetically and then heal illness. Like Puységur, Barberin believed in the ability of somnambulists to see into their own bodies and those of others to diagnose and prescribe remedies. But in the mystically oriented environment of masonic Lyon, even more was expected: magnetic somnambulists were also consulted for advice about problems of family life, difficulties in daily affairs, the lot of the dead, the meaning of heaven and hell, the nature of the future life, and every other matter of curiosity.

Another Freemason of Lyon, Antoine Esmonin, marquis de Dampierre, found himself in agreement with Barberin's approach to magnetization. In his *Réflexions impartiales sur le magnétisme animal* (1784) he deplored the convulsive crises of orthodox mesmerism, which he considered embarrassing and obscene for the patient and flattering for the magnetizer. Pointing out the advantages of Barberin's noncontact technique to produce a somnambulistic crisis, he expressed his belief that Barberin had succeeded in magnetizing at great distances, even from one village to another. Dampierre also declared that at Lyon conclusive experiments with clairvoyant medical diagnosis had been carried out in conditions that precluded fraud (pp. 8ff).[7]

Puységur was himself quite closely associated with the mystical philosophers of the day. In 1785 he organized a Society of Harmony in Strasbourg. His regiment had been stationed in that city, and the comte de Lutzelbourg had taken advantage of his presence by asking him, on behalf of a group of Freemasons, to teach them about animal magnetism. Puységur accepted the invitation and went on to establish an association, called the Société Harmonique des Amis Réunis, whose purpose was to train magnetizers, treat patients, and report the results of its work. The Strasbourg society was extremely

7. For more about masonic magnetism in Lyon see Andry 1924, Viatte 1928, Amadou 1971, Chevallier 1974, Ladrey 1976, and Rausky 1977.

active, setting up treatment centers in various localities and publishing three volumes of reports (Society harmonique 1786, 1787a, 1787b, 1789). Dampierre was one of those who had benefited from this activity.

The Strasbourg society embodied the spirit of Puységur's approach to animal magnetism. Most of its members were from the aristocracy, and they were required to treat their patients free of charge. Puységur's emphasis on good will and the spirituality of the magnetic operation fired the society to great heights of dedication. The mystical, masonic ambience of the society was a congenial environment for Puységur's healing "apostolate." He delivered a series of talks to the new society and helped set up its regulations, based on those adopted by the magnetic Society of Harmony at Nancy and by the regiment at Metz. After Puységur's seventh and concluding instructional talk, the comte de Lutzelbourg commented: "Well, my dear Puységur, this is a real initiation, like our masonic induction ceremonies. Isn't animal magnetism a long-lost truth? And isn't the clarity achieved in the state of magnetic somnambulism the light to which we aspire?" (Puységur 1820, p. 142).

Puységur was enthusiastic about the Strasbourg society and the support he found there.[8] He mentioned with appreciation the masonic Martinists at Lyon, the work of Cagliostro in Paris and Rome, and the alliance of Freemasonry with the followers of Swedenborg in Germany and Sweden. He clearly felt a kinship with these metaphysical systems and believed them to be quite compatible with his version of Mesmer's animal magnetism (Puységur 1820, p. 101).

Given this philosophic compatibility, it is not surprising that the Swedenborgian society of Stockholm, called the Exegetic and Philanthropic Society, became heavily involved with magnetic somnambulism and set out to become an associate of the Strasbourg society. Founded in 1786, the Stockholm society had as its purpose the promotion of the teaching of the famous Swedish seer Emanuel Swedenborg (1688–1772), whose visions revealed a world of spirits, good and bad, which secretly influence the conduct of ordinary human life. On the lookout for philosophies complementary to those of their teacher, the society was impressed by the animal-magnetic literature of France, particularly that produced by the Strasbourg group. In 1789 the Stockholm society sent the Strasbourg society a *Lettre sur la seule explication satisfaisante des phénomènes du magnétisme animal et du somnambulisme. . . .* The letter had a preface addressed to King Gustavus III and was published under the royal imprint. Its subject was "the cure of diseases by means of Magnetism, and its

---

8. Puységur (1820) wrote that when he departed from Strasbourg in 1791 the society was still vital. The following year, however, it was dispersed under political pressures (p. 144).

adjunct, Somniloquism" and "the practical use and application of this science in everyday life" (Bush 1847, pp. 258, 259). The society states its position up front:

We believe that those systems which have their foundation in mere physical causes, as "La Psycologie [sic] Sacrée de Lyon," are quite inadequate to explain how those singular effects take place and are produced, which Magnetism and Somnambulism present. It seems to be impossible fully and rationally to explain them, unless we once and for all, and without shrinking from the shafts of ridicule, take it for granted that spiritual beings exert an influence upon the organs of the invalid during the time that the power of Magnetism has produced a partial cessation of the functions of the soul, and that these spiritual agents, in virtue of the higher degrees of knowledge which they possess, originate these wonderful and otherwise inexplicable phenomena. (Bush 1847, pp. 261–62)

The Stockholm society here took a stand in favor of a spiritistic interpretation of the phenomena of magnetic somnambulism. This was a new step in the evolution of explanatory philosophies of animal magnetism. All beneficial effects were in the last analysis attributed to the work of wise and helpful spirits who, upon induction of magnetic sleep, were given the opportunity to act. This viewpoint derived from a broader Swedenborgian framework that saw human beings as a kind of battleground between good and evil spirits. Disease was attributed to the action of an evil spirit in the body of the patient. Healing was brought about by the action of a good spirit to eject the evil influence. The good spirit may even take possession of the body of the patient and speak through him. This, said the Stockholm society, is what happens in certain somnambulistic states.

This view of disease and healing somnambulism is spelled out in the letter:

That state of the Somnambulist, during the magnetic sleep, may be called ecstatic, in order by that epithet to indicate a suspension of the functions of the will and of the understanding, in the exercise of which, man's ordinary being, or esse, consists. Such a state plainly demonstrates, that what is said or done through the sleeper's organs, is not the act of his soul, but of some other being, who has taken possession of his organs, and operates through them. So long as the magnetized person exhibits painful paroxysms, such as convulsions, &c., it is a sign that the spirit of disease, which certainly cannot be a benevolent being, is still present; but this spirit has no power to speak through the organs of the patient, unless he is fully possessed; . . . So soon as the magnetized person begins to talk in his sleep, it is a sure sign, that a spiritual being and friendly to the person (as being his guardian-angel or good genius, and possessing the same measure of goodness and wisdom with the patient) has succeeded if not entirely to remove the disease or rather the spirit of the disease, has at least in so far rebuked its influence, that he, who is a benevolent being, is able to speak and act through means of the patient's organs, and to give suitable advice to those present to promote his recovery; as also to impart information on all subjects which do not transcend his own knowledge. (Bush 1847, pp. 268–69)

The view that outside spiritual agents are essential to account for magnetic healing explicitly contradicts the "psychology of Lyon," which, although spiritual in orientation, sees the action of the soul of the healer on the soul of the patient as a sufficient explanation for what occurs.

In their brief position paper the Stockholm Swedenborgians laid out the classical spiritistic stance with regard to dissociated psychological phenomena: the phenomena of magnetic somnambulism must involve mental acts that do not take place in the consciousness of the somnambulist. It was not sufficient, in the society's opinion, to say, as Puységur and the Freemasons did, that magnetic somnambulism proved the spirituality of the soul; one must determine *how* the information used by somnambulists is obtained and processed. The Stockholm society said that the wisdom involved in diagnosis and healing requires an intellectual, spiritual agency beyond the somnambulist himself. Since the somnambulist does not experience himself as acquiring this wisdom through any identifiable conscious process—through learning, reasoning, or experience—the mental processing must take place in another intelligent agent: a beneficent spirit.

If the Strasbourg society was going to take up this challenge, it would have to search for an arena of mental processing that was neither the mind of an external entity (spirit) nor the conscious mind of the somnambulist. In other words, they would have to posit mental acts that occur within human beings but outside of and unavailable to consciousness. Needless to say, thinking about magnetic sleep had not yet developed to such a sophisticated stage. It would be much later, when a new form of spiritism (American spiritualism with its table-turning phenomena) met with a more advanced form of animal magnetism (magnetic and hypnotic experiments with double consciousness), that this mystery at the heart of magnetic sleep could be solved.

Chapter 5

Magnetic Sleep and Psychotherapy

The theory of animal magnetism stated that all bodies exert an influence over one another by the action of animal magnetic fluid. With this concept, Mesmer built a strong social component into his theory with direct implications for psychology, politics, and education. The mutual influence of magnetist and patient involved a kind of sympathy which Mesmer called the "sixth sense." Cures wrought by animal magnetism always had this empathic component strongly at the forefront.

By placing empathy at the center of magnetic treatment, Mesmer became an innovator in the field of psychology. He stated that animal magnetism could cure nervous disorders directly and other disorders indirectly. This assumed that the nervous system, along with its psychological aspects, was always involved in the successful application of animal magnetism.

Through his personal charisma Mesmer was able to set up a powerful empathic relationship with his patients, as is exemplified in the strong emotional bond he established with Maria Paradis. But although Mesmer may have been a "natural" psychologist and used psychological principles in applying magnetic healing, he did not develop an identifiable psychological theory. The first psychological investigation of magnetic phenomena and the formulation of a primitive magnetic psychotherapy must be attributed to the marquis de Puységur and those who were influenced by his work.[1]

Rapport and Love

Puységur, as we have seen, wrote about the "intimate rapport" that exists between magnetist and somnambulist and that, he believed, produced a remarkable communication between the two. Tardy de Montravel also perceived the central place of intimate rapport in magnetic somnambulism. He compared the nervous systems of the magnetizer and the magnetized to two

---

1. The term "psychotherapy" was first used in 1887 (Pivnicki 1969). It seems legitimate, though, to speak of the evolution of a magnetic "psychotherapy" if that word is used in its root meaning: nurturing of the spirit or soul.

musical instruments in such "harmonic rapport" that a chord played on one creates a corresponding chord in the other. The rapport between magnetizer and somnambulist—the result of the action of sympathy found everywhere in nature—allows two beings that have the same physical constitution to modify each other's magnetic fluids. Tardy de Montravel referred to harmonic rapport as a kind of platonic love between magnetizer and patient, a union of souls in which the somnambulist establishes a strong dependency on the magnetizer. This dependency was an advantage for the healing process and in no way detrimental to good morals (pp. 69–71).

The notion of harmonic rapport and love as essential to the healing process was developed even further by Charles de Villers (1767–1815), a friend and onetime aide-de-camp to the marquis de Puységur and a member of the Society of Harmony of the Metz artillery regiment at Strasbourg. He evolved a highly philosophical theory of animal magnetism and somnambulism which he used in a novel, *Le magnétiseur amoureux* (1787).[2] The title refers both to the magnetizer-hero, Valcourt, who is in love with the daughter of the household in which the novel is set, and to the central theme of the book, the affection and "cordiality" the magnetizer must exercise toward his patient in order to cure effectively. Although the book is written in the form of a romance, it is as much a theoretical treatise on animal magnetism as it is a work of fiction.

Le magnétiseur amoureux showed that Villers did not believe that a physical agent (magnetic fluid) was involved in animal magnetism. For him animal magnetism was the work of the soul, a spiritual entity that makes use of the will to bring about the desired effects. "My soul, which is the first principle of movement, can convey its effects to another being. If that being is organized, it unites itself to the principle of movement it finds there: that is how I produce a salutary effect on a patient" (1824, 1:121). Healing, said Villers, is produced through sympathy and harmonic rapport, an "amalgam of souls" in which the magnetizer is essentially active and the somnambulist essentially passive. The

---

2. *Le magnétiseur amoureux* had an unusual publication history. Villers sent copies of the book to a few of his friends, including Puységur. The rest were ordered destroyed by the minister of police. Today only one copy of the 1787 version is known to exist; it is housed in the public library at Besançon, where Villers was living when he wrote it. In 1824 Puységur, presumably using the copy given him by Villers, decided to issue a new edition, in two volumes. As he stated in his preface, he made some changes, but left the work principally intact. He then, however, ordered the entire edition to be destroyed—perhaps for fear that the title could give a bad name to animal magnetism—so that copies of this second edition are also extremely rare.

soul of the magnetizer is "identified" with that of the somnambulist, leading to a communication of thoughts and feelings (1:68–70, 128–56 passim).[3]

Villers's description of the communication of images from the imagination of the magnetizer to the somnambulist is one aspect of what is today called "suggestion." The other aspect of suggestion, which Villers called "inspiration," he uses to explain how someone may be induced to fall into a somnambulistic state when touching a magnetized tree or a magnetic *baquet*. For Villers it was not magnetic fluid that produced the effect but the inspiration the person feels because he has formed the expectation that somnambulism will result from these experiences (pp. 166–67).

Healing is brought about principally through the conveyance of actions of the soul of the magnetizer to that of the patient. "This conveyance can only happen if the soul desires it and knows about it. Therefore, will and thought produce the conveyance of the soul. . . . Now our soul exercises the faculties of will and thought; so it can also convey itself to another being. This conveyance is accomplished through thought. So when I think of some exterior object, some man, for instance, my soul produces a real effect on this man" (pp. 123–24).

With his denial of a physical agent and his emphasis on the accomplishment of animal magnetism through the union of souls, Villers moves far from the views of Mesmer, who downplayed the personal in the relationship between magnetizer and patient. Villers pushed the more psychologically interactive approach of Puységur further than anyone else in the first decades of the animal magnetic tradition. He stated that in order to heal, the magnetizer must have "moral affection" or "sympathy," brought about through a feeling of "cordiality." He equated this moral affection with what had been termed "rapport" in magnetic writings. In magnetic healing, patients put their trust in the magnetizer and open themselves completely to his influence. He in turn exercises a familial benevolence toward them that is the means by which the healing action takes place:

> [The two souls] unite so much more effectively insofar as they are of the same ilk, that is, insofar as I have many moral affections in common

3. This "identification" was offered to explain how the somnambulist is able to carry out the commands of the magnetizer without a word being spoken. It was also meant to explain how the magnetizer can create hallucinations in the somnambulist through the use of his imagination. If, for example, he presents the somnambulist with an object and suggests it is a fragrant rose, the magnetizer's memory recalls to his imagination the experience of smelling a rose; this is in turn picked up and amplified by the somnambulist, who then believes he is smelling a rose (1:156–57).

with this person. Now what is the moral affection, the tint of the soul, most marked in the patient? It is to be healed. It is necessary, therefore, that I also have this will to heal in order to act as efficaciously as possible on him. . . . So I bear in myself the power to heal my fellow men. The most sublime part of my being is devoted to this undertaking. And it is in the sentiment of the most sweet beneficence that my friend is certain to find a remedy for his ills. If a tender father sees his son or his wife suffering, it is his own substance that goes to cure them; it is the sweet action of his tenderness that communicates life to their organs. And, once health has been reestablished in them, their souls, imprinted by that of the father or spouse, leap forward towards it, being drawn there by a sympathy that they neither can nor want to withdraw. . . . My love for my fellow men is my guide. . . . The action of thought, which has too often been applied to harm, now has the means to be more useful. And when a compassionate soul recognizes its faculties and makes use of them, it will say: "I will" and ills vanish before the words pronounced interiorly and with energy. What an inexhaustible source of joy for the magnetizer! (p. 146)

Villers thus constructed a system based on a strong psychological interaction between magnetizer and patient. Both had to be emotionally involved in the magnetic exchange for it to be salubrious.

The Psychotherapy of Alexandre Hébert

Puységur, perhaps more than any other early magnetizer, put these principles of the healing relationship into practice. He believed the "intimate rapport" between patient and magnetizer to be most delicate, and he perceived both the benefits and the dangers of such a relationship. His humane and respectful treatment of his subjects was acknowledged by all and held as an ideal by generations of magnetizers.

Puységur insisted that a treatment once started must be followed through to the end, no matter what the inconvenience to the magnetizer. This extraordinary commitment led him to conduct the first case of magnetic psychotherapy, that of Alexandre Hébert. His account of this case (Puységur 1812, 1813) shows that many of the basic elements of psychotherapy that were explicitly established by the end of the nineteenth century were already present in Puységur's treatment.

Alexandre, a boy of twelve and a half, was suffering from paroxysms of rage in which he was a danger to both himself and those around him. He experienced severe headaches and would fall into fits of weeping and moaning while hitting his head against the wall, sometimes even attempting to throw himself out of windows. If someone touched him while he was in this condition he became violent, thrashing around and biting anyone who tried to restrain him.

The boy had been subject to mental problems from age four. At seven he underwent surgery of the skull to relieve "pressure" in his head. At that time, apparently, part of his brain was surgically removed, but the operation had no beneficial effect.

Alexandre's parents had sought medical help for their son for years, but to no avail. In June 1812 they brought him to stay with the curé at the church in Buzancy, and it was there that he came to the attention of Puységur. He magnetized Alexandre, who immediately closed his eyes and became perfectly peaceful for about fifteen minutes. Puységur repeated this treatment for four days. On the fifth day the boy fell into a sleep state and Puységur, believing that he was now somnambulistic, began to question him in an attempt to learn what was causing his condition and what course it would take. The somnambulistic Alexandre was at first pessimistic about any healing, but, in response to Puységur's inquiries, began to prescribe the means that should be used to bring about his cure (Puységur 1812, pp. 9–18).

The principal means of cure, the somnambulistic boy said, was magnetization, applied no less than every other day. The first time Puységur neglected to keep to the timetable Alexandre fell into a raging fit of extraordinary intensity. He screamed, pounded the furniture, tried to throw himself out the window, and generally terrified everyone else in the house (1812, pp. 18–24). Puységur was called and in a few minutes was able to place Alexandre in a somnambulistic state and question him about the attack. Alexandre said it was a consequence of Puységur's neglect of the prescribed schedule. After this conversation, Puységur brought Alexandre back to his waking state, in which he now was peaceful. A short time later the boy had two more violent fits (at one point four men were needed to restrain him). Puységur asked him what was to be done, insisting that Alexandre himself make an effort to provide some solution. The boy replied that magnetizing was required. Puységur pointed out that he could not always be with the boy and asked why Alexandre was tranquil at the moment, while being questioned, even though he was not magnetized. Alexandre answered, "Because you are at my side." Reflecting on this reply, Puységur decided to remain in the same room with Alexandre that night so that both the boy and the household could get some rest (pp. 34–37).

This was a turning point in the therapy. From then until the end of the treatment, Puységur kept the boy with him constantly. Alexandre was moved to his home, and a bed was set up in the marquis's own bedroom. The marquis had to be with Alexandre every night and was parted from him for only short periods during the day. Puységur decided that when he had to travel he would take the boy with him. For a period of some months he more or less devoted his life to Alexandre's cure.

Rapport and Transference

Puységur had a great sympathy for Alexandre and showed unusual patience with his demands. It is clear that through Puységur's devotion to Alexandre, the boy was able to experience the attention and care that had previously been lacking in his life.

The treatment of Alexandre Hébert shows that Puységur was a man of remarkable humanity and kindness. He keenly felt the inconvenience of the situation—"the obstacles of every kind that it was necessary for me to submit to in order to have the time available during the cruel illness of this child"— but made up his mind to accept it. In his journal he records with extraordinary precision the working of his own emotions and conscience that motivated his perseverence in the treatment:

> What misfortunes could result from the least inadvertence on my part, both with regard to my small patient and with regard to those who might find themselves without me, in those moments when the boy was subject to dizziness or madness! And even if I could foresee all and prevent all, such constraint, such clashing of interests is not for me—not to be able to be in charge of even one of my trips or my evenings. . . . If I had not been of aid to this poor little boy when, in his first headache attacks, he wanted to throw himself against the walls, or bite those who tried to prevent him, and if, as a result of his furors, he had ended up dead in his fit of rage so terrible to see and so difficult to imagine—I would have undoubtedly been grieved.

Puységur felt that once it was started he had no choice but to pursue the treatment to the end. He sometimes wished he had been a mere observer on the sidelines, but fate had not so ordained. He was Alexandre's sole source of help and "answerable to the tribunal of my conscience for all the good that I do for this child and all the evil that I can avert." He summed up simply: "In any case I did, if not what was best, at least what I thought was best to do. And this gives me a great deal of peace" (Puységur 1812, pp. 43–45).

In the relationship between Alexandre Hébert and Puységur we see what psychodynamic therapy calls the transference. The boy's powerful connection to Puységur, characterized by strong trust and dependency along with episodes of petulance and angry reproach, is depicted again and again throughout the account of Alexandre's treatment. That there was a countertransference on the part of Puységur is also clear. He was to some extent prey to Alexandre's manipulations, and he often candidly described his pain in watching the suffering endured by the boy. But Puységur was not without insight into the situation. In fact we see, in his description of the magnetic rapport between himself and Alexandre, Puységur's seminal grasp of the notion of transference. One day Puységur returned home to find that Alexandre had been taken to the parsonage at Buzancy in a catatonic state: hardly breathing, eyes fixed, responding to no one. When Puységur went to magnetize the boy, to his surprise he could not produce somnambulism: "After two or three minutes, however, I heard him pronounce with some difficulty the words: 'Mama, Mama . . . at Soissons.' 'Has your mother magnetized you?' I asked him. 'At Soissons, Mama . . . never to heal.' I immediately judged him to be in a kind of delirium; and whether or not he had been magnetized by his mother, I very quickly made the resolution to break this new rapport and to dominate his senses and his imagination. I did not speak a word to him but continued to magnetize him strongly. After four or five minutes he woke up and was astonished to find himself in a strange house. I immediately asked him if his mother or others had touched [magnetized] him at Soissons. There was no response" (1812, pp. 47–48).

Puységur then discovered that the boy had overheard him talking about his concern for his daughter's safety in a recent storm, and this had thrown Alexandre into a state of disturbance. Although we are not given enough information to speculate about the emotional connection between the overheard conversation and Alexandre's relationship with his mother at Soissons, we see how important Puységur considered such psychological associations to be. In his view, somehow the boy had been thrown back into magnetic rapport with his mother and this was harmful to him. Puységur recognized the therapeutic value of shifting this disturbing connection over to the therapist, thus replacing a chaotic somnambulistic state with an orderly one centered around himself.

Mental Disturbance as Disordered Somnambulism

This insight led Puységur to a general theory about the nature of mental disturbances. The theory was not based wholly on his experience with Alexandre, however. As background, he described another case that demonstrated the same feature of disturbing rapport. In 1790 he had treated an artillery soldier of seventeen or eighteen with animal magnetism. The soldier was just as susceptible to somnambulism as was Alexandre, but his treatment differed in that a young woman assisted Puységur with the magnetizing.[4] In discussing this case, Puységur insisted that the relationship between the soldier and the woman was completely above board. The young woman was forced by political events to leave France before treatment was complete. Immediately upon her departure the soldier fell into a state of depression. Hoping to uncover the reason for his condition, Puységur repeatedly placed the soldier in a state of somnambulism:

> At last he said to me . . . that he had been in an extraordinary state for many days. He did things, went through the exercises, ate with his comrades at the inn—all this without revealing the inner illness that was consuming him. For he was alive, so to speak, without living. He could not remember anything he saw or heard while in this sad state. Asked in the magnetic state about the cause of this madness and the means to draw him out of it, he said to me: "Madame de *** magnetized me with you. She alone is the cause of the illness I suffer." I asked him, "Are you in love with her?" "Oh," he replied with an offended air, "Do not even speak that word, Sir; it does not give you the proper notion of the sentiment that connects me with Madame de ***. I assure you that my senses are not at all involved in the sentiment she inspires in me. The care she has shown me is so pure that I blush at any desire that could shock her purity and scandalize her angelic virtue. But this rapport that her tender interest and my gratitude have established between us is, however disengaged from the senses, nonetheless more and more durable. You help me, no doubt. Your magnetism supports me. But you only deal with half of my life. The other half, that which constitutes my being, that which alone I value, is with Madame de *** . . . only her will can bring about a separation." (pp. 51–52)

Puységur decided to write to the woman with instructions about how to release the soldier from his rapport with her, and he sent the young man to find her and present her with the letter. Eventually, the soldier did find Madame de ***, who placed him in a state of magnetic sleep and severed the magnetic rapport between them. He awakened in his normal state, free of all depression

4. Throughout the history of mesmerism it was not uncommon to find women magnetizers. Pearson (1970, p. 10) mentions that female magnetizers were at work in London in 1790. Suzette Labrousse, a prophet and mystic of the era of the French Revolution, performed several magnetic cures, and the duchesse de Bourbon and Madame de Krüdener, from the same period, were well known as mesmerists (Darnton 1968, pp. 129–49). Deleuze took it for granted that women magnetizers would be available if needed (Deleuze 1850, pp. 107-8), and in mid-nineteenth-century America "clairvoyant physicians," most of them women, used a form of spiritistic mesmerism to heal disease (Braude 1989, p. 146).

and confusion. Puységur then pronounced this momentous conclusion: "The example of this young artillery officer and that of little Alexandre, as well as many other patients I have had occasion to observe, justifies fully my assertion that most insanity is nothing but disordered somnambulism. This is also the opinion of Mesmer" (p. 54).

Puységur believed that this theory of mental illness was supported by a number of phenomena observed in Alexandre and others. The first was sleepwalking, the state of natural somnambulism that often occurs spontaneously in children and more rarely in adults. Alexandre was subject to frequent episodes of sleepwalking (and the related phenomenon of sleep-talking) throughout his illness. At first Puységur worried about this, but Alexandre, in the state of magnetic somnambulism, assured him that these episodes were not only unavoidable but were actually beneficial to him (pp. 27, 40–41). As we have seen, Puységur considered sleepwalking, or natural somnambulism, to be essentially the same as magnetic somnambulism, except that in natural somnambulism the sleeper is in rapport with no one, whereas in magnetic somnambulism the sleeper is in rapport with the magnetizer. From this Puységur concluded that the treatment of the insane (those subject to disordered somnambulism) with magnetic somnambulism counteracts the "disorder" of chaotic rapport by making the magnetizer the center and focus of rapport. By repeatedly establishing this state of ordered somnambulism, the disorder of the insane is cured.

Puységur believed that another confirmation for his theory of disordered somnambulism came from dreams. Talking with Alexandre about his dreams became an important part of the boy's therapy. At first Puységur believed that the disturbing dreams to which Alexandre was subject were harmful. But Alexandre, in the state of magnetic somnambulism, assured Puységur that dreaming actually brought him relief from his illness. Gradually Puységur came to accept this surprising revelation and learned from the material revealed in the boy's dreams. Among other things, his dreams exposed the underlying relationships between Alexandre, his mother, and his siblings. Although Puységur lacked a conceptual frame to account for the meaning of dreams, he made good use of what he could see (pp. 37, 84–85, 88; 1813 1:35–39). He believed that he perceived in Alexandre's dreams the same elements the boy would act out in his waking "insane" states. From this he drew these shrewd conclusions: "But if dreams are little fits of nocturnal madness, is not madness really only a dream more or less prolonged into the waking state? Do not these similarities between the fallacious illusions of our dreams and the distressful manifestations of madness, which learned physiologists and skillful physicians have noted before me, serve to support my opinion that the insane, maniacs, the frenzied, and mad people are simply deranged or disordered somnambulists?" (1813, 1:39–40).

Puységur's profound but simple notion was that the somnambulism of the mentally disturbed was disordered precisely because the patient was unwittingly caught in a magnetic rapport with someone no longer present (Alexandre's mother, the soldier's Madame de ***). Fits of "madness" were merely periods of spontaneous somnambulism in which rapport was momentarily reestablished with an absent object. The somnambulism was disordered either because of some disturbance in the original rapport or because it removed the person from contact with reality. This state of affairs could be cured by breaking the hidden, disturbing rapport and replacing it by rapport with the magnetizer. Repeatedly establishing a stabilizing rapport with the magnetizer gave the patient a point of reference for present reality while divesting the hidden rapport of its strength.

Puységur's theory of mental disturbance filled out his picture of the psychotherapeutic elements much later called transference. Seeing the need to establish a strong state of rapport with Alexandre, Puységur kept the boy with him constantly for months. Alexandre's transference toward Puységur sometimes produced bizarre effects (see, e.g., Puységur 1813, 1:58–59). At times the magnetizer became personally distressed by Alexandre's behavior and state of mind—evidence of what Freud later termed "countertransference" (Freud 1964c, p. 144). And Puységur was quite honest about his struggle with his own feelings. There is perhaps no more candid account of the phenomenon of countertransference to be found until Freud.

Divided Consciousness

Again and again in Puységur's account of his treatment of Alexandre Hébert, one is struck by the sharp contrast between the boy in the state of magnetic somnambulism and in his normal waking state. In his normal state, Alexandre had severe problems remembering even the simplest things, a condition that made it nearly impossible for him to read, write, or apply himself in a concentrated way to any task (1812, p. 42). He was also subject to seizures and was generally slow to understand the realities of life. In the magnetic state, he experienced no problems of memory, never had seizures, and was extremely perceptive and acute in his understanding of reality. The contrast of the two states of consciousness is reminiscent of the alteration of personality Puységur noted much earlier in his treatment of Victor Race.

The alteration in the ability to remember is very striking. In his normal state Alexandre could remember almost nothing of his life from day to day, but when magnetized he could recall even the smallest details of his experiences since age four. Puységur was concerned that this memory defect might persist even after Alexandre was cured, and he did not understand why somnambulism totally removed the problem (1813, 1:76–77). Like most somnambulists, Alexandre while in a state of magnetic somnambulism was able to remember the events of previous somnambulistic episodes, but lost all memory of them when awake. Puységur often kept Alexandre in the magnetic state for long periods of time, during which the boy, though somnambulistic, would seem to onlookers to be functioning normally. It bothered Alexandre when other people would refer to things he had seen or done while somnambulistic but for which he retained no memory in his waking state; he would feel that he was being mocked when people were amused at his lack of memory.

Alexandre's change in personality characteristics while in magnetic somnambulism was as striking as his alteration in memory. Normally a quiet, uncertain, sometimes petulant, sometimes violent boy, Alexandre would become articulate, self-assured, clearheaded, and calm in the somnambulistic state. The sole exception to this was described by Puységur. He wrote that one day, after having his small drink of brandy (self-prescribed by the boy in his somnambulistic state), Alexandre became calmer and more reasonable than ever before. This mood carried over to the next morning. En route from Paris to Buzancy "Alexandre without ceasing entertained me with his little reflections and observations. . . . He occupied himself with very ordinary subjects, such as the events of his childhood, the inner details of his family and of his education. All this he expressed with such sense and so little infantilism that I often looked at him with astonishment and listened with the same. So, I told myself, a mere drop of brandy has been able to produce such a rapid and marvelous change!" (1812, pp. 86–87).

It may be that what Puységur saw in Alexandre was not a remarkably lucid normal state at all but a spontaneously produced state of magnetic sleep. For when he inquired of Alexandre, later in the magnetic state, why he had been so changed, the boy responded that two things contributed to "the good state I am in: the movement of the coach and the brandy of the day before." Perhaps the effect of the alcohol and the monotonous motion of the carriage (along with Alexandre's conditioning for magnetic somnambulism in Puységur's presence) combined to produce an unnoted somnambulistic state. Whatever one makes of this incident, the contrast of mood and mental clarity between magnetic and normal states was profound. Virtually all of the therapeutic work accomplished by Puységur with Alexandre was carried out while the boy was somnambulistic. In that state he had insight and information not available to him in the waking state.

Alexandre Hébert was eventually cured of most of his major symptoms. He ceased to have fits of rage, in which he would bite and bash his head, three months after Puységur started working with him. Five months later, all other symptoms (including agitation and headaches) ceased and he was then able to read and write a bit and apply himself to simple tasks. His memory, however, was never restored to normal functioning—just as he himself, when somnambulistic, had predicted (1813, 1:72–75; 2:117–19).

The Elements of Magnetic Psychotherapy

Recognizing that physical cures often necessitate dealing with psychological issues in the ill, the early magnetists found themselves attempting to carry out a rudimentary psychotherapy with some of their patients. From early accounts of magnetic somnambulism, particularly those by Puységur and his followers, the elements of a magnetic psychotherapy clearly emerge. In those early years no complete system was formulated by any one practitioner, yet all the ingredients of an overall theory were present. Although the elements of magnetic psychotherapy were in place by 1820, the principles embedded there would not find full expression until after 1880.

There are six basic elements of magnetic psychotherapy. They involve the recognition of (1) a second consciousness accessible in magnetic sleep, (2) the fact that this second consciousness often exhibits qualities uncharacteristic of the waking person, (3) the presence of two distinct streams of memory, with the waking person being unable to recall the events occurring in magnetic sleep, (4) the accessibility of painful secrets in the state of magnetic sleep, (5) a view of mental disturbance as "disordered somnambulism," and (6) the importance of establishing a therapeutic rapport between magnetizer and patient to correct that disordered somnambulism.

The recognition of divided consciousness is the keystone of magnetic psychotherapy. It is the second stream of consciousness, revealed in magnetic sleep, that makes psychotherapy possible, for it provides access to hidden thoughts and creates a special connection with the magnetic psychotherapist. Often magnetic somnambulism was characterized by greater objectivity about oneself and a remarkably clear-sighted perspective on one's problems. Puységur made effective use of this state in his psychotherapeutic work with Victor Race, and by the time he encountered Alexandre Hébert he had for some time considered it his principal therapeutic tool.

It has long been known that emotional disturbances can result from keeping painful secrets hidden (see Ellenberger 1966, pp. 29–42). Over the ages various techniques for revealing and dealing with such secrets have been devised. With magnetic somnambulism, a new means of resolution became available. In the very first recorded case of magnetic sleep, Victor Race revealed a painful secret while somnambulistic, enabling Puységur to help him with a difficult personal problem. In 1786 Count Lutzelbourg confirmed this phenomenon in a fascinating note:

> The greatest obstacles opposing the cure of patients come either from a temperament worn by the abuse of remedies or from painful secrets that make the physical treatment insufficient. A lucid somnambulist had been obliged by me to confide a sorrow of a most singular kind. He had had an intimate liaison with a person of the same sex whom he loved to the point of folly in his ordinary state and in whom he had complete confidence, never concealing anything from him. In the state of crisis [magnetic somnambulism] he saw clearly that his so-called friend abused his confidence, betrayed his secrets, thwarted his projects, and even destroyed his reputation. He revealed to me the means for procuring proofs of this and prescribed what I ought to do to clarify things, while recommending circumspection and caution in using these means. Everything turned out exactly that way, but after a long period of time. I had the great good fortune of overcoming the obstacle that had made all my efforts useless and the cure impossible. And at the same time I had the pleasure of dealing with two different and opposed individuals, of whom the one was timid, pliable, and even credulous to an excess; while the other was clairvoyant, firm, and judged men and things according to their just value. (p. 47)

In this paragraph, apparently one of the earliest accounts of psychotherapy with a homosexual patient, Lutzelbourg presented a clear outline of the psychotherapeutic process. He declared that sometimes it is necessary to deal with psychological matters to obtain a physical cure; that this involves uncovering hidden, painful secrets; that these secrets, unknown in the ordinary state, can be revealed in the somnambulistic state; that the contrast between the two states is like that between two different persons; and that once the hidden material is dealt with, the cure is possible.

In the early decades of animal magnetism the view of emotional disturbance and madness as disordered somnambulism was most clearly expressed in Puységur's treatment of Alexandre Hébert. The recognition of the place of dreams in magnetic psychotherapy seems to have been limited to that case, and it would be many years before this early insight would be pursued. By contrast, the therapeutic importance of the relationship between magnetizer and somnambulist was accepted by nearly all magnetic practitioners. The literature tends to concentrate on the experience of the somnambulist, paying little attention to that of the magnetizer. Puységur's emphasis on the crucial importance of good will on the part of the magnetizer, Tardy de Montravel's concepts of sympathy and harmonic rapport, and Villers's notion of cordiality do draw attention to the mutuality of the magnetic therapeutic experience. Only Puységur, however, in his description of his involvement with Alexandre Hébert, gives any hint of how a magnetizer's subjectivity might enter into the dynamism of the therapeutic relationship.

Paradigms of Mental Illness

Mental disorders like that experienced by Alexandre Hébert have always existed and have been interpreted in a variety of ways. The interpretation has determined the therapy.

Until relatively recently in the history of human thought, when a person was subject to unaccountable thoughts, feelings, or impulses, the source was thought to be some outside entity—the devil, an evil spirit, or a witch—intruding into the consciousness of the afflicted person and causing him or her to think strange things and have weird visions. The extreme form of this harassment was possession, a state in which the intruding consciousness seized control from the victim's ordinary consciousness and took charge of the body. This paradigm for disturbed mental functioning was employed to account for everything from "bad thoughts" to madness and was long considered an adequate and complete explanation. This framework for explaining mental disorders can be called the *intrusion paradigm*. The intrusion paradigm implies that all mental aberration is of supernatural origin. The remedy therefore has to be applied by those skilled in spiritual matters—the shaman, the sorcerer, and the priest-exorcist. History has shown that these workers have at times been able to help their clients greatly.

Later a second paradigm arose for understanding the origin of disturbances in human consciousness, one that recognized a natural cause rooted in imbalances of the physical organism. Although the foundations for this paradigm already existed in ancient healing practices, it came into its own in the western world only in the sixteenth century, heralded by the appearance of two works: the *Occulta naturae miracula* of Levinus Lemnius (1505–1568), which appeared in 1559, and the *De praestigiis daemonum* of Johann Weyer (1515–1588), first published in 1562. Lemnius and Weyer believed that mental dysfunction was not a spiritual or moral problem but a physical one, that victims of disturbances of consciousness were suffering not from the intrusion of spirits or demons but from a bodily malady that could be corrected medically.

Lemnius contended that disturbances of consciousness were the result of vehement cerebral stimulation, and the cause should therefore be sought in the brain and the humors that affect it. Weyer, speaking of the persecuted witches of his day, asserted that they were the victims of crazy imagination caused by drugs or poisons they had ingested, or humoral imbalances brought about through other means. Weyer denounced those who resorted to exorcism or magic to cure these conditions, claiming that they were even more deluded than those they treated.

Both Lemnius and Weyer believed that the treatment of disturbances of consciousness must be medical, first counteracting the distortions of the imagination, then correcting the underlying physical problem with the conventional tools of medicine. Since this explanatory framework holds that mental aberrations are the result of a malfunctioning organism, it can fittingly be called the *organic paradigm*.

From the sixteenth century the organic paradigm steadily gained ground over the intrusion paradigm. Its exponents pictured themselves in a noble battle with the benighted superstitions of primitive thought and as promoting the inevitable triumph of reason. They believed that science would eventually provide the answers, that it was only a matter of time until medical research would uncover the physiological sources of all disturbances of consciousness.

Mesmer's discovery of animal magnetism effectively gave birth to a third paradigm. Mesmer himself had no intention of challenging the organic paradigm. Quite the contrary; as the rational medical practitioner opposing the superstitions of the exorcist Gassner, he saw himself as the prototypical champion of the organic paradigm, combatting the unenlightened ignorance of the intrusion paradigm.

The discovery of magnetic sleep by Puységur, however, introduced a radically new view of the human psyche and opened up a fresh vista of psychological inquiry.[4] Magnetic sleep revealed that consciousness was divided and that there exists in human beings a second consciousness quite distinct from normal, everyday consciousness. This second consciousness in some cases displays personality characteristics unlike those of the waking self in taste, value judgments, and mental acuity. The second consciousness has its own memory chain, with continuity of memory and identity from one episode of magnetic sleep to the next, but it is separated from the person's ordinary consciousness by a memory barrier, and the two consciousnesses are often sharply distinquished, as though there were, as Puységur put it, "two different existences."

The importance of this discovery cannot be overestimated. The newfound second consciousness became the ground for a third paradigm for explaining mental disorders, based on the interplay between ordinary consciousness and the usually hidden second consciousness that is revealed in magnetic sleep. According to the *alternate-consciousness paradigm,* humans are divided beings. We have our ordinary consciousness, which we normally identify as ourselves, and a second, alternate consciousness that reveals itself in the magnetic trance and can seem quite alien to ordinary consciousness. The feeling of alienation is due in part to the memory barrier and in part to the fact that a distinct sense of identity is often present in the second consciousness. This alienation is the basis on which the alternate-consciousness paradigm explains mental disorders, for the second consciousness may develop thoughts or emotions very different from and even opposed to those of the ordinary self, causing one to think, feel, and act in uncharacteristic ways.

In the intrusion paradigm the feeling of alienness is automatically interpreted as an invading presence from the outside which is the source of the unaccountable thoughts, feelings, or impulses. In the organic paradigm the source is the body. In the alternate-consciousness paradigm the source is a dissociated part of the mind itself. The feeling of alienness results from a condition of inner division that, in normal circumstances, prevents one's ordinary consciousness from being aware of one's second consciousness.

The discovery of magnetic sleep thus led to the formulation of a revolutionary view of the human mind. Although the full implications of this discovery would not be developed until the latter part of the nineteenth century, the basic elements were there from the beginning. The alternate-consciousness paradigm made it possible to conceive of a subconscious or unconscious life in human beings and thereby provided the foundation for all modern psychotherapies that recognize the existence of a hidden arena of dynamic mental activity.[5]

5. The credit for first clearly recognizing this lineage goes to Ellenberger, whose *Discovery of the Unconscious* (1970) is a pioneering work of great originality.

Chapter 6

Love, Sexuality, and Magnetic Rapport

The basic elements of a magnetic psychotherapy were established in the very first decades of animal magnetism. Two of them, the uncovering of painful secrets in the state of magnetic sleep and the establishment of therapeutic rapport between magnetizer and patient, brought with them delicate issues.[1] Was the exposure of secrets desirable or hazardous? Did magnetic rapport involve excessive influence of the magnetizer over the subject? Could that relationship foster an unhealthy sentimental or even sexual attachment between the two? Could a magnetizer use somnambulism to sexually overpower an unwilling subject? The fate of a magnetic psychotherapy, especially at a time when the magnetizers were mostly men and the patients predominantly women, hung on the answers.

The Exposure of Secrets

Although the exposure of painful secrets was noted by a few magnetizers as a powerful therapeutic tool, it was ignored by most. Puységur discovered with Victor Race that secrets that one would never reveal when awake could be revealed in magnetic sleep, and Lutzelbourg used the somnambulistic revelation of secrets to deal effectively with the depressed state of one of his magnetic subjects.

Of course, not all forgotten memories retrieved in magnetic sleep were painful. Deleuze pointed out that an enhanced ability to remember past experience—a long-ago conversation, a pleasant childhood smell—was common among somnambulists.

He was also aware of the therapeutic value of revealing secret shame, the therapeutic usefulness of suggestion, and the responsibility of the magnetizer to maintain confidentiality:

> If your somnambulist has mental troubles which aggravate his malady, seek with him the means of easing them. You will console him,

1. The other elements of magnetic therapy—recognizing a second consciousness and its altered personality traits, dealing with two memory streams, and correcting disordered somnambulism—shall be discussed in chapters 14 and 15.

and profit by his confidence to relieve his anxieties, and destroy the cause. If he has any inclinations which you disapprove of, employ your ascendancy in vanquishing them. You must avoid, most carefully, penetrating into the secrets of your somnambulist, when it is not evidently useful to him to have these secrets known to you. I need not add that, if he tells you any thing which he would not have told you in the ordinary state, you will never permit yourself to impart it to any person, not even to your most intimate friend. (1850, p. 83)

Rostan (1825) was concerned that the mesmerist might use knowledge of this accessibility of memories against the somnambulist: "Might not a magnetizer rob the subject of important secrets and use them to his advantage? Do we not know that the well-being of families is often dependent upon a secret pertaining to particular circumstances? One family might be concealing its origins, another its fortune, another the illness of one of its members, still another an ambitious project, etc. Could not the disclosure of one of these secrets ruin an entire family?" (pp. 38–39). Demarquay and Giraud-Teulon (1860), in their study of somnambulism, feared that the revelation of embarrassing secrets might have its dangers, even given a trustworthy operator: "A local woman, hypnotized and questioned under conditions analogous to those mentioned in the previous observation, during the state of sleep became loquacious. In response to our scientific curiosity she began to bring forward confidences aimed at satisfying a totally different kind of curiosity, and *so grave were these confidences, so dangerous for her* that, afraid for this patient and struck by our responsibility so fatefully engaged, we hastened to awaken the unfortunate author of this excessively frank information" (p. 33).

A Special Connection

When Puységur initially described an "intimate rapport" between himself and his first magnetized subjects, he referred to one of the most extraordinary features of magnetic sleep. Rapport was a special connection between operator and subject in which the magnetized person was aware only of the magnetizer and could discern even his hidden thoughts, showed a strong dependency on the magnetizer, and therefore was ready to carry out the suggestions and commands of the magnetizer.[2]

---

2. Puységur (1820) compared the rapport manifested in magnetic somnambulism to that demonstrated in mineral magnetism between an iron rod and a magnetic needle: the somnambulist will obey the magnetizer as the needle obeys the rod (p. 2). In the state of rapport, he believed, the subject is so completely concentrated on the magnetizer that he or she has no relationship with anyone or anything else (Puységur 1811, p. 43).

The power of rapport was used (and abused) in magnetic practice with full awareness over the hundred years following Mesmer's discovery. For some practitioners, the phenomena flowing from this special connection between magnetizer and subject were little more than a matter of curiosity. For other practitioners, rapport was an important part of the healing process, providing a sense of confidence and security for the patient and in other ways enhancing the magnetic effects.

As Puységur became more experienced, he recognized that the rapport between magnetizer and patient was a powerful psychotherapeutic tool. He viewed mental disturbance as "disordered somnambulism," a state in which the patient, without realizing it, was in magnetic rapport with someone no longer present. Treatment involved breaking this hidden, disturbing rapport and replacing it by rapport with the magnetizer.

Dependency and the Sexual Link

Puységur himself agonized over the dependency created by magnetic rapport. He feared that the dependency of the subject might become a burden to the magnetizer and also that the magnetizer might abuse his power over the subject. The problem of excessive dependency predated Puységur and magnetic sleep. Even a casual reading of Mesmer's (1779) description of his treatment of Maria Theresa Paradis raises serious questions about the nature of the emotional forces unleashed in the application of animal-magnetic procedures.

The power of the feelings of this eighteen-year-old girl toward her hero-healer are not hard to imagine. In her parents' haste to remove their daughter from proximity to Mesmer one may perceive their dimly conscious misgivings about the emotional bond that was developing between doctor and patient. When Maria Theresa chose to remain in Mesmer's house rather than accede to her parents' demands, they saw that bond as a threat to familial ties and reacted violently. There are some hints in Mesmer's account that the parents may even have accused him of unspecified sexual improprieties.

The problem of the emotional bond that can be generated between magnetizer and magnetized arose frequently in the literature of animal magnetism.[3]

3. The dangers of dependence on the mesmerist were also noted in contemporary fiction. The German writer and musician E. T. A. Hoffmann (1776–1822), for example, was fascinated by animal magnetism and often incorporated a mesmeric theme into his tales. He warned of the dangers inherent in the mesmerist's power over his subject. In his stories mesmerists could, among other things, control the content of their subjects' dreams. The power Hoffmann attributed to the mesmerist made animal magnetism a "dangerous instrument," but also a fascinating literary theme. Nathaniel Hawthorne (1804–1864) was also intrigued by mesmerism, espe-

At times that bond was seen as an integral part of a healing process that provided much-needed security for the patient. But just as often the dependency was viewed with alarm. Again and again the theme of dependency and control in mesmerism reduced itself to the power of a male magnetizer over a female subject, and in the last analysis it becomes an issue of sexual passion. From the earliest years, concern that the magnetizer might sexually misuse the mesmeric relationship was expressed by both practitioners and opponents of mesmerism. The first lengthy treatment of the issue was by the members of the Franklin commission.

The Franklin Commission

In its public report on animal magnetism (1786) the Franklin commission had concluded that all the effects of animal magnetism could be attributed to touching, imagination, and imitation. In its secret report to the king of France, which was published only in 1800, Bailly pointed out that women, who comprise the majority of magnetized individuals who fall into a convulsive crisis, are particularly susceptible to these three influences. They are extremely imaginative; their constitutions make them very responsive to touch, which affects them throughout their bodies; and they are highly prone to imitation.

Men, Bailly wrote, exercise "natural empire" over women, which attaches women to men and stirs them emotionally. Since it is "always men who magnetize women," there is real potential for abuse. Many women who seek a magnetic cure are not really ill but are simply bored or looking for entertainment. On the other hand, some ailing women retain their freshness, sensibility, and charm—their sexuality—which can act on the physician. And since prolonged touch, communicated heat, and deep gazes—all of which are involved in magnetization—are ways in which affection between men and women are expressed, Bailly concluded that there is mutual danger to magnetizer and magnetized.

The intimacy of the magnetizing process came across in Bailly's description of the technique involved:

> Ordinarily the magnetizer has the knees of the woman gripped between his own; the knees and all the parts below are therefore in contact. The hand is applied to the diaphragm area and sometimes lower, over the

cially by the control he believed the mesmerist could exert over young women, enslaving their wills and passions. For him the influence of the magnetizer was both psychological and sexual (Schneck 1954, Kaplan 1975, Tatar 1978).

ovaries. So touch is being applied to many areas at once, and in the vicinity of some of the most sensitive parts of the body. . . . All physical impressions are instantaneously shared and one would expect the mutual attraction of the sexes to be at its height. It is not surprising that the senses are inflamed. The imagination, which is active, fills the whole machine [body] with a kind of confusion. It suppresses judgment and diverts attention, so that women cannot give an account of what they are experiencing and are ignorant of their state. (pp. 148–49)

Bailly then described the resulting "crisis," drawing a barely hidden parallel with sexual excitation and orgasm:

When this kind of crisis is coming, the face reddens and the eyes become ardent—this is a sign from nature that desire is present. One notices that the woman lowers her head and puts her hand in front of her eyes to cover them (her habitual modesty is awakened and causes her to want to hide herself). Now the crisis continues and the eye is troubled. This is an unequivocal sign of a total disorder of the senses. This disorder is not perceived by the woman but is obvious to the medical observer. When this sign occurs, the pupils become moist and breathing becomes shallow and uneven. Then convulsions occur, with sudden and short movements of the arms and legs or the whole body. With lively and sensitive women, a convulsion often occurs as the final degree and the termination of the sweetest emotions. This state is followed by a languor, a weakness, and a sort of sleep of the senses which is the rest needed after a strong agitation. (pp. 149–50)

Bailly pointed out, almost with surprise, that women feel no guilt about this experience and are quite ready to repeat it. Although many women do not experience these effects, many of those who do are not aware that the cause is sexual excitement. Some, to be sure, do become aware of what is happening to them and withdraw from magnetic treatment. But the problem is that the more innocent the woman is, the less likely she is to realize the nature of the experience. Bailly summed it up:

The magnetic treatment cannot but be dangerous for morals. In undertaking to heal illnesses that require a long treatment, one stimulates agreeable, sweet emotions that are missed afterwards and are sought to have again because they have a natural charm for us and contribute physically to our happiness. But that does not make them any less reprehensible morally, and in fact they are all the more dangerous since they can easily create a sweet habit. . . . Exposed to this danger, strong women will move away, but weak women can lose their morals and their health. (pp. 150–51)

When a police lieutenant who was present at some of the commission's investigations asked D'Eslon whether a woman being magnetized or in crisis could be sexually abused by the magnetizer, D'Eslon answered that she could be and added that this is why only men of high character should be permitted to practice magnetism. This was the reason D'Eslon's "crisis rooms" were open to the public. D'Eslon's answer hardly put the commissioners' minds at ease. They reasoned that if the possibility of abuse was real, sooner or later it was going to happen. If a physician was working privately with a woman, spending many hours alone with her, the opportunity for sexual abuse would present itself over and over. It seemed to the commissioners that for a magnetizer in such a situation to refrain from taking advantage of his subject would require a virtue beyond human capacity.

Puységur and Tardy de Montravel

The potential for abuse also concerned the discoverer of magnetic sleep. The marquis de Puységur had fears that the "intimate rapport" between operator and subject could lead to serious problems. He first voiced his concern in a letter about Victor Race, where he wondered "whether one can will evil as well as good" when using animal magnetism. Puységur lamented of ever resolving on his own the question of the potential for abuse and so turned to his patients for guidance. "They all assure me that in this state they maintain their judgment and their reason, and they add that they quickly perceive any bad intentions which one might have towards them; for then their health would suffer and they would wake up on the spot. Despite this, I cannot put blind confidence in this solution. . . . I am always left with inquietude concerning the abuse to which one may subject the discovery of this greatest good that exists" (1784, pp. 39–41).

Although the subject is not explicitly mentioned here, it is clear that the abuse he is particularly concerned about is sexual, with the typically male magnetizer taking sexual advantage of the typically female somnambulist. Puységur put one of his female somnambulists to the test:

One day I questioned a woman in the magnetic state about the extent of the power I could exercise over her. I was going to force her (without even speaking to her) through jests to hit me with a flyswatter she had at hand. I said to her: "Since you are obliged to beat me—me who has helped you—it seems likely that if I absolutely willed it I could do what I want with you: make you undress, for example, etc. . . ." She said to me, "Not at all, sir. It would not be the same. What you ask me to do [the beating] does not seem to be a good thing. I would resist it for a long time. But since it is a joke, in the end I would give in, because you will it absolutely. But as for the other matter, you would never be able to force me to take off my underclothing. My shoes, my hat—that would be as you please. But beyond that you would obtain nothing." (pp. 182–83)

Puységur eventually concluded that the power of the magnetizer over the somnambulist was limited to matters related to the good health of the subject. Beyond those borders, he believed, that power ceased.

Tardy de Montravel also emphasized the central place of the will in producing salutary magnetic effects and hence insisted on the magnetizer's will to do good. He believed that "platonic love" could be explained in magnetic terms. As a matter of fact, he saw magnetic rapport as identical with the state of sympathy that is at the heart of platonic love. He recognized that this view gave the detractors of animal magnetism the opportunity to charge magnetizers with actions contrary to good morals, making their female subjects slavishly dependent on them and subject to lascivious advances (pp. 70–71).

But Tardy de Montravel insisted that platonic love could never sully the purity of a magnetic subject because a somnambulist's moral instincts remain intact while in the state of magnetic sleep. Although awareness that comes through the external senses is stifled, that which is due to the internal sense, the sixth sense, remains in force, and it is here that one's moral values reside: "The moment the exterior senses take over from above, the moment a moral love and sympathy is displaced by the excitation of a physical love, somnambulism ought to cease and will cease, the internal sense, shocked by the external senses, will be extinguished and come to a waking state. And I am not surprised at the response given by so many female somnambulists on the subject: 'If you would try to do any dishonorable thing, anything contrary to my principles, you would do me great harm and I would immediately awaken'" (pp. 71–72).

Sounding the Alarm

Although Puységur and Tardy de Montravel may have put their own doubts to rest concerning the possibility of the sexual abuse of a woman by a male magnetizer, others were not so easily satisfied. In fact, in the undated *Lettre à l'intendant de Soissons* (publ. 1800) an anonymous writer penned his criticism of Puységur himself (described as a "young man") for the type of magnetic stroking he used to heal a seventeen-year-old girl suffering from anemia. The author noted that Puységur "placed his hands on her head, her brow, her neck; he gently tickled her nostrils, pressed on her breast in a manner that her nipples would have to have felt a slight rubbing; he closed her two eyes with his fingers; he moved an iron rod he had in his hand across her face, guiding it from the top of her head to the bottom of her breasts" (pp. 160–61). The letter strongly hinted that these touches along with the other passes involved in Puységur's technique excited the girl sexually (p. 163) and suggested that it was nearly impossible to believe that such operations could do otherwise:

> Will not a young man who takes a young woman of nervous disposition into his arms, rubbing the vertebrae, the diaphragm, the stomach, the nipples of the breasts, the navel area, cause a revolution . . . in the subject he is massaging? Surely he will. If the imagination is struck by fear and hope, or by a sensual delirium that no word in our language can express, then the subject is experiencing a dangerous abandon, and the physician will be able with guilty daring . . . I have to stop here! . . . This sect, embraced among voluptuous or credulous women—surely it is an evil. (pp. 166–67)

The British writer John Martin voiced similar, if less vehement, fears in his *Animal Magnetism Examined* (1790). He wrote that mesmerists had on occasion argued for keeping the techniques of animal magnetism secret "because a man who seemed only to play with a woman might throw her into a crisis, and behave improperly" (p. 19). Martin agreed that such abuse was possible, asking: "If a woman in a *common* crisis, has her understanding deranged, if the most usual symptoms are wildness of eyes, distortion of features, attended with convulsions and violent agitations, who could be tempted to behave improperly to such a woman? If in a *luminous* crisis, a Lady appears to be in a soft sweet sleep, temptation, to a stranger, may be stirring; but are Magnetists in danger? . . . can they . . . behave impertinently to such a Lady, in such a situation? They can" (pp. 19–20).[4]

---

4. Although the dire warnings voiced by the anonymous letter writer and Martin were heeded by few, there was abroad in the 1780s a popular notion that mesmerism had a mysterious sexual power. In a play called *Animal Magnetism,* published in 1789 and performed in both Dublin and London, the author, Mrs. Inchbald, used the notion that animal magnetism could be employed as a force to compel love. The comedy features two young couples whose love wins out over various obstacles through a cunning plan involving animal magnetism: the discovery of a certain "Doctor Mystery," who represented Mesmer. Doctor Mystery does not himself appear in the play. One of the young men passes himself off as this "magnetizing doctor" and makes extravagant claims for the power of magnetic fluid, including the production of *disease.* When a doubting local physician asks Doctor Mystery, "Pray, Doctor, is it true what they report, that he, who is once in possession of your art, can, if he pleases, make every woman who comes

Although he had no doubt about the reality of the phenomena of animal magnetism, A. Lombard thought that its benefits were outweighed by its drawbacks. In *Les dangers du magnétisme animal, et l'importance d'en arrêter la propagation vulgaire* (1819), he stated that women should stay away from animal magnetism altogether. They should not themselves magnetize people because "they have a constitution more frail than ours, and magnetism, in lavishing nervous fluid, very much exhausts them and leaves them no energy for fulfilling their sacred duties. Pregnancy and its concomitants, suckling, the periodic vicissitudes to which their health is bound, the ills which almost always occur when their fertility leaves them—these states require all the strength nature gives them" (pp. 99–100). According to Lombard, the modesty required of women dictates that they not practice animal magnetism, and certainly never on a man.

As to whether a woman should be magnetized, Lombard laments that in healing the body, magnetism may make the soul ill. The patient, aware of the magnetizer's intense care and affection, may be moved to respond in kind: "For a young woman with a virginal heart . . . it would be difficult to resist an emotion of esteem and gratitude and not overstep proper boundaries." A second danger is that the magnetizer may exercise his will over the patient to take sexual liberties with her. Lombard illustrates this risk by a decorous example:

A virtuous and profound young man had a young somnambulist. He wanted to experiment with his rational discoveries and see whether magnetism was dangerous. One night he instructed her. The next day, when the young woman was placed in somnambulism, she became agitated and her face showed the embarrassment of conquered modesty. When asked how she was during the night, she answered, in a tone that was troubled and full of pain: very badly. Overcoming her natural shyness she strongly reproached her magnetizer and begged him not to abuse his power, if he respected her life. I do not allow myself to repeat this experience; suffice it to say that this and many other things are possible. (pp. 101–2)

---

near him, in love with him?" (p. 12), the bogus mesmerizer assures him that it is true. The physician then confesses his love for a young woman who does not return his feelings, and asks whether Doctor Mystery can help him. Doctor Mystery responds: "Then this little wand shall cause in her sentiments, the very reverse. In this, is a magnet which shall change her disposition—Take it . . . and while you keep it, she will be constrained to love you with the most ardent passion" (p. 13). Not surprisingly, the physician is tricked and the young woman winds up with the man she really loves—all because of the sexual power popularly attributed to the animal magnetism of Doctor Mystery.

A more moderate view was expressed by J. J. Virey in the *Dictionnaire des sciences médicales* (1818). Virey believed that animal magnetism could be used beneficially but cautioned that forces were at work that could be dangerous. It was his opinion that convulsions and hysterical symptoms sometimes produced in women through animal magnetism were "the result of nervous emotions naturally produced, either by the imagination or through affections present between various individuals, and especially those that emanate from sexual relationships" (pp. 23–24). The connection that is often sought between magnetizer and female subject is the kind that is properly found between husband and wife—and this is a problem.

In 1825, in an article on animal magnetism written for the *Dictionnaire de Médecine,* Rostan perceived sexual dangers arising from the intimacy of the situation and the amnesia that regularly accompanied the somnambulistic state:

The magnetized person is in a state of absolute dependence on the magnetizer and his will is for the most part hers. Should she want to oppose her magnetizer, he can remove from her the faculty to act or speak. . . . What terrible consequences can come from this total power. What woman, what girl could be sure of leaving without having been reached by the hands of her magnetizer, who would act all the more confidently because on awakening all memory of what had occurred will have been entirely effaced. . . . The somnambulist develops towards her magnetizer a gratitude, an attachment without boundaries. She will freely follow him as a dog follows his master. The path is not long from there to a true passion. . . . How does one resist repeated touches, tender looks, daily meetings, protestations of interest on the part of one and gratitude on the part of the other. It is not possible. The intimacy is established . . . one can foresee what will follow. (pp. 38–39)

In 1839 Gabriel Lafont-Gouzi, who considered magnetic somnambulism a "monstrous situation," wrote that the practice of animal magnetism could not be tolerated and gave the following reasons: "The magnetic action directly and of itself attenuates reason, health, dignity, the will, and the liberty of those who are magnetized. . . . It inevitably gives magnetizers the means, the opportunity, and the pretext to satisfy their views, their interests, their passions, and even their foolish or criminal inclinations. . . . The magnetic effects can greatly surpass the intention and honest aim of the magnetizer" (p. 83). Lafont-Gouzi, a professor of medicine at Toulouse, called for a law prohibiting the teaching and practice of animal magnetism, which he considered "unhealthy, immoral, and subversive to the rights of man" (p. 136).

In 1845 *The Confessions of a Magnetiser, Being an Expose of Animal Magnetism* was published anonymously. The author, a former mesmerizer who called himself "R.S.," presented the problem of sex and control in animal magnetic practice as an issue of power:

> I am one of those who believe that there is a certain will within the cultivation of every powerful mind, which may be so exercised over the *dormant* powers of another, as to render the object or person, for the time being, subject to the authority or wish of the active agent. For instance, if you, kind reader, were willing to subject yourself to the power of Magnetism, you must first seat yourself, acquire an agreeable and quiet state of mind, be perfectly willing, or rather exert no will against the process, and your mind or brain becomes dormant, while the active mind of the *Mesmeriser* is exerted to its utmost power and capacity, and thus takes possession of your will, as it were, through the agency of the nervo vital fluid or invisible electricity, and becoming for the time and until that agency shall cease, the master of the body of the subject, its own mind being no longer active. (pp. 5–6)

R.S. believed that mesmerism created a state of "sympathy" between operator and subject that caused the subject to share the magnetizer's intimate thoughts and sentiments. He confessed that he had misused this sympathy when he was a practicing magnetizer, specifically in mesmerizing the "gentler sex":

> Reader, let me tell you that to be placed opposite a young and lovely female, who has subjected herself to the process for the purpose of effecting a cure of some nervous affection or otherwise, to look into her gentle eyes, soft and beaming with confidence and trust, is singular entrancing. You assume her hands, which are clasped in your own, you look intently upon the pupils of her eyes, which as the power becomes more and more visible in her person, evince the tenderest regard, until they close in dreamy and as it were spiritual affection.—Then is her mind all your own, and she will evince the most tender solicitude and care for your good. . . . Self is entirely swallowed up in the earnest regard that actuates the subject, and somnambulists will stop at no point beyond which they may afford you pleasure should you indicate it by thought or word. (pp. 9–10)

R.S. believed that the danger of abuse in this situation was great. He said that he knew a magnetizer who bragged that he could choose a wife from among his patients whenever he pleased. R.S. himself had felt his power over his female subjects and discovered that mesmerism could even be used to create affection where there had been anger. He described being at a social gathering and meeting a woman whom he had once seriously offended. He begged her forgiveness, but she rejected every attempt on his part to repair the rift. R.S. then suggested to the woman that he would be able to change her feelings if she would give him a few minutes of her time. He magnetized her on the spot and left her in the trance state for half an hour, during which time he mentally directed thoughts of kindness and affection toward her. Upon awakening, her attitude was entirely changed and she apologized for her coldness (pp. 10–11).

R.S. offered a stern warning to any woman considering magnetic treatment:

> There is at this writing, a man in this city, a practitioner in the art of Magnetism, whose pretended sanctity of character, would lead us to deem him worthy of trust, but it is to be deeply regretted that he has proved himself to exert his power in the art mainly for vile and sensual purposes. I cannot refrain from warning young females and even married ladies not to trust themselves alone with practitioners who are comparative strangers to them, for did I feel at liberty to reveal some startling facts with which I am conversant, the community would be thoroughly awakened to the danger of permitting publicly the exercise and employment of this agency. The ruin and utter destruction of many a domestic circle must eventually follow, and in one case I already know the greatest unhappiness to exist. (pp. 11–12)

R.S. stated that he had not himself actually engaged in a sinful act with a subject, but he reproached himself for an abuse of a more subtle kind. He felt that he had exercised the magnetic art not so much for the good of the client as for the pleasure it afforded him to feel such affection from his female subjects and such control over their emotions (p. 12).

As the practice of animal magnetism became more standardized and manuals of practice began to appear, mesmerizers made frequent mention of the sexual dangers and the precautions that should be taken when a male magnetizer treated a female patient. Although it was generally assumed that the healing practitioner was a man of virtue, he was admonished to beware of situations that might develop in good faith but lead to dangerous consequences.

Deleuze (1813) pointed out that even the best of gifts could be corrupted and that magnetism was no exception. He warned that a mother should never leave her daughter alone to be magnetized by a young man, no matter how high an opinion she might have of him. Also, a wife should always be magnetized in the presence of her husband when a male magnetizer was involved. He warned the magnetizer to be on guard against any feeling that developed within him other than the desire to heal and relieve suffering. In regard to deliberate and premeditated seduction, he stated: "As to the possibility of abusing magnetism and making it a means of seduction, I will not even speak about that. A man who makes himself guilty of such a crime would be an object of horror to society" (1:204).

Beyond the danger of direct sexual interaction, Deleuze wrote about the "tender attachment" a patient can feel toward the magnetizer. He cited an example from his own experience:

> For some time my health had been bad, and a young woman obliged me by magnetizing me in a chain that included her parents, some friends, and two or three ill persons. When she touched me I fell into a light sleep that lasted the length of the séance. After ten or twelve days I became aware that she had inspired me with a special affection and that I was involuntarily consumed by the feeling. Fifteen days later I felt well and we ended the treatment. Afterwards the impression that she had made on me gradually diminished and I saw her as I had before, with a respectful attachment but without emotion. In recounting this I must say that during the time when her image was constantly with my spirit, I never had a thought that I could not acknowledge without blushing— whether it is because the affections produced by magnetism are disengaged from the senses or whether my friendship and my place in the family removed every reprehensible idea from me, I do not know. (1:204–5)

Just how the magnetizer might lessen the chance that a problematic attachment would occur Deleuze described in his *Instruction pratique* (1825):

> When a man is desired to magnetize a woman who is ill, he ought to avoid whatever may wound the most scrupulous modesty, or cause the least embarrassment, and even whatever might to a spectator seem improper. He will not place himself directly in front of the person whom he intends to magnetize; he will not request her to look at him; he will merely ask her to abandon herself entirely to the influence of the action; he will take her thumbs during some moments, and he will then make

passes at a distance, without touching her. It is unnecessary to observe
that some one of the family, or a female friend, ought always to be
present. (Deleuze 1850, p. 161)

Not all magnetic practitioners seemed to grasp the gravity and complexity
of the problem as clearly as Deleuze did. Thomas Pyne (1849), expressing
unbounded confidence in human nature, wrote: "No good person would be
magnetised any more than he or she would converse with those in whom
confidence could not be placed" (p. 75). William Gregory (1851) admitted
that "no power is known which cannot be perverted and abused by human
depravity" (p. 46) and yet claimed: "In general, the moral perceptions and
feelings of the somnambulist are exalted and strengthened in the sleep, and he
generally exhibits a profound aversion for all that is bad, false, and mean. In
vain might we try, in many cases at least, to induce the subject to violate
confidence, or to betray a secret which he has learned in his sleeping
state. . . . Were we capable of trying to persuade the sleeper to do a bad
action, we should soon discover that he is awake to moral obligations, and
usually much more so than in his ordinary waking condition" (p. 48).

One might expect that J.J.A. Ricard, in his *Physiologie et hygiène du
magnétiseur* (1844), would have followed Deleuze's lead, offering prudent
advice about contact between male magnetizers and female patients, but he
did not. Although Ricard spent a great deal of space debating which of the four
temperaments (sanguine, nervous, lymphatic, and bilious) was most suited to
the magnetizer, emphasized the importance of strength of will in magnetic
work, and pointed out how bad breath and smelly feet on the part of the
magnetizer would imperil the treatment, he hardly seemed to be aware that
emotions exchanged between operator and subject might be a problem. He
wrote about the necessity of a parental attitude in the magnetizer, recommended that reciprocal affection should exist between the two parties, and
even described how magnetic effects occurring between the opposite sexes are
most powerful, but he said nothing about the boundaries that should be
observed in these cases.

The magnetizer Alphonse Teste showed more psychological insight into
the magnetic relationship. In a chapter titled "On the Necessity for Morality in
Magnetizers" in his *Manuel pratique de magnétisme animal* (1840), he undertook to "boldly expose abuses which, to the shame of humanity, I consider to
be too real" (1843, p. 307). He proclaimed as fact the "absolute unlimited
power" of the magnetizer over his subject and warned that this power levies a
moral responsibility on the practitioner that he "may not always prove himself
worthy of." Even if the magnetizer does not overtly abuse this power, it may

happen that the somnambulist "penetrates your most secret desires, associates her soul with all the emotions of your soul, and, without perceiving that she only obeys your will, anticipates even your most secret intentions." That is, she may work his will for him. "And though she still retain the discernment of right and wrong, she belongs body and soul to him, if he is base and dastardly enough to abuse such power. . . . Yes, that is true, unfortunately too true; magnetism may produce between two persons of different sexes a profound, extreme, and insurmountable attachment" (pp. 397–98, 401). Teste's solution to this problem was to recommend that only physicians practice animal magnetism. He believed that, although anyone may be subject to such temptation, physicians as a professional group were the least vulnerable.

Although there were many anecdotal allusions to the sexual abuse of magnetized women, no clearly documented cases are to be found until the 1860s. The best-known early case was one that came to trial in 1865. On March 31, 1865, a twenty-five-year-old vagrant named Timothée Castellan, who had been roaming the French countryside, arrived at the village of Guiols. He was taken into a household where a twenty-six-year-old woman, Joséphine, lived with her father. He secretly made magnetic passes behind her back and acted in other bizarre ways, finally succeeding in putting her into a somnambulistic state. He then had sexual relations with her. From then on she seems to have been continually under his power, and the two of them left the village together. For some time he succeeded in maintaining his control over Joséphine by means of magnetic passes, forcing her to stay with him and sleep with him. Eventually someone noticed that the young woman was acting peculiarly, seeming to be in a kind of trance state. She was not able to escape from Castellan until he was detained by conversation with some hunters; she used the opportunity to run away and succeeded in returning home. Castellan was charged and brought to court. At the trial, expert testimony agreed that Joséphine had lost her free will through the induction of magnetic somnambulism and that Castellan was guilty of rape. He was sentenced to twelve years' hard labor (Tardieu 1878, pp. 92–101).

Suggestion

Puységur's fears about the possibility of sexual misuse of the magnetic situation were based on his experience with suggestibility in somnambulists. Although the term "suggestion" was not employed in its psychological sense until about 1820, the reality of the phenomenon was noted very early. Puységur first encountered suggestibility in his magnetic treatment of Victor Race. In Puységur's account, he "inspired" in Victor's mind ideas that caused both perceptions and actions (Victor imagined himself to be shooting for a prize and danced as though he were at a party) determined by Puységur. The king's commissioners had noted similar phenomena and described them in terms of imagination and imitation.

Virtually every magnetizer after Puységur who induced magnetic sleep was aware of the reality of magnetic suggestion and the control that "intimate rapport" between magnetizer and somnambulist could produce. Much was written about the role of the "will" of the magnetizer and the subjugation of the somnambulist. Practitioners were often both enamored with and awed by the power they could exercise over magnetic subjects. To a degree, the somnambulist was seen as an extension of the magnetizer and capable of operating only according to his will. In these early years magnetizers concentrated their attention on the actions and physiological states that magnetic suggestion could induce. Using suggestion to create hallucinations was less investigated and would become more important only in later decades.

Post-somnambulistic suggestion was also widely recognized among magnetizers. One of the early enthusiasts of magnetic somnambulism was de Mouilleseaux (1787), who proposed the establishment of a journal to publish articles on the systematic and scientific study of animal magnetism (a journal that never saw the light of day). His treatise contained, among other things, a succinct and informative list of the phenomena of magnetic somnambulism. Among them are the following:

> To attend to, respond to, and obey (in the crisis) the commands, the signs and even the thoughts of the magnetizer. . . . To execute (in the natural state), the will of the magnetizer—the execution occurring through a kind of irresistible impulse, without the somnambulist being able to give a reason for the action and without the somnambulist realizing that the command for the action was given during the crisis— whether the effect of this will takes place right away or is delayed by some days. (p. 83)

In this rather awkward passage, Mouilleseaux provided the earliest known reference to post-somnambulistic (or later "post-hypnotic") suggestion.[5] The command of the magnetizer is given to the subject while he or she is in "crisis" or a state of magnetic sleep, but the execution takes place afterwards, in the "natural" or waking state. The subject in the waking state knows nothing of

5. In the same section, Mouilleseaux stated that one of the characteristics of magnetic somnambulism was that it could be self-induced (1787, p. 84). This is probably the earliest reference in the literature to self-mesmerization (later "autohypnosis").

the command, and when the act is carried out, he or she can give no reason for the action.

The power of post-somnambulistic suggestions was noted by many magnetizers. In his *Instruction pratique* Deleuze (1825) advised the magnetic healer to make direct suggestions to the somnambulist about what to eat and how to care for himself. When in the normal state, the patient would then carry out those suggestions without realizing he was doing so (Deleuze 1850, p. 118). Bertrand (1823) noted that if the magnetizer said to a somnambulist, "Come to see me on such and such a day at such and such a time," the subject would execute that command even though he had no memory of it and was not able to account for his action (pp. 298–99; see also Gregory 1851, pp. 110–11). Teste (1845) also used post-somnambulistic suggestions to inculcate healthy actions (pp. 431, 435), and Jules Charpignon (1841) was able to create a complex post-somnambulistic hallucination that persisted for two days after awakening (p. 82).

This extension of the power of the will over the magnetic subject beyond the state of somnambulism added to concern about the possible abuse of the somnambulistic subject and raised the problem of the subject's responsibility for actions performed as a result of suggestions from the magnetizer. Both issues had obvious legal implications, and cases began coming into the courts around the middle of the nineteenth century.[6]

Although concerns about control and sexual abuse of somnambulists continued to haunt those involved in the practice of animal magnetism, they were never sufficient to imperil its use. Because those concerns were voiced not only by critics of mesmerism but also by its advocates, the cautions and safeguards recommended seem to have provided a fairly effective curb on potential misuse and helped insure that the good of the patient would remain the primary concern in the practice of magnetic healing.

6. Among the principal nineteenth-century works dealing with this aspect are Despine 1868; Mesnet 1874, 1894; Tardieu 1878; Brouardel 1879; Ladame 1882, 1887; Liégeois 1884, 1889; and Gilles de la Tourette 1887. A full treatment of the medico-legal aspects of artificial somnambulism is beyond the scope of this work; the reader is directed to the excellent study *Hypnosis, Will, and Memory* (1988), by Jean-Roch Laurence and Campbell Perry.

Part II

Magnetic Medicine

Chapter 7

Magnetic Healing

From its very beginnings animal magnetism was first and foremost a system of healing. Puységur discovered that healing was not merely a physical process but also involved the mind and soul of both healer and healed. Along the way, he opened to investigation a new world within the human psyche and set the stage for the development of a system of healing mental ills that was revolutionary.

In contrast to Mesmer's physicalist approach, Puységur insisted that healing also involved the psychological. Although after Puységur's discovery the psychological (and parapsychological) aspects of the healing process were of great interest to magnetizers, their main concern remained the cure and amelioration of physical ills. For decades Puységur was almost alone in his interest in psychological healing. The truth is that for the most part magnetic practitioners were unable to distinguish between psychological and physiological healing. So while their theoretical explanations of their work often incorporated psychological elements indicated by Puységur, their understanding of the treatment of mental disorders did not reflect the progress Puységur had made in this area. This state of affairs is evident in the literature of animal magnetism in the decades following its discovery.

Mialle's Compendium of Cases Treated by Animal Magnetism

Detailed descriptions of the treatment of illnesses by animal magnetism can be found throughout the mesmeric literature of the early nineteenth century. The vastness of this literature, however, made it difficult for those interested in learning magnetic healing to perceive the patterns of treatment and appreciate the range of cures performed. Students of magnetism were greatly helped when, in 1826, Simon Mialle published a two-volume collection of cures effected through animal magnetism. Mialle combed hundreds of books, pamphlets, and articles to compile nearly twelve hundred pages summarizing treatments and results. This valuable source contains both an index and a list of works consulted, labeled pro, con, or neutral in regard to animal magnetism. Writing at a time of renewed controversy about animal magnetism (see chapter 9), Mialle hoped, among other things, to induce his readers to look objectively at the accounts given by magnetizers and judge for themselves. For that reason he often provided extensive quotations from his sources.

Mialle himself had been healed of a "coughing of blood" by the marquis de Puységur. After this successful encounter with magnetic healing, Mialle began to study animal magnetism. This led to his examination of the literature and eventual cataloguing of cures. Mialle discovered that many terms had been used for animal magnetism over the decades, but the reality designated by the terms was remarkably consistent.[1] In his compendium, he often quoted the words of the original account. However, where he presented his own summary he used the standard terms "animal magnetism" and "magnetize."

At the beginning of each case description, Mialle stated the illness treated, the names of the patient and magnetizer(s), and the magnetic technique used. He called the usual technique of magnetic passes made with the hands "immediate magnetism." Other frequently mentioned techniques were the use of the magnetic *baquet,* the magnetic chain, and the magnetized tree. Some of the cases presented involved magnetic somnambulism; others did not. Where somnambulism was involved, the case description tended to be lengthier because it usually included some dialogue between patient and magnetizer. Some of the accounts were brief—only a few lines—and did not follow up the condition of the patient after treatment. For example, in 1789 Catherine Montavon Werner was treated with "immediate magnetism" at Strasbourg by a magnetizer named Waldt. Mialle tells us: "After a difficult childbirth experience, this woman was beset by a strong pain on the left side. She had endured it for nineteen years, and ordinary medicine had been unable to cure her. Monsieur Waldt magnetized her. After three days the pains ceased" (1:142).

Elsewhere Mialle went into much greater detail. Thus he reports that in 1818 Dubois Maillard was treated at Nantes under the direction of a magnetizer named Segretier:

> [Maillard], a seaman by profession, was shipwrecked on the coast of Africa and was battered by the sea for many hours. A six-month illness

1. Mialle found that animal magnetism was called mesmerism, puységurism, barbarinism, noctambulism, somniloquism, sympathism, onirabanism, phantasiexoussism, and hypnology. The last term is an interesting anticipation of Braid's later nomenclature, "hypnotism." To magnetize was called to touch, to mesmerize, to actuate, to de-organize, to sympathize, to give a crisis, and to put into crisis, etc. Magnetic operators were called mesmerists, touchers, magnetizers, directors, onirexitists, phantasiexoussists, and hypnologists. Somnambulists were labeled noctambulists, crisiacs, crisiologues, somniloquists, époptists, hypnologues, hypnoscopists, onirobatists, and oniroscopists, etc. (Mialle 1826, p. xxxix n.)

followed. Just beginning to recover, and racked by a continual cough, he returned to France and there, during maneuvers, was struck by a drive wheel on his right side. The blow was so violent that he fainted and then vomited a great deal of blood. Lacking a surgeon, he did not heal and had only water for medicine for the rest of his passage. Arriving in his homeland . . . he worsened day by day. Then his wife and his mother went by boat to Monsieur Segretier, who had already refused many times to see the man, since he believed him beyond help. The way the man looked, his color, the stench of his breath, the perpetual cough accompanied by a prolonged wheeze—all this simply confirmed him in his belief. He had the man placed under one of his magnetized trees and attached him to it with ropes, placing one as a local remedy on the side on which he had been injured. And, turning the man's head to one side, he magnetized him at the top of his arm three or four times for an hour and a half. The action had an immediate effect, and the pains returned to the point of making the patient sweat. All of his expectorations were infected. Because it was impossible for the man to come often for treatment, and considering his weakness, the distance of the place, and the bad effect the man had on other patients, Monsieur Segretier gave him a magnetized bottle, fitted with an iron rod at the end of a cord. He said he should take a spoonful of watercress juice in two spoonfuls of syrup of Althéa every day in the evening, on an empty stomach, etc. He said he should return after eight days, if he had the strength. The bottle procured him a good sleep and the next day he vomited a panful of clotted blood. He felt much better and improved through the whole week so that he was able to keep the appointment. His change of health was already so remarkable that Monsieur Segretier began to envision a speedier and more complete cure than he had thought possible. From then, Dubois began to spend all day at the *baquet*. He continued to cough up pus until all the congestion was expelled. At the end of two months, he took three baths and his cure, which had not at first seemed possible, was complete. Since then this man has regained his weight and his strength. He went back to work, and before the grape season he came to thank his magnetizer for having returned him to life and to his family. (1:200–202)

Sometimes healing was accompanied by somnambulism and somnambulistic diagnosis. A simple case of this type involved a resident of Nantes in 1817: "For twenty years Mlle Brunelière suffered from headaches, which many physicians attributed to an abscess. She became somnambulistic the first time she was magnetized and said that in three days she would discharge through her nose a congested deposit that had accumulated in her head. Her prediction was exactly fulfilled and from that time on she has been in good health" (1:195).

Mesmer himself believed that animal magnetism would be particularly effective with hysteria or hysteria-related conditions, and this proved to be true. His treatment of hysterical blindness in Maria Theresa Paradis in Vienna showed signs of success but was aborted by family interference. A number of magnetizers since then reported cures of similar cases of blindness, as well as hysterically produced deafness, paralysis, convulsions, and other related conditions.

Mialle described a case that involved hysteria with somnambulism treated at Aschaffenbourg in Bavaria in 1817 by the surgeon Herr Zahn. The subject was a young woman between eighteen and twenty years of age:

> Monsieur Zahn undertook the treatment of this young woman in the month of July. She had been ill from age fifteen. Physicians had treated her for hysteria, but none of their remedies had worked. Her condition was complicated by natural somnambulism. In this state the patient conversed with her parents and read with eyes closed. She was considered insane and feared to be epileptic, since she had terrible convulsions. From the first [magnetic] passes she became somnambulistic and accurately answered all questions. A remarkable clairvoyance developed, but her treatment was very difficult. Often Monsieur Zahn had to spend three or four hours with her and bear with annoyances of many kinds. Nevertheless, his courage, patience, and perseverance triumphed over all obstacles, and he had the good fortune to successfully conclude the cure of this person towards the end of the year [in six months]. . . . Her only medicine was magnetized water, and it bothered her to drink anything else after her cure. (1:492–94)

Convulsions of various kinds were often cured through animal magnetism. In most of these cases their hysterical origin seems fairly certain. Puységur himself successfully cured fifty-two-year-old Catherine Bauz of this condition in 1784: "Monsieur de Puységur, being in Strasbourg, was invited to magnetize a woman who for twenty-one years had been subject to convulsions that would occur many times a week. She was magnetized for only twenty-one days, during which time the convulsions occurred only once, ending after five minutes of magnetizing. Being obliged to return home . . . to aid her ailing husband, her good health continued. On two different occasions she has written to Monsieur de Puységur to confirm her cure" (1:125).

There were, however, successful cures of convulsions that were not hysterically produced—infantile convulsions, for instance. For those who believed that animal magnetism cured merely through imagination and suggestion, successes recorded in treating infants had to be accounted for. D'Eslon had earlier recorded such a case. Mialle described the successful treatment of infantile convulsions by Puységur in 1809. The child, Honorine, was ten or twelve months old and had been in a state of violent convulsions for some hours. Medicines normally employed in such a case had proven useless, and the desperate parents sought out Puységur to ask him to apply animal magnetism. Puységur wrote:

> When I came to the home of Madame Mo***, I saw a scene of great sorrow. The little Honorine, eyes open and fixed, was rigid and unmoving. Her parents, mournful and silent around her, seemed to be simply waiting for the moment of her last sigh. Without saying anything to them, and without even asking anew for their consent, I took the little Honorine into my arms along with the pillow on which she was lying, and I sat down and placed her on my knees. Then, without thinking about it and without thinking about anything that was going on around me, I concentrated entirely on magnetizing this child, with the single volition of producing the effect that would be most salutary for her. After some minutes I thought I perceived a return of respiration. I placed a hand on her heart and sensed feeble beating. Second by second, I announced each of the consoling remarks that I was making, but without becoming distracted and as though I was speaking to myself. My profound meditation imposed a silence which, in the sorrowful suspense of the moment, no one dared break. Then all of a sudden we heard the reassuring noise of an abundant evacuation. I expressed my joy, and then, without uncovering or looking at the little one, I continued with even more energy to exercise my magnetic action. The immediate happy result was a general relaxation of the muscles and the cessation of the convulsive state. After less than a half hour, I had the sweet satisfaction of returning this beautiful little one, now freed from the danger that had been menacing her, to the arms of her mother. (1811, pp. 71–72)

Theories of Treatment

Mesmerizers persisted in trying to account for the cures and ameliorations brought about through animal magnetism—even though many said, with Puységur, that they did not completely understand how and why animal magnetism worked. Magnetic theorists fall into five groups: those who emphasized the action of the magnetic fluid (orthodox followers of Mesmer); those who stressed the importance of the magnetizer's good will (followers of Puységur); those who believed the healing action to be mainly spiritual (followers of Barberin); those who related magnetic action to the concept of sympathy (especially Hufeland); and those who emphasized the importance of suggestion (Faria, Hénin de Cuvillers, and Bertrand).

Orthodox Mesmerism

After the turn of the century there were few who limited themselves to the orthodox magnetic technique of Mesmer.[2] Puységur's discovery of magnetic sleep and his use of somnambulists in the healing process so strongly influenced practitioners that few restricted themselves to the magnetic passes and healing aids (such as the *baquet*) recommended by Mesmer. Most magnetizers employed these procedures in conjunction with diagnosis and treatment prescriptions obtained from patients in the state of "lucid" magnetic sleep.

Perhaps only Mesmer himself maintained a truly orthodox approach, resisting the tide of interest in artificial somnambulism as an adjunct to treatment. His view of the theory of animal magnetism remained basically unchanged from 1775 until his death in 1815. Mesmer did, however, elaborate on the basic principles laid down in his *Mémoires* of 1779 and 1799 and especially in his master work, written with the assistance of Karl Wolfart, *Mesmerismus. Oder System der Wechselwirkungen* (1814). Here he reaffirmed his view of animal magnetism as a physically based phenomenon of nature and stated certain implications for family and social life that he believed necessarily followed from that view.

In 1811 the Berlin Academy of Science, learning to their surprise that Mesmer was still alive and living in Meersburg on the Bodensee (in southern Germany), sent Karl Christian Wolfart (1778–1832) to ascertain Mesmer's

2. At least one practitioner who would perhaps qualify as a totally orthodox magnetizer was Dr. Georg Christophe Würtz (1756–1823). He was a personal friend of Mesmer who published a prospectus for a new theoretical and practical course on animal magnetism (Würtz 1787), in which he deplored the use being made of magnetic somnambulists in healing. From letters written to Mesmer from Versailles from 1805 to 1809 (Kerner 1856, pp. 127–37), it is clear that his support of Mesmer was still complete. Another old friend of Mesmer's, P. J. Bachelier d'Agès, although apparently not a practitioner, continued to write about animal magnetism in orthodox fashion. Bachelier d'Agès had assisted Mesmer in the construction of his *Mémoire* of 1779 and supported him over the years. A man of the Revolution and an author of a treatise on finances and the French constitution, Bachelier d'Agès wrote a political work that incorporated the theory of animal magnetism and the socio-political implications drawn from it principally by Bergasse (Bachelier d'Agès 1800).

views about animal magnetism. In September 1812 Wolfart arrived at Meers-burg, where he spent about a month discussing animal magnetism with Mesmer and helping him to put together a new treatise on the subject. In *Mesmerismus* Wolfart is listed as "editor." Oddly enough, the work was compiled from notes written by Mesmer in French and translated by Wolfart into Mesmer's native German.[3]

With the controversy concerning the control of the propagation of information about technique in the past, Mesmer wrote quite freely in *Mesmerismus* about both theory and its practical implications. As before, he took as his starting point the general notion of magnetism, positing a universal "nature-magnetism" that affects all things. In living things, universal magnetism takes the form of animal magnetism. Just as it had been discovered that one could produce artificial mineral magnets, so also Mesmer had discovered that he could directly affect the animal magnetism of the human organism.

At this point Mesmer introduced a new term to designate the concentrated magnetism of the living body. He called it an "invisible fire," stating that it could be distinguished from the general nature-magnetism in the same way that fire is distinguished from warmth. Invisible fire is not ordinary fire, vibration, light, mineral magnetism, or electricity. It is more subtle than all these and is what acts on the nerves of the body and brings the organism into harmony with nature. Invisible fire can affect all bodies, "besouled and un-besouled." That is why animal magnetism can be applied to all living things and can be stored in nonliving substances, such as water or silk. The magnetizer, using his conscious will, can transmit this fire by touch, but he can also convey it by pointing his hand or a conductor, or by a glance (Schott 1982, pp. 111–12).

When he magnetized, Mesmer wrote, he concentrated and built up this invisible fire in his being to such a degree that it could be called forth and applied to others. The magnetizer taps an inner energy or movement, so that to magnetize is "to direct and share this fire through a kind of overflow or discharge of this movement. The overflow acts through direct contact, or by pointing with the extremities or 'pole' of an individual who possesses this power or this fire, or through one's intention or thought. Since all organic substances are penetrated by this streaming flow, they are capable of receiving this 'tone' and being magnetized, just as every substance penetrated by air can be a conductor of sonic resonance" (p. 117).

There are several ways to concentrate and strengthen the action of invisible

3. Wolfart wrote his own commentary on *Mesmerismus* preceded by a brief history of Mesmer's work (Wolfart 1815).

fire or animal magnetism, including combining a number of organic bodies in the magnetic action (e.g., the magnetic chain); using hard substances, such as metals, that "accelerate" the action through their concentration of matter; using bodies that have an "inner movement," such as electrified substances, animals, trees, or plants; or concentrating thought or will (pp. 113–14).

The magnetizer, in applying the invisible fire, assists nature by affecting the very life principle of the patient: "The basic essence of life in man, the life principle, consists in his share of the life-fire which he received at the beginning of his life and which is maintained and nourished through the influx of the universal-movement. This life-fire . . . maintains and regulates the performance of the organism. . . . Total loss of movement or life-fire is death" (p. 163).

According to Mesmer, life consists in a balance between movement and rest. Health exists when all the parts of the organism can operate without hindrance—when there is complete order and harmony. Sickness is a state of disharmony, resulting in a hindrance of function. Healing is the reestablishment of the order and harmony that has been lost. There is no healing without a crisis, "the striving of Nature against the illness" (p. 170). The art of healing, therefore, is knowing how to bring on the crisis, facilitate it, and guide its development. Once healing has taken place, no more crises are experienced and, in fact, the patient is no longer subject to the magnetic action (pp. 168–73).

The Tradition of Puységur

Magnetic practice in France in the early nineteenth century was dominated by Puységur's emphasis on somnambulism and its uses in healing. When animal magnetism spread to the German-speaking world in the 1780s, it came almost exclusively in the Puységuran form. Similarly, after the seminars of John Bell and the classes of John de Mainauduc had faded, the new wave of animal magnetism in England, from the late 1820s on, centered around the production of magnetic sleep.

Puységur was an ardent supporter of Mesmer and never had anything but praise for the discoverer of animal magnetism. Though maintaining all the basic features of magnetic practice prescribed by Mesmer, he gave his personal stamp to that practice by ascribing less importance to the magnetic fluid and more to the application of a "will to do good" and, most important, through his discovery of magnetic sleep and his recognition of its usefulness in the healing process.

In the early 1800s, probably the best spokesmen for animal magnetic healing in the tradition of Puységur were August Roullier and Joseph Deleuze, whose manuals for magnetic healing provided a good summary of the theory of animal magnetism according to Puységur for that time. In *Exposition physiologique des phénomènes du magnétisme animal et du somnambulisme* (1817) Roullier credited Mesmer with discovering certain fundamental principles of magnetic theory which hold their own against all criticism, such as the doctrine of the magnetic fluid. Deleuze also accepted the reality of the fluid. In his *Instruction pratique sur le magnétisme animal* (1825) he insisted that if a magnetizer is to be effective, he must know about the fluid's movements in the living body and how to control its flow from his body to that of the patient. Roullier and Deleuze left Mesmer far behind, however, when dealing with magnetic somnambulism. Although Mesmer discussed artificial somnambulism in *Mesmerismus* (pp. 172, 198–212), he did not appreciate its usefulness and certainly did not give it a central position. Roullier (1817) wrote, as had Tardy de Montravel some decades earlier, that somnambulists could see the magnetic fluid and describe its movements (p. 3), and both he and Deleuze emphasized the important uses that could be made of this kind of somnambulistic clairvoyance.

In his *Instruction pratique,* Deleuze summarized the nature of magnetic somnambulism:

> Magnetic somnambulism . . . is a mode of existence during which the person who is in it appears to be asleep. If his magnetizer speaks to him, he answers without waking; he can also execute various movements, and when he returns to the natural state, he retains no remembrance of what has passed. His eyes are closed; he generally understands only those who are put in communication with him. The external organs of sense are all, or nearly all, asleep; and yet he experiences sensations, but by another means. There is roused in him an internal sense, which is perhaps the center of the others, or a sort of instinct, which enlightens him in respect to his own preservation. He is subject to the influence of his magnetizer, and this influence may be either useful or injurious, according to the disposition and the conduct of the magnetizer. (Deleuze 1850, pp. 68–69)

In good Puységuran fashion Roullier stated the importance of the operator's will in magnetic somnambulism: "Everyone in a crisis of lucid somnambulism senses the will of the magnetizer and more or less submits himself to it in everything that is not harmful or contrary to the ideas of justice and truth. But this salutary dependence is only established through a well-founded confidence in the morality of the magnetizer and his sincere desire to do good. No one can be magnetized despite themselves—much less be placed in a state of somnambulism" (pp. 94–95).

In the tradition of Puységur, the importance of somnambulism for the healing process consists in three things: it puts the ill person in a calm state in which the body can better heal itself; it enables some somnambulists to diagnose their own illnesses, or those of others, and prescribe remedies; and it enables some somnambulists to foresee the course of an illness and predict its conclusion (Roullier 1817, pp. 88ff; Deleuze 1850, pp. 68ff).

Deleuze also believed that the somnambulistic state provides an opportunity to impose helpful suggestions on the patient to aid the course of the treatment:

> Somnambulists perfectly abstracted, whose interior faculties have acquired great energy, are often found in a frame of mind of which you might avail yourself advantageously to make them follow a course of regimen, or to make them do things useful for them, but contrary to their habits and inclinations. The magnetizer can, after it has been mutually agreed upon, impress upon them, while in the somnambulistic state, an idea or a determination which will influence them in the natural state, without their knowing the cause. For instance, the magnetizer will say to the somnambulist, "You will return home at such an hour; you will not go this evening to the theater; you will clothe yourself in such a manner; you will take your medicines without being obstinate; you will take no liquor; you will drink no coffee; you will occupy yourself no longer in such a thing; you will drive away such a fear; you will forget such a thing." The somnambulist will be naturally induced to do what has been thus prescribed. He will recollect it without suspecting it to be anything more than a recollection of what you have ordered for his benefit; he will have a desire for what you have advised him, and a dislike for what you have interdicted. Take advantage of this empire of your will, and of this concert with him, solely for the benefit of your patient. (Deleuze 1850, pp. 88–89)

Deleuze had grasped the idea of post-somnambulistic (post-hypnotic) suggestion. He knew that when the patient in the waking state responded to implanted suggestions, he would not realize that the impetus for his action or feeling had come from the somnambulistic session, and he appreciated the power of this kind of suggestion.

In the tradition of Puységur, somnambulistic clairvoyance was more or less taken for granted. This faculty was always considered completely natural, and some practitioners believed it could be explained in terms of purely mechanical laws. There were, however, magnetizers who saw the magnetic operation as a basically spiritual one. These operators followed in the tradition of the chevalier de Barberin, on the one hand, and the Stockholm Exegetical Society, on the other (see chapter 4).

The Spiritualists

Although Puységur emphasized the importance of the will, he also recognized the central place of touch and other physical actions. The spiritualist magnetizers, however, held that after establishing rapport, touch had no further purpose to serve. They taught that will and thought were everything, stressing mental intention and prayer in the magnetic process (Deleuze 1813, 1:96).

Deleuze was frankly puzzled by the magnetic theory of the spiritualists. He believed in the immateriality of the soul, but was convinced that in our world one soul always operates on another through material means. He was also bothered by the fact that some of the spiritualists had turned their somnambulists into oracles whose pronouncements on spiritual and mystical matters were taken as infallible truth (1:98, 235ff).

In all his writings about spiritualist magnetizers, Deleuze never made it clear whom specifically he was talking about. It seems safe to say, though, that he was referring principally to those magnetizers in Germany who were followers of both Mesmer and the Swedish seer Emanuel Swedenborg.

In Germany the chief theoretician for the spiritualist magnetizers of the early nineteenth century was Johann Heinrich Jung-Stilling (1740–1817). Born to pious parents in Westphalia, he dreamed of being ordained a preacher. Instead he became a physician and eventually professor at the universities of Heidelburg and Marburg. He was first exposed to animal magnetism by two friends who were important participants in the German magnetic movement in the late eighteenth century: Arnold Wienholt and Johann Böckmann.

Jung-Stilling's main work in the area of spiritualistic magnetism was *Theorie der Geister-Kunde* (1808). Following the teachings of Swedenborg, he believed it could be shown through empirical evidence, derived largely from the visions of somnambulists, that human beings dwell in two worlds, the material world and the world of spirits. He claimed that since somnambulists, in ecstatic vision, have direct access to the spiritual world and can bring forward verifiable information not available through ordinary means, we have direct and irrefutable proof of the reality of that world.

For Jung-Stilling, human beings are composed of three elements: the immortal spirit, the ethereal fluid, which is the principle of corporeal life, and the material body. Together the spirit and ethereal fluid make up the human soul, which is capable of existing apart from the material body. Through magnetization, the soul becomes free from its usual adherence to the brain and nervous system. In that state it may perceive the world without the use of bodily senses, communicate directly with other souls in this world, or communicate with spirits not of this world. The more completely it is detached from the body, the more it is capable of communicating with that other world. Less perfectly detached souls may be subject to imagined apparitions and delusions (Jung-Stilling 1854, pp. 27–54).

It is not hard to see why Deleuze objected to the spiritualist view of animal-magnetic action. If the soul is basically independent of the body and in magnetic somnambulism operates separately from the body, physical action becomes more or less superfluous. Not only that, the emphasis on communication with the spiritual world takes the magnetizer in a direction very different from that conceived by Mesmer or Puységur. He becomes more a midwife for mysticism than a healer of physical ills. He concentrates more on the supernatural than the natural and can easily be led into what Deleuze considered dangerous excesses produced by the fertile imaginations of somnambulists. Nevertheless, the spiritualists continued to thrive and, as we shall see, eventually returned to France in force.

Sympathy

In 1789 Johann Rahn (1749–1812) wrote a work entitled *Über Sympathie und Magnetismus* examining animal magnetism in the light of a concept of sympathy deriving largely from Renaissance medicine. Rahn stated that he discovered nothing basically new in animal magnetism, finding it to be simply a rehash of the old notion of a world spirit that also permeates the human body. This universal world spirit, he said, produces not only the general mutual influence between all bodies, heavenly and earthly, but also a special sympathy between one man and another, which is, in the last analysis, the bond between body and soul.

Some years later, Friedrich Hufeland (1774–1839) took up the concept of sympathy, using it as the basis for a sweeping, Romantically oriented criticism of the Enlightenment view of science:

> Up until now it has been the custom of physical scientists to separate living things from dead things and make only the latter the object of their observations and investigations. They were misled by a false notion of life that sees organic bodies as isolated, existing by themselves and out of connection with nature in general, as though they were examining a separated essence, and they abandoned the investigation of

their physical relationships to the physicians, who for the most part share this one-sided view with them. It directly followed . . . that each single body could only be studied in connection or relationship with others if one zealously endeavored to examine, through a physical and chemical investigation, the way, as natural bodies, they acted on one another in terms of mass and quantity. (Hufeland 1811, pp. iii-iv)

Hufeland strongly objected to this and said it was time to arrive at a comprehensive theory that took the unifying interconnections of nature into account. This is the reason he considered animal magnetism to be of great importance, since it offered a starting point for a truly unified theory of nature, revealing its hidden power and laws.

Hufeland believed that each individual being is a limited manifestation of nature as a totality. All things are involved in a coordinated developmental process and necessarily affect one another as they go through this process. Just as heavenly bodies are bound together by gravity and enjoy an individual unity by weight, so also all living things are bound together as a whole and enjoy an inner unity as an organism. This is what Hufeland calls "sympathy" (pp. 5–6).

Although human beings participate unreflectively in the unconscious developmental process of nature and, because they belong to the same species, experience a special sympathetic bond with each other, at the same time they have the special faculty of rational consciousness, which allows them to stand apart from the process, reflecting on the nature of the world and viewing it with a kind of objectivity. Hufeland taught that the invisible bonds that unite all things in sympathy act at a distance. Physical contact is unnecessary, for sympathy is not a mechanical action but a dynamic tension. This means that the influence of living things extends beyond the limits of their bodies.

Human beings have a unique capacity to affect the way sympathy operates among them. That faculty is exercised in the application of animal magnetism. Through the magnetic art, a healthy individual can induce a sympathetic response in an ill person that may restore him to health:

The phenomena that accompany animal magnetism show us most palpably how the sympathetic action of a living body acts on an ill person. In essence, these phenomena are nothing but the action of the sympathy that exists within individuals expressed in a more visible outward form. Every individual organism that acts positively on another polarizes it, and, pulling it (as it were) into its sphere, magnetizes it. For that reason, the notion of animal magnetism should not be limited to that sympathetic relationship (commonly called "rapport") brought about through artificial manipulation. The conditions in which it occurs are the same as those in which sympathy operates, namely through a definite grade of difference between two distinct subjects. . . . The phenomena of animal magnetism are more remarkable and striking when the positive principle of life predominates in one individual and his sphere of action has a greater intensity; and when, at the same time, the other subject, because of the "disorganization" of his subjective sphere, is weakened and receptive to sympathetic action. (pp. 98–99)

With his notion of animal magnetism as a dramatic manifestation of the bonds of sympathy present in nature, Hufeland brings the Romantic nature philosophy of his time to bear on the theory of animal magnetism. As we will see in chapter 10, this Romantic view of animal magnetism continues to be a strong theme in the German-speaking world, finding its principal expression in the writings of Eschenmayer and Ennemoser.

Suggestion

Although the beginnings of hypnotic suggestion and suggestive therapeutics are commonly traced to the 1880s, they actually predate that period by many decades. Mouilleseaux had a notion of postmagnetic (hypnotic) suggestion, and Deleuze incorporated into his healing technique suggestions that would affect the patient after the magnetic session. There were, however, three men who developed the notion of magnetic suggestion far beyond these passing references: the priest José Custodio de Faria (1755–1819), the historian Etienne Félix Hénin de Cuvillers (1755–1841), and the physician Alexandre Bertrand (1795–1831).

Faria, Portuguese by birth, spent much of his adult life in France. He eventually became a member of the Société Médicale of Marseille and taught at a number of universities. Although it is not clear that he ever received formal instruction in animal magnetism, there is little doubt that he was inspired by the works of the marquis de Puységur, and it is known that by 1802 Faria was himself practicing animal magnetism (see Dalgado 1906; Moniz 1925). By 1813 Faria had developed a unique view of animal magnetism and was teaching a public course on "lucid sleep," his term for magnetic somnambulism. In 1819 he published his theory in *De la cause du sommeil lucide*.[4]

Because his conception of animal magnetism was so different from the views of both Mesmer and Puységur, Faria developed a new vocabulary.

---

4. For more detailed biographical information on Faria, see Dalgado 1906b and Moniz 1925.

Besides "lucid sleep," which he defined as natural somnambulism "developed with art, guided with wisdom, and cultivated with caution" (p. 22), he used "concentration" for "animal magnetism" and *épopte* for "somnambulist" (pp. 39–40).

Faria contended that the true cause of the phenomena of animal magnetism was psychological. This did not mean that he believed they were simply the result of imagination. But he did teach that, contrary to Mesmer's view, there is no magnetic fluid and that, contrary to Puységur's view, the will of the magnetizer is not involved. In other words, for Faria there is no external agent that produces the effects. Rather, the magnetizer makes use of suggestion to allow the subject to put himself into a state of lucid sleep, although the subject has the false conviction that he needs an external agent to accomplish this. Through the induction of lucid sleep the subject is brought into a state of inner concentration whereby the soul is disengaged from the external senses (pp. 74–82).[5]

The technique Faria used when publicly inducing lucid sleep demonstrated the importance and power of suggestion: "In advance I make myself aware of who, according to the external signs I notice, has the required disposition for occasional concentration [ordinary lucid sleep]. I seat them comfortably and energetically pronounce the word 'sleep.' Or I show them my open hand, at some distance, and have them fix it with their gaze, not turning their eyes aside and not resisting the urge to blink. In the first case, I tell them to close their eyes, and I always say that when I forcibly pronounce the command to sleep they will feel a trembling all over and will fall asleep" (pp. 192–93).

Faria believed that the healing powers of lucid sleep are due to the powerful effects of suggestions from the operator. He called these healing suggestions "effects announced in advance" and insisted that they were effective whether made in lucid sleep or in the waking state. In fact, Faria taught that all the phenomena of lucid sleep could be brought about in the waking state, since the powers exercised by the subject are always there (pp. 51, 78).

Among the powers of suggestion Faria included the ability to affect the functioning of the body. He claimed that through post-sleep suggestion he was able to cause patients, upon awakening, to have a menstrual flow, to vomit, or to experience other physical movements needed for their healing. Faria considered suggestions so powerful that "at a word one can cause healthy *époptes* to become ill, and render ill *époptes* healthy" (p. 269).

5. Excellent discussions of Faria's theory and technique will be found in Dalgado's introduction to the 1906 version of *De la cause du sommeil lucide* (Faria 1906) and in his article "Braidisme et Fariisme" (Dalgado 1906a).

In the year following the appearance of *De la cause du sommeil lucide,* a new periodical devoted to animal magnetism appeared. The *Archives du magnétisme animal* was intended to supplant *Bibliothèque du magnétisme animal,* which had earlier replaced *Annales du magnétisme animal.* In his introduction to the first volume (1820) the editor, Etienne Félix Hénin de Cuvillers, proclaimed that the *Archives* would be a forum for the expression of all views on animal magnetism, but in fact it turned out to be principally a vehicle for his own theories, and by the third volume he openly acknowledged that it had ceased to be a periodical in the usual sense.

Hénin de Cuvillers divided views about animal magnetism into three categories: those of the "magnetists," who naively and uncritically accept the reality of magnetic fluid; those of the "physiologists," who reject the notion of the fluid and for the most part all the phenomena of magnetism; and those of investigators who agree with him in rejecting the fluid but accepting the phenomena that can be produced by the imagination.

Endorsing the notion of a magnetic fluid, said Hénin de Cuvillers, leads to belief in such absurdities as "action at a distance" and an "occult" power exercised by a few over the many. He substituted a theory of imagination, a term that in his usage was the equivalent of suggestion (1822, pp. 362ff). Since the term "animal magnetism" embodies the notion of a magnetic fluid, Hénin de Cuvillers substituted his own nomenclature. He built one set of terms around the Greek word for sleep: *hypnos.* Somnambulists were called *hypnoscopes* (those who see while asleep), *hypnologues* (those who talk during sleep), and *hypnobates* (those who walk during sleep); the process of producing somnambulists was called *hypnoscopie;* and *hypnocratie* referred to the power that produced the effects of *hypnoscopie.* He also adopted certain words that were already in use to describe effects related to the sleep state, such as *hypnologie* and *hypnotique* (*Archives* 1:131–34, 198; *Archives* 3:43–44).[6] Using another Greek word for sleep, Hénin de Cuvillers established an equivalent vocabulary—*oniroscopes, onirobates,* etc.—and used the two sets of terms interchangeably (*Archives* 1:134; *Archives* 3:43–45). In 1822 he invented yet another set of terms, this one based on the Greek word for imagination. The result was an extremely awkward vocabulary which, in French, included the words *phantasiésouxie, phantasiéxousisme, phantasiéxousique,* and *phantasiésouxiquement* (*Archives* 5:53ff). What he at-

---

6. Hénin de Cuvillers's original terms include *hypnotisme, hypnotiseur,* and *hypnotiste* (Hénin de Cuvillers 1821, p. 33). With this nomenclature, he anticipated the terminology of "hypnotism" developed by Braid in 1842 and 1843 (see chapter 8).

tempted to do with this final nomenclature was to express his concept of the phenomena of animal magnetism and the reason it works:

> After all, it is truly the imagination that always produces the pretended magnetic phenomena. . . . One can say, without compromising oneself, that one believes in *phantasiéxousisme* or in *phantasiésousie,* that is to say in the power of the imagination. Who would dare to doubt this astonishing power of the imagination, so well known in antiquity as well as modern times? . . . Animal magnetism has since Mesmer been abandoned into the hands of men who are, for the most part, not very enlightened by philosophy, and especially the science of physiology . . . they continue to accept the existence of a pretended animal magnetic fluid that operates immaterially and absolutely independently of the senses and the imagination. It is time that the truth bring enlightenment to this chaos of absurd and mystical ideas that ignorance and obstinacy seems to always want to perpetuate. (*Archives* 5:55–57)

Hénin de Cuvillers was strongly critical of the writings of the magnetists, stating that, lacking a critical sense, they grouped genuine phenomena with patently absurd phenomena and called them all magnetism (pp. 46–47). Always seeking the sensational, they on the one hand exaggerated their experiences to absurd proportions and on the other explained purely natural phenomena on the basis of a nonexistent fluid. He attributed great power to the imagination and the "faith" stirred up by the circumstances of magnetic treatment (*Archives* 5:97ff). He contended that even animals are capable of having their imaginations affected by magnetic procedures (5:55ff). He believed that it was imagination that produced artificial somnambulism, with its special sensitivities, and that "in certain circumstances the imagination can control all the faculties and all the passions" (p. 184). He insisted that authentic cures are often effected by the imagination, and that accounts from antiquity confirm that this power is real. He also said that many times the ill person is overwhelmed by negative thoughts, and when the magnetic ritual ignites new hope, a sudden, striking, but short-lived improvement may result (p. 369).

For Hénin de Cuvillers, imagination, fired by the expectation of a particular outcome, was the sole cause of successful magnetic action. The magnetizer works in a state of self-delusion, believing that he produces the magnetic phenomena through the action of the fluid directed by the force of his will. His success is real, however, because the patient shares the same delusion, so that his imagination works strongly to produce the desired results. Hénin de Cuvillers called for recognition of the true source of the power of animal magnetism. But if an aspiring magnetizer should agree with him that all was the power of the imagination, how was he to carry out his healing work? Hénin de Cuvillers never tackled the problem of how to excite the power of the imagination without using the conventional magnetic procedures. While telling the magnetizer what not to do, he did not suggest an alternative. That is probably why his influence on magnetic practitioners was never great.

A few years later, Alexandre Bertrand, a highly regarded magnetizer and physician, published *Du magnétisme animal en France* (1826), a work of history that sent shock waves through the community of magnetizers, for it was an about-face from his much praised *Traité du somnambulisme,* which had appeared in 1823. Bertrand's preface, entitled "How the author came to realize that animal magnetism does not exist," denied the reality of animal magnetic fluid and presented a totally different explanation for the phenomena of animal magnetism. He stated that he accepted the phenomena of animal magnetism as real. On the other hand, he had come, through close observation of many cases, first to doubt and then to deny the accepted explanation for those phenomena. When he began magnetizing, the striking results he obtained convinced him that a new agent was at work—the magnetic fluid of Mesmer. Many of his somnambulists confirmed this view by stating that they could perceive the fluid emanating from his fingers, and he saw no reason to deny the existence of one more "imponderable fluid" among the many posited by scientists of the day (pp. i–xi).

But certain anomalies in the phenomena began to give him doubts. He came to realize, for instance, that one could not accept as truth everything somnambulists said, for there were many direct contradictions in the pronouncements of somnambulists from various backgrounds, and they could not all be right. Also, he noted that persons instructed to sit under a certain magnetized tree or hold a magnetized article would fall into a state of somnambulism even when they mistakenly sat under the wrong tree or held an unmagnetized article (pp. x–xix). The only conclusion he could draw from this evidence was that the phenomena of animal magnetism, genuine and beneficial as they may be, were due not to a magnetic fluid directed by magnetizer to subject but to the command, overtly or subtly expressed, that the magnetizer conveys to the subject. Without realizing it, the magnetizer and subject cooperate psychologically to produce the effect. The effect is suggested by the words or actions of the magnetizer or by the context of the encounter, which creates the expectation of certain phenomena.

The Theory of Petetin

Jacques Henri Desire Petetin (1744–1808) was a physician of Lyon who developed a special interest in the treatment of hysterics. In 1787 he wrote a work on his experiments with somnambulistic hysterics, *Mémoire sur la découverte des phénomènes que présentent la catalepsie et le somnambulisme,* in which he stated his belief that certain hysterics spontaneously enter somnambulistic states and that magnetic somnambulism could be better understood through the experiences of these patients. Petetin also believed that hysterical symptoms were produced by an accumulation of "electrical fluid" in the brain and the stomach area. It was his contention that somnambulists were particularly sensitive to electrical fluid and that this sensitivity made them clairvoyant, enabling them to discern objects concealed in opaque containers. Like Tardy de Montravel before him, Petetin thought that somnambulists were capable of seeing and hearing from the stomach area, and he believed that experiments he carried out with hysterical somnambulists confirmed that belief.[7]

According to Petetin, the application of animal magnetism excites the attention and the imagination, thus producing a concentration of electrical fluid in the brain and an irritation of the nerve fibers. The result is the state of induced hysterical somnambulism. The view of magnetic sleep as an artificially produced state of hysteria was reintroduced much later by Charcot.

Petetin evolved a theory of action of electrical fluid within the nervous system to replace that of Mesmer. From this theory, Petetin developed a method of treating hysteria and illnesses arising from that condition. Although he used some conventional approaches, his principal technique was to apply "artificial electricity" to remove the imbalances of electrical fluid in the body. In 1805 Petetin published a new treatise on the subject, *Electricité animal,* further developing the ideas stated in the earlier work.[8]

Petetin considered himself an electrical and galvanic experimenter rather than an animal magnetizer. His writings on the action of the electrical fluid in the body and his views on hysterical somnambulism, however, tie him to the animal magnetic theorists of his day.

7. Bertrand criticized Petetin's work for its naivete, claiming that Petetin unwittingly created an epidemic of cataleptic hysteria and engineered the production of the symptoms, which he then used to confirm his theory (Bertrand 1826, pp. 254–55).

8. There is a second edition of this work that appeared posthumously (Petetin 1808). It contains additional material including a long note on Petetin's life and writings, a letter written to Petetin in 1808, and a lecture delivered by Petetin in 1807. Petetin also wrote a more technical manual in the same area in 1802 titled *Nouveau méchanisme de l'électricité.*

Techniques and Conditions of Treatment

If someone wanted to take up the practice of animal magnetism in the early nineteenth century, there was no lack of opportunity to learn how to do it. Unlike the days when Mesmer claimed absolute control over the teaching of the art, those interested in magnetic practice could now find copious explanations of theory and technique in books, journals, and manuals. As we have seen, by this time even Mesmer was willing to get down to the brass tacks of magnetic technique.

A number of journals specializing in magnetic practice made their appearance in the first decades of the nineteenth century. In France the most important was *Annales du magnétisme animal,* which, under various titles, ran from 1814 to 1823. Edited in its first years by Alexandre Sarrazin de Montferrier (1792–1863), who went by the pseudonym of A. de Lausanne, its articles were well written and knowledgeable in both the history and practice of animal magnetism. In Germany the most influential magnetic periodical was Karl Christian Wolfart's *Jahrbücher für den Lebens-Magnetismus oder neues Askläpieion.* It ran from 1818 to 1822 and contained a wealth of practical information.[9]

Among the manuals published in the same period, the most important were those written by Alexandre Sarrazin de Montferrier (from this point referred to as Lausanne) and Joseph Deleuze. Lausanne's two-volume *Des principes et des procédés du magnétisme animal* (1819) was probably the most thorough treatment of magnetic practice ever published, and Deleuze's *Instruction pratique sur le magnétisme animal* (1825) was without question the most popular.

Lausanne

Lausanne's approach shows the influence of the three main magnetic schools of the 1780s: those of Mesmer, D'Eslon, and Puységur. He was able to synthesize the main elements of these schools in a comprehensive form, augment his synthesis with original methods deriving from his own experience, and present the resulting system in a cogent way to the public.

---

9. The *Jahrbücher* was successor to Wolfart's *Askläpieion: Allgemeines medicinisch-chirurgisches Wochenblatt* that appeared for only two years (1811–1812). Other German periodicals on animal magnetism in the first four decades of the nineteenth century were *Archiv für den thierischen Magnetismus* (1817–1824), *Blätter für höhere Wahrheit* (1818–1827), *Zeitschrift für psychische Aerzte* (1818–1819), and *Blätter aus Prevorst* (1831–1839). In France, besides the *Annales* there were *Le propagateur du magnétisme animal* (1827–1828), *Le révéla-teur* (1837), and *Journal du magnétisme animal* (1839–1842).

Lausanne had an appreciation for Mesmer and his groundbreaking work, but he also saw what he believed to be his limitations. Chief among them was Mesmer's inability to recognize the importance of the influence of the "moral" or psychological on the physical. He stated that Mesmer failed to give weight to the way the patient's mental state affects his body, and ignored the importance of such factors as suggestion and imagination in producing the cure. Lausanne learned this from D'Eslon, from whom he had his first lessons in animal magnetism, and who treated Lausanne's own son for an illness (Montferrier 1819, 1:3, 132, 182n). Lausanne also held that Mesmer did not perceive the importance of the magnetizer's will in the magnetic procedure, a factor recognized clearly by Puységur. It is the will that directs the magnetic fluid, said Lausanne, and without a knowledge of that fact the magnetizer cannot be effective (pp. 24–27).

Convinced of the importance of the magnetizer's psychological participation, Lausanne recommended that he steadfastly concentrate on the patient and fix his attention on the healing action. In this connection, Lausanne proposed mental images for the magnetizer to hold in his mind to make the healing more effective.

Another central element in Lausanne's system was his belief that the magnetizer was capable of using his natural instincts to sense the nature and location of the illness in his patient. Although he was not the first to mention this idea, he was the first to describe how that instinct functions and to advise in detail how it could be most beneficially applied.

Lausanne began his treatment sessions with the conventional magnetic pose. Seated face to face with the patient, the magnetizer placed his legs against the outside of the patient's legs, foot against foot and knee against knee. In that position, he put his hands on the shoulders of the patient, envisioned the nerve plexus behind the stomach, and directed the first movement of fluid to that area. Or, as an alternative, he might concentrate on the principal parts of the body, one after the other, from the head downward. After this, he made passes slowly from shoulders to stomach, with the hands a few inches from the patient's body. Lausanne believed that at the conclusion of this action he should be in rapport with the patient.

Once in rapport, Lausanne's next step was to discover the "seat of the illness" in the patient. To do this he mentally concentrated on the various parts of the patient's body in turn, directing magnetic fluid to each. He alternated his attention between the body part of the patient and the corresponding part of his own body, attempting to sense if there was some impediment or deflection of the magnetic current. Any aberration from an easy flow back and forth indicated an illness.

From this it is clear that Lausanne demanded an unusual degree of involvement with the patient, a concentration that could brook no distraction. He also demanded that the magnetizer be in a state of good health when doing healing work—for two reasons. First, he explained, the sensing of the seat of the illness depends upon sensing any interruption of the magnetic current, so the magnetizer himself has to be free of illness to make an accurate diagnosis. Second, the magnetizer directs his own magnetic fluid into the body of the patient, so that any disorder in his own body will produce a similar disorder in the corresponding part of the patient's body. Lausanne held it as a principle that an unhealthy magnetizer produces bad effects, and he believed that D'Eslon's "pernicious habit of allowing himself to be magnetized daily by people suffering from the gravest illnesses" was the reason for his death at age thirty-six (Montferrier 1819, 1:59n). For similar reasons, Lausanne insisted that during the magnetic operations the patient must be passive. If the patient were active or resistant and directed his own magnetic fluid into the magnetizer, the result would be detrimental to the health of the magnetic operator.

Central to Lausanne's magnetic technique was the notion of "sensing the currents." In this he left Mesmer far behind, because although Mesmer wrote of a current of fluid that was set up between the magnetizer's hand and the patient, he did nothing to clarify how the magnetizer should use that current. Lausanne stated that he was personally acquainted with many of Mesmer's disciples, and none of them knew anything about sensing the currents.

In his instructions, Lausanne detailed the technique for perceiving and interpreting the fluidic currents. He noted that they vary subtly according to the type of illness involved, its location, and its severity. Placing his hands at the area of illness, the magnetizer may feel the emanation of a breeze. He may then experience heat, coldness, tickling, tingling, numbness, quivering, or sensations of wavelike motion. Each of these sensations tells the magnetizer something about the illness. Lausanne gave examples of how to interpret the sensations, but he seemed to believe that experience would be the magnetizer's best teacher.

Lausanne held that it is not just the hands that sense the currents. Once in magnetic rapport with the patient, the magnetizer also experiences sensations in those parts of his body corresponding to the affected part of the patient, especially the internal organs. Lausanne believed that this could happen because of the effects of sympathy. In this way the magnetizer may experience the patient's symptoms (pain, for example) in the corresponding part of his own body. This is what Lausanne called a distinct and direct sensing of the illness (p. 108).

Lausanne used all of the experiences summed up under the term "sensing the currents"—surface sensations (such as tingling), sympathetic feelings (such as pain), and "drawing" feelings—as sources of information for treatment. He counseled that the magnetizer should use this sensing ability to explore the body of the patient to diagnose the illness and determine the proper healing procedure. He described the process of exploration in systematic detail, stating that to discover the seat of more elusive illnesses the drawing sensations were particularly effective (pp. 155ff).

When Lausanne wrote his manual of magnetic practice, he had been working as a mesmeric healer for thirty-five years. He had developed a remarkably consistent synthesis of all the main magnetic traditions while adding original touches of his own. No one before him had so clearly detailed a system of magnetic procedure, and no one who followed added anything essentially new.

Joseph Deleuze

Although Lausanne's *Des principes et des procédés* was the most complete magnetic manual, it was not the most popular. The most widely read compendium of magnetic practice was by Joseph Philip Francis Deleuze (1753–1835). Highly regarded for his kindness and integrity, Deleuze practiced his mesmeric art for half a century without charge to his patients. Although he learned from Mesmer, Lausanne, and others, his technique was based on the approach of Puységur, which he elaborated into a comprehensive system of healing.

Deleuze was not just a healer. He was also a scholar and a man of science who developed an international reputation as an expert on plants. He was the first real historian of animal magnetism, writing a two-volume account (Deleuze 1813) that remains to this day an indispensable source of information for scholars. His other works on animal magnetism (Deleuze 1817, 1819, 1821, 1825, 1826, 1828, 1834) were written in defense of its genuineness and utility through a period of revitalized interest and renewed controversy in France.

Deleuze was introduced to animal magnetism in 1785 when he read the account of the cures performed at Buzancy (Cloquet 1784). At first he believed that the descriptions of cures were untrue, fabrications intended to ridicule animal magnetism. But when he heard that a friend had succeeded in producing somnambulism in a female patient through animal magnetism, he decided to visit him and investigate the matter. His friend demonstrated the effects on his patient by having Deleuze and others present form a magnetic chain with the woman. She fell asleep in a few minutes, but Deleuze's scientific observations were cut short when he himself entered a somnambulistic state a little while later. Much affected by this striking and very personal demonstration, Deleuze learned the technique of magnetic induction and returned home to experiment with it on some ill persons in the vicinity of his home. He was, in his own words, "careful not to excite their imaginations," for he wanted to satisfy himself that the effects were really due to animal magnetism and not suggestion. So he magnetized them "under various pretexts" and obtained beneficial results. Shortly afterwards he attempted to convince a friend of the efficacy of animal magnetism and was rewarded with remarkable success: "He introduced to me a young woman who had been sick seven years. She constantly suffered great pain and was very bloated. She also had a local swelling externally because of a notable enlargement of the spleen, which she showed to us. She was not able to walk or lie down. I succeeded in removing the obstruction. Circulation was restored, the swelling gradually disappeared, and she was enabled to attend to her customary duties" (Deleuze 1850, p. x).

Deleuze pursued his experimentation with animal magnetism for two years. Then in 1787 he traveled to Paris and threw himself into the study of the liberal arts, concentrating on botany. In 1798 he was appointed assistant naturalist of the Garden of Plants in Paris and in 1802 became the first secretary of the Association of the Museum of Natural History. All this time, however, Deleuze continued his work with animal magnetism, becoming increasingly proficient as a magnetizer.

In 1813 he published the two-volume *Histoire critique du magnétisme animal*, which was followed twelve years later by his manual *Instruction pratique sur le magnétisme animal* (1825), a highly popular work that went through many editions and was translated into English and German. The *Instruction pratique* provided a wealth of detail about technique and conditions for magnetic healing, and it became the basis for the magnetic practice of many mesmerizers in the decades to follow. In the matter of magnetic procedures, Deleuze was far from dogmatic. He realized that there was a great diversity of practice and did not intend to change well-established routines. Rather, he wrote for those who needed guidance in arriving at a predictable and efficacious approach.

Deleuze shunned working with patients who were not determined to make the most of the opportunity, or whose family or advisers were going to become obstacles to a smooth treatment. Although Deleuze insisted on having at least one witness present, he did not want to have people idly standing around at the sessions, so he suggested that anyone who was in the room with magnetizer and patient should unite himself with the magnetizer in the intention to do good for the patient.

In regard to the magnetic procedure proper, Deleuze adopted the standard position and posture, saying that the magnetizer should face the patient with the patient's knees and feet between his own. Then, wrote Deleuze, having placed himself in a self-collected state, the magnetizer should grasp the two thumbs of the patient, one in each hand, and hold them until he perceives an "equal degree of heat between your thumbs and his." The next step was for the magnetizer to put his hands on the shoulders of the patient for about a minute and then draw them along the arms to the fingertips, only lightly touching the patient. This pass was to be repeated five or six times, sweeping the hands off to the side at the end of each pass. The hands were then to be placed on the top of the patient's head for a moment and moved down the front, at a distance of about two inches from the body, to the stomach, pausing there for two minutes. Continuing the downward movement of the hands, the magnetizer was to bring them to the knees and then the feet, if convenient. Similar passes could then be performed over the back surface of the patient's body, and more abbreviated passes could be applied to various parts as the magnetizer saw fit (Deleuze 1850, pp. 28–30).

The session would be terminated by special passes to take away the effects of the magnetic action. These were conducted down the whole body to the feet and beyond, with the magnetizer shaking his fingers at the end of each pass. To these were added horizontal sweeping passes made at a distance of four or five inches in front of the face and the stomach.

Deleuze went into great detail about the form and direction of the passes, their distance from the body, the muscular force to be used, the images to keep in mind, and the attitudes to be assumed. He also presented alternative methods to be used if, for instance, the patient is in bed.

Much of Deleuze's handbook was taken up with prescriptions for how to treat specific conditions magnetically. The mode of treatment seems to have been arrived at through a combination of deductions derived from the magnetic theory of illness and experience by trial and error.

To aid the practitioner, Deleuze included detailed information about "accessories" to be used to increase the magnetic action (Deleuze 1850, pp. 56–67). First in importance he lists magnetized water, which could be used in a variety of ways. Deleuze also stated that food could be magnetized and used medicinally. He claimed that he knew of someone who was allergic to milk but had no trouble with magnetized milk. He also recommended the use of "magnetic reservoirs" such as the *baquet* and the magnetic chain.

Looking at animal magnetism from the point of view of the patient, Deleuze made suggestions about choosing a magnetizer. He believed that the most desirable thing would be to have a magnetizer in your own family: "The ties of blood contribute, by a physical sympathy, to establish a communication. The confidence and friendship which exist between a husband and his wife, between a mother and her daughter, and between near relations, have already produced that affection and that devotedness which ought to unite the magnetizer to the somnambulist, and which authorize the continuance of these sentiments when the treatment has ceased" (pp. 106–7).

Deleuze stated that, all else being equal, the best magnetizer for a woman is her husband, and vice versa. But since often a magnetizer will not be found within the family, Deleuze further advised that women be magnetized by women: modesty, he asserted, prevents male magnetizers from using the same posture and physical operations with a woman as with a man; the lively affection felt by a female somnambulist for a male magnetizer may not seem proper; the female patient can discuss certain female symptoms more easily with a woman than with a man; and crises may induce spasmodic movements in women that it is not proper for a man to view (pp. 107–8). Moreover, for Deleuze, deleterious gossip was not a matter to ignore: "It is almost impossible, especially in a small town, for a man to come each day and pass an hour with a woman without people's perceiving it, and discovering the reason. Then inquisitive persons ask the magnetizer many questions which embarrass him; and, if the disease be not a very severe one, the incredulous will indulge in ill-placed pleasantries. Indiscreet persons will talk to the patient about the method she has chosen to pursue, and give her inquietude" (p. 109).

For Deleuze, the choice of the magnetizer was important for another reason. He considered it a fact that not all magnetizers were equally effective and that there was a difference in power from magnetizer to magnetizer. How could this difference be accounted for? Seven factors explain it: the force of the will; the capacity of the attention; the direction of the will (it should be consistent, uniform, and tranquil, and never directed by mere curiosity); the strength of belief in the power of animal magnetism; the magnetizer's confidence in his or her own ability to use that power; the benevolence of the magnetizer's attention; and the physical constitution and health of the magnetizer (Deleuze 1813, 1:127–28).

From an overall view of Deleuze's writings, it becomes clear that he saw the magnetizer as a lay person rather than a medical man. He believed that the practice of animal magnetism ought to become so widespread and so well understood by the populace at large that the number of people capable of providing good magnetic treatment would be very great. He did point out, however, the importance of informing one's physician when beginning magnetic treatment, so that the physician would not misinterpret symptoms that arise from magnetic crises, and the magnetizer would not do anything contra-indicated by medical concerns. He envisioned a close, harmonious working relationship between physician and magnetizer, with both using their skill to the patient's benefit (Deleuze 1850, pp. 106–59).

Deleuze pointed out that recently physicians and medical students attached to hospitals had begun to experiment on their patients with animal magnetism. He saw a problem in that they tended to choose "young women or young girls attacked with nervous diseases, because they believe them more susceptible and more likely to present curious phenomena." Being physicians and accustomed to touch patients as a matter of course, it did not occur to them to use the precautions he had already spoken about in regard to matters of modesty. Deleuze believed they ought to "distrust themselves; to dread equally the impressions which they can experience, and those they can produce" (p. 162).

Deleuze's manual was easy to read but not simplistic. In its small space, Deleuze was able to provide the learning magnetizer with a wealth of helpful information for his practice. Outside of France, its influence was particularly great in the United States, where it went through a number of editions, to which were attached longer and longer appendixes of American cases. It was probably because of its popularity in America that no other manual of comparable quality was ever published there.

Animal Magnetism as a Surgical Analgesic

When Puységur began his magnetic practice, the first opportunity to use his newly acquired art was on the daughter of his estate manager who was suffer-ing from a painful toothache. It took only ten minutes of magnetizing to remove the pain. Immediately after that, the wife of his watchman came to him suffering from the same problem; Puységur also cured her in short order (Puységur 1784, pp. 27–28). Puységur's brother, the comte Jacques Maxime de Puységur, also a magnetizer of some note, who published a summary of his cures (J. M. Puységur, 1784), described how he relieved a soldier of the pain of a sprained ankle. After two minutes of magnetizing, the young man was able to walk without pain.

Subsequent magnetic healers repeatedly discovered that pain could be reduced or eliminated with the application of magnetic passes. Although animal magnetism was routinely used to relieve pain, it took some decades before things were taken to their next logical step: the application to surgery.[10]

10. Treatises on the subject of mesmeric analgesia include Podmore 1909, Rosen 1946, Brunn 1954, Chertok 1979 (one chapter), and Gravitz 1988.

Early Surgical Applications

Probably the first reference to the use of animal magnetism as a surgical analgesic is in the *Annales du magnétisme animal*. The account described a female magnetic somnambulist, skilled at diagnosing diseases in women, who developed an abscess below her left breast. She insisted that an operation had to be performed to remove the danger posed by an abscess so close to the heart, but refused to allow a surgeon to be called. Instead she herself carried out the lancing. The somnambulist made two 2-inch incisions in the shape of a cross over the abscess, drained it of its contents, and dressed the wound. On another occasion this same woman, while somnambulistic, successfully and apparently without pain split open an abscess on her tonsils (*Annales,* no. 35, pp. 193ff).

In his *De la cause du sommeil lucide* (1819), Faria indicated that the surgical use of animal magnetism was fairly well known. He wrote that in more profound somnambulistic states people can be insensitive even to "grave incisions, wounds, and amputations" (pp. 241–42). In his practical manual, Deleuze too mentioned that in certain cases painless surgery could be performed under magnetic anesthesia (Deleuze 1850, p. 90).

In 1821 the baron Jules Du Potet (1796–1881) arranged a demonstration of physical insensibility at the Hôtel-Dieu in Paris. The surgeon Joseph Claude Récamier (1774–1856) attended and performed two surgical procedures, one to relieve a buildup of fluid in the abdomen of a young woman and the other to treat sciatica in the thigh of a man. The patients were rendered somnambulistic and experienced no pain during the operations, but suffered greatly upon awakening (Du Potet 1826, pp. 92–93). Although Récamier found the technique successful in this case, he did not consider it suitable for general use by physicians.

A more serious surgical operation was performed on a patient in the state of magnetic somnambulism in 1829. Mme Plantin, sixty-four years of age, suffered from a cancerous breast. She had been treated for some months with animal magnetism by her physician, Monsieur Chapelain, but had found no relief. In the process he discovered that Mme Plantin was a very good magnetic subject and would readily fall into a state of somnambulism in which she remained alert but lost all physical sensibility. It became clear that the only way to help the woman was to remove the breast surgically, and it was decided to attempt the operation while the woman was somnambulistic. Chapelain magnetized her many times to prepare her for the operation. In her normal state the coming surgery terrified her, but when somnambulistic she viewed it with complete calm. The operation was performed by the surgeon Jules Germain Cloquet (1790–1883) on April 16, 1829. The surgery took only about ten minutes, and during that time the somnambulistic woman spoke in a relaxed fashion with the surgeon. She experienced no pain, no emotion, no change in respiration, and no alteration in skin color. Upon awakening she was unaware of what had occurred (Foissac 1833, pp. 156–158).[11]

Surgery in the United States and Great Britain

Writing in the *Boston Medical and Surgical Journal* of June 1836, Dr. Benjamin H. West reported magnetic experiments carried out on a twelve-year-old girl who had been afflicted with epilepsy since her sixth year. The magnetizer was B. F. Bugard, a French teacher in Boston who, West was careful to note, was not an exploiter of magnetism, but used his magnetic powers "merely for the benefit of his fellow creatures, and in philosophic investigation" (p. 350).[12]

The girl had been magnetized thirteen times in the presence of doctors Ware, Lewis, and Glover, all of Boston. Bugard noted that she had no sense of feeling while magnetized. He told Ware that the decision had been made to take advantage of this insensibility to extract a decayed molar. Ware believed the girl would not remain asleep during the operation, and decided to attend. The actual extraction was performed by a Dr. Harwood in the presence of Professor Treadwell of Harvard, doctors Ware, Lewis, and Lodge, Mr. A. D. Parker, and certain "medical students." During the operation no elevation of pulse was observed, but the girl's face was flushed and a look indicative of pain was noted. West stated that during the whole time the girl was apparently sound asleep. The operation was begun ten minutes after the initiation of magnetic passes, and the girl was awakened twenty-five minutes later. Upon questioning, she maintained that she had been asleep, that she only knew the tooth had been pulled by the space she could feel with her tongue and the taste of blood, and that there had been no pain (pp. 350–51). Subsequently a number of painless extractions of teeth from mesmerized patients were reported in the American press (Gravitz 1988, p. 205).

Meanwhile, other uses for mesmeric insensibility to pain were starting to

11. A report on this operation was presented before the Section on Surgery of the French Academy of Medicine (Gravis 1988, p. 202). During the ensuing discussion, Dominique Jean Larrey stated that he had often used animal magnetism on the battlefield to eliminate pain in dealing with wounds (Brunn 1954, p. 338). Charpignon (1841), before citing the Plantin surgery, described an operation performed by the doctor Filassier to remove a neck tumor accomplished without pain. He did not give the date of the operation (Charpignon 1841, p. 229).

12. Since the *Boston Medical and Surgical Journal* had previously reported extensively on Poyen's lectures in Boston, and since no other magnetic teachers were mentioned, it is safe to assume that Bugard learned his technique from Poyen.

be developed among American medical doctors. In 1837 a Dr. Cutter of Nashua, New Hampshire, a follower of Poyen, acted as both magnetizer and physician to assist a woman who was having a difficult childbirth. William Baker Fahnstock used magnetic somnambulism to aid in deliveries in a number of cases starting in 1844 (Fahnstock 1869).

Apparently taking their cue from the dentists, American physicians turned their attention to painless surgical operations through magnetic anesthesia. In 1843 B. F. Edwards of Alton, Illinois, removed a tumor from the left side of a patient's face; W. P. Shattuck of Lowell, Massachusetts, excised a tumor from the shoulder, an unnamed physician in Missouri did likewise, and George B. Rich of Bangor, Maine, stated that a tumor removed from his patient's shoulder measured approximately two inches. In the same year Josiah Deane, also of Bangor, performed a painless amputation of the leg. In July 1843 Albert T. Wheelock removed a nasal polyp from a patient who was magnetized by Phineas Parkhurst Quimby (1802–1866), a medical mesmerist whose ideas about healing later influenced Mary Baker Eddy, the founder of Christian Science. Neck tumors were removed by Professor Ackley of Cleveland and J. V. Bodinier of New York in 1845. And in the same year Louis A. Dugas of the Georgia Medical College used mesmerism in the removal of the cancerous right breast of a forty-seven-year-old woman (Gravitz 1988, pp. 205–6).

In England, in 1838 the noted British physician John Elliotson (1791–1868) carried out an experiment with one of his more famous mesmeric subjects, Elizabeth Okey. According to his own account, he witnessed the insertion of a seton in her neck from behind without her becoming aware of it. He claimed that this procedure, conducted without pain, constituted the first surgical intervention under mesmeric anesthesia in Britain.

Elliotson described other medical uses of animal-magnetic anesthesia carried out shortly after this episode, including the replacement of the dressing on a skin eruption on the head of a young boy, the successful painless extraction of the molar of a young woman, and the complex extraction of the molar of a young Frenchman in 1841. The man was placed in a mesmeric trance by the magnetizer Lafontaine and the procedure carried out with no sign from the patient of awareness or distress (Elliotson 1843, pp. 39–40). In the summer of 1841 Elliotson set out to convince medical practitioners to use mesmerism as an anesthetic for removing teeth. The response was a number of successful tooth extractions, reported in Elliotson's pamphlet on the surgical use of animal magnetism (Elliotson 1843, pp. 41ff).

The issue of mesmerism as a surgical anesthetic came to a crisis with an amputation performed in 1842.[13] A forty-two-year-old laborer named James Wombell suffered from a neglected disease of the left knee which had progressed to an extreme state of degeneration. When the man's surgeon, W. Squire Ward, judged that the leg had to be amputated, it was decided to use mesmerism in the hope that the pain could be reduced or eliminated. A magnetizer, William Topham, mesmerized Wombell for four minutes and induced magnetic sleep. He then continued to mesmerize the man for another fifteen minutes and concluded by laying two fingers of each hand on Wombell's eyes, believing that this would serve to further deepen the trance. Then Topham began the amputation. Wombell's breathing remained steady and his body quiet as the first incisions were made. Due to certain complications in the condition of the leg, Ward discovered that he had to carry out more lengthy procedures than he had first intended. Yet "notwithstanding all this, the patient's sleep continued as profound as ever. *The PLACID look of his countenance never changed for AN INSTANT*; his whole frame rested, *uncontrolled*, in *perfect stillness* and repose; *not a muscle was seen to twitch*. To end the operation, including the sawing of the bone, securing the arteries, and applying the bandages, occupying a period of upwards of twenty minutes, he *lay like a statue*" (Elliotson 1843, p. 5). When questioned on awakening, Wombell said he did not know anything about what happened after he was put to sleep and he felt no pain during that period. Topham and Ward also used mesmerism for the changing of dressings, which rendered that process painless to Wombell. Within three weeks the man was at home and recovering well.

The amputation was reported to the Royal Medical and Chirurgical Society, but it was not well received. Some claimed that the patient was an impostor who had been trained not to show pain. Elliotson was astounded at the hostile reaction and answered with the treatise *Numerous Cases of Surgical Operations without Pain in the Mesmeric State* (1843). Through numerous articles published in his mesmeric journal *The Zoist*, Elliotson continued to promote the use of animal magnetism as a safe and effective means for producing anesthesia.

James Esdaile

The most prolific practitioner of mesmeric anesthesia was James Esdaile (1808–1859). After graduating from the University of Edinburgh medical school in 1830, Esdaile began to practice medicine in Calcutta, India. He

---

13. The original account was published in a small pamphlet (Topham and Ward, 1842).

showed himself to be a surgeon of great skill and in 1847 was appointed
surgeon to the government of India. In 1845 Esdaile came across some of
Elliotson's writings on mesmerism. He was impressed by what he read and
decided to try to use magnetic sleep as an analgesic for his own patients.

Esdaile performed his first mesmeric surgery on April 4, 1845. Continuing
to experiment with mesmeric analgesia, he sent a report to *The Zoist* on
January 22, 1846, detailing seventy-three cases of surgery carried out while
the patient was in a state of magnetic sleep. These included an arm amputation, a breast removal, the extraction of a tumor from the jaw, and the removal
of scrotal tumors weighing from eight to eighty pounds (Rosen 1946, p. 540).

In the same year Esdaile published his *Mesmerism in India and Its Practical Application in Surgery and Medicine*. In this work he listed the seventy-
three surgical operations depicted in *The Zoist* and described eighteen "medical cases" cured by mesmerism, including nervous headache, tic douloureux,
spasmodic colic, chronic and acute inflammation of the eye, lumbago,
sciatica, and palsy of the arm. Probably no other book on mesmerism gives
the reader such graphic descriptions of the practical use of animal magnetism,
such a sense of being there while the surgical and medical procedures are
performed.

Esdaile did not spare the details as he depicted his treatment of more and
more severe cases. His graphic and gory description of the painless removal of
a tumor of the jaw (Esdaile 1846, pp. 147–50) had to be impressive to even the
most serious skeptic. And his emphasis on the successful use of mesmeric
anesthesia in postoperative procedures made surgeons sit up and take notice.

In 1852 Esdaile wrote a second book on animal magnetism: *Natural and
Mesmeric Clairvoyance, with the Practical Application of Mesmerism in
Surgery and Medicine*. Here he republished some of the material contained in
Mesmerism in India and attempted to promote further the use of mesmerism in
England by adding material he considered to be convincing to the British
mind. He described how he had, since 1846, been promoted to the position of
presidency-surgeon by the governor general of India—specifically in recognition of his work with mesmerism. He pointed out the keen interest in
mesmerism evinced by the government of India, noting that it appointed a
committee to report on his mesmeric operations, established an experimental
hospital supported by public funds to aid in mesmeric experimentation and
practice, and supported him in introducing mesmeric practices into the hospitals of India.[14] Esdaile noted that such a thing was, unfortunately, unlikely to

14. The committee eventually published a report (*Report* 1846).

occur in England, where "I find the ever-beginning never-ending task of *putting down Mesmerism* still hopelessly going on" (Esdaile 1852a, p. xi).

For all the success enjoyed by Esdaile and others in using mesmerism to carry out painless surgery,[15] mesmeric analgesia never received widespread acceptance.[16] This was not simply because of the stubborn rejection of mesmeric phenomena bemoaned by Esdaile in his second book. For just as mesmeric analgesia was being developed and established as a viable surgical procedure, another approach to painless operations was being discovered: chemical anesthesia. This new development, more than anything else, doomed mesmeric anesthesia to obscurity in the annals of medical history.

Chemical Anesthesia

American dentists, who were the first to popularize the use of mesmerism as an analgesia, also brought chemical anesthesia into its own. There had been some experimentation with the use of chemical anesthesia on animals by the British surgeon Henry Hill Hickman (1800–1830) in 1824, but its first use on humans seems to have been in the United States, where Dr. Elijah Pope of Rochester, New York, applied ether to a patient for the abstraction of a tooth in 1842. In the same year, the physician Crawford W. Long (1815–1878) of Jefferson, Georgia, used ether as an anesthetic for removing tumors from the neck of a friend, James M. Venable (Keys 1963, pp. 19, 22).

These were isolated incidents, however, and it was only in 1844 that chemical anesthesia began to be used more regularly. On December 10 a dentist, Dr. Horace Wells (1815–1848), attended a demonstration by Gardner Quincy Colton of the effects of nitrous oxide gas. One man while under the

15. Podmore summarized Esdaile's success: "During Esdaile's six years' practice in India he performed, on patients rendered insensible by the operation of Mesmerism, no fewer than 261 serious operations, besides a large number of minor cases. Of the serious operations two hundred consisted in the removal of scrotal tumours varying from 10 lbs. to 103 lbs. in weight. In these two hundred cases there were only sixteen deaths, though the mortality from the removal of similar tumours had previously been 40 or 50 percent. There were besides several cases of amputations, removal of cancerous and other tumours, &c." (Podmore 1909, p. 140).

16. A number of significant surgical operations performed under magnetic sleep were carried out at Cherbourg by a Dr. Loisel from 1845 to 1847. They were reported in the *Phare de la Manche*, the *Journal de Cherbourg*, and a number of short monographs (Loisel 1845a, 1845b, 1846, *Magnétisme* 1847). They were also described in Du Potet's *Traité complet du magnétisme animal*, 3d ed. (1856). Among the procedures performed were the amputation of the arm of a seventeen-year-old girl, the removal of cancerous glands from the neck of a thirty-year-old woman, and the removal of a tumor of the jaw from a fifteen-year-old girl. The operations, most painful when performed without an analgesic, were accomplished apparently without the slightest discomfort.

influence of the gas accidentally injured himself rather severely, but he experienced no pain. This gave Wells the idea that nitrous oxide might be used to carry out painless tooth extractions. He enlisted Colton's aid in applying the gas for the extraction of one of Wells's own teeth. The painless operation was carried out by a Dr. Riggs, and Wells began using the gas regularly in his dental practice (Thatcher 1953, pp. 7–9).

On September 30, 1846, the American dentist William T. G. Morton (1819–1868) used the inhalation of ether to produce unconsciousness and then painlessly extract a tooth. Morton, a former dental partner of Wells, then continued to employ ether in dental surgery and assisted physicians in using it for other surgical operations (Gies 1848, pp. 6–7).

Chloroform was introduced as an anesthetic inhalant by James Y. Simpson in November 1847 and thereafter was widely used. Ether and chloroform were commercially produced, easily handled (being liquid in form), and very effective in producing unconsciousness. For that reason they quickly supplanted nitrous oxide as the anesthetic of choice for most dentists and physicians. Nitrous oxide did make a comeback among dentists, however, when in 1863 Colton, with the aid of some dentists, conducted experiments demonstrating the effectiveness of that inhalant for extracting teeth (Gies 1848, pp. 7–8).

Because of the complexity of events surrounding the first use of nitrous oxide and ether as anesthetics, there has been an ongoing controversy about who was the true discoverer of chemical anesthesia. Although Long used ether for surgery in 1842, he did not make this known to medical colleagues at that time and only published his findings in 1849. Wells first employed nitrous oxide in 1844 and immediately tried to promote its surgical use through a public demonstration in Boston. Because of technical problems the demonstration was unsuccessful and the use of nitrous oxide did not catch on as he had hoped. But Wells's early public promotion of this anesthetic gas seems to establish him as the discoverer of chemical anesthesia (Gies 1948, pp. 8–9).

The availability of the three chemical inhalants and their ease of use made the widespread practice of chemical anesthesia in general and dental surgery inevitable. These factors gave it an advantage over mesmerism, also developing as a surgical analgesia in the mid-1840s. Nevertheless, magnetic sleep did continue to be used in some surgical operations for many years, thanks in part to James Braid's "hypnotism," a new, "purified" form of mesmerism.

Hypnotic Analgesia

In 1842 James Braid attempted to revolutionize thinking about animal magnetism and its phenomena, explaining it as a psychologically induced psycho-physiological state and renaming it "hypnotism." At the time he was only partly successful in his attempt to convince the public and medical coworkers in Britain of his views. But in 1859 the French medical world was alerted to Braid's work and responded with enthusiasm.

Among the mesmeric phenomena that Braid accepted into his concept of hypnotism was induced analgesia. In his first major work on hypnotism, written in 1843, he stated: "I am quite satisfied that hypnotism is capable of throwing a patient into that state in which he shall be entirely unconscious of the pain of a surgical operation, or of greatly moderating it, according to the time allowed and mode of management resorted to. Thus, I have myself extracted teeth from six patients under this influence without pain, and to some others with so little pain, that they did not know a tooth had been extracted" (Braid 1899, p. 310). In this Braid was keeping pace with the practitioners of animal magnetism who, in 1842 and 1843, were discovering its usefulness in dental surgery.

In *Hypnotic Therapeutics* Braid described his successful use of hypnotism in other minor operations (Braid 1853b, pp. 9–10). In the same work he gave credit to the mesmerists (particularly Esdaile) for their surgical successes and claimed that ether and chloroform, though certain and speedy in their results, were not as useful as hypnotism in the treatment of certain medical conditions (Braid 1853b, p. 10).

In 1858 Eugène Azam, professor of medicine at Bordeaux, read Braid's masterwork, *Neurypnology*, and was impressed. Late in 1859 he acquainted the surgeon Paul Broca with his reflections on hypnotism, and Broca immediately applied what he had heard to a case he was treating. A twenty-four-year-old woman suffering from an abscess of the anus was hypnotized by concentrating on a bright object. Broca then lanced the abscess, and when the woman was awakened from hypnotism she remembered nothing of the operation. On December 5, 1859, Broca presented his findings to the Académie des Sciences, and his report was immediately published (Broca 1859).

Azam himself reported on hypnotism in the *Archives générales de médecine* of January 1860. He strongly urged the use of hypnotism and emphasized its importance for surgery. He stated that although he did not believe that hypnotic analgesia could replace the use of chloroform in every case, it should be investigated as a practicable and desirable surgical analgesic, especially because of the benign effects of hypnotism on the body of the patient compared to the "brutal" effects of chemical inhalants (Azam 1860, pp. 13–14).

Later in 1860 a treatise on the surgical use of hypnotism was published by Jean Nicolas Demarquay and M. A. Giraud-Teulon. Taking note of the optimism of Broca and Azam about the potential of hypnotism as an analgesic, these two surgeons decided to make experiments of their own. Their results were less than encouraging. Through eighteen trials they found only one patient who became insensible to a significant degree, and that patient they considered to be hysterical. From their investigations they concluded that magnetic sleep, magnetic somnambulism, and hypnotism were the same phenomenon manifested in varying degrees, and that their efficacy was dependent on a condition of hysteria or near-hysteria in the patient (Demarquay and Giraud-Teulon 1860).

After animal magnetism had enjoyed sixty years of widespread medical use and many documented successes, Mesmer's dream that it would be accepted by the medical establishment was still not realized. Its fulfillment was impeded by the close alliance of animal magnetism to phenomena that many savants considered too improbable to believe, and to cures that many medical people thought too extravagant to credit. But there were some physicians who, while remaining skeptical, nonetheless were not prepared to dismiss mesmeric phenomena altogether. One of these was James Braid of Manchester. He attempted to separate what he saw as the chaff from the wheat and single-handedly initiated a new chapter in the history of animal magnetism.

Chapter 8

Animal Magnetism and Hypnotism

As magnetic healing continued to be practiced successfully and as animal magnetism enjoyed a brief flash of attention as an analgesic, practitioners were still trying to explain how it all worked. In the process, the psychological aspects of mesmerism—particularly suggestion—received growing attention. Then in England a major attempt to develop a totally psychological understanding of mesmeric phenomena was set in motion with James Braid's introduction of the notion of hypnotism. Braid's approach evolved in the context of a thriving magnetic practice in that country.

Animal Magnetism in Great Britain

Although there had been some mesmeric activity in England in the decades following Mesmer's discovery, the practice of animal magnetism there really came into its own in the 1830s. The man most responsible for this renewal of interest was Richard Chevenix, a chemist and mineralogist from Ireland who had been instructed in the traditions of Puységur and Faria. In 1829 Chevenix attempted to demonstrate the medical value of animal magnetism through experiments conducted at Saint Thomas's Hospital in London. The results were published in the *London Medical and Physiological Journal* (March, June, August, and October 1829) and noted with disfavor by the rival medical journal *The Lancet* (June 13, 1929). A talented young physician named John Elliotson (1791–1868) witnessed Chevenix's experiments and was impressed, although he had some reservations.[1] He later conveyed his impressions in his book *Human Physiology* (1840).

In 1830 William Newnham, writing under the patronage of the bishop of Winchester, published an *Essay on Superstition,* in which he attempted to formulate physiological and psychological explanations for supernatural phenomena that were embarrassing the Anglican church. In the process, he presented an appreciative discussion of animal magnetism as a naturalistic explanation for some of them.

---

1. Much of the biographical information about Elliotson presented in this chapter is drawn from Fred Kaplan 1982, pp. vii–xv.

In 1833 John Campbell Colquhoun published an English translation of the favorable report on animal magnetism produced by the French Royal Academy of Medicine in 1831. In a lengthy introduction he delineated the history and theory of animal magnetism. In 1836 Colquhoun produced what he called a "second edition" of this work under the title *Isis Revelata*. In fact it was so thoroughly revised and augmented that it was really a new book. The introduction to the 1833 publication was expanded from one hundred to nearly six hundred pages, and the translation of the report was merely an appendix. This was by far the most exhaustive study of animal magnetism to appear to that date and marks an important point in the development of interest in the subject in England.

In 1838 Colquhoun issued a pamphlet entitled *Hints on Animal Magnetism* in which he answered critics of *Isis Revelata*—among them John Elliotson. Then, in 1851, he published his final apology for animal magnetism, the two-volume *History of Magic, Witchcraft, and Animal Magnetism,* in which he attempted to show that the "establishment" of each age tends to link new scientific discoveries to witchcraft and superstition and reject them outright. Just as the great discoveries of Galileo, Harvey, and others were initially rejected on these grounds and eventually proved beneficial to mankind, so it would be with animal magnetism.

The decisive event in the revival of British interest in animal magnetism was the arrival in London in 1837 of the baron Du Potet, who had come from Paris to teach animal magnetism. Du Potet was invited by John Elliotson to demonstrate the medical uses of animal magnetism at the University Hospital in London, where Elliotson was professor of principles and practices of medicine. These demonstrations continued until the managing committee of the hospital voiced its objection to them. Du Potet then moved his practice to his apartments in Cavendish Square (Du Potet 1838, p. 3).

Elliotson himself then began to conduct experiments at North London Hospital based on those of Du Potet. Among the observers of these experiments was Herbert Mayo, professor of comparative anatomy at the Royal College of Surgeons. Mayo was convinced that at least some of the phenomena were genuine and eventually pursued mesmeric experiments of his own. He published his conclusions in his *Letters on the Truths Contained in Popular Superstitions, with an Account of Mesmerism* (1851). Mayo's thought on the subject evolved from an initial skepticism in 1837 to acceptance of most of the more extraordinary phenomena by 1851.

Elliotson expressed his enthusiasm for Du Potet in an article in *The Lancet* in September 1837 (Kaplan 1982, pp. vii–xv). Because of Elliotson's considerable reputation, *The Lancet* reluctantly printed this article and others he wrote on mesmeric practice throughout 1837 and 1838—along with many articles rejecting his findings. Elliotson insisted that the fluid of animal magnetism was a physical reality and that its effects were susceptible to scientific observation. But this view was unpalatable to most physicians, and when problems and errors in some of Elliotson's experiments were pointed out, he lost much of the support he first enjoyed.

Late in 1838 the Council of University College decided to put an end to the practice of mesmerism at its hospital. Elliotson resigned in protest and founded the London Mesmeric Infirmary, which treated patients with mesmeric techniques. He continued to insist on the scientific nature of his mesmeric experiments and strongly opposed those who connected animal magnetism with the occult.

One of Elliotson's most important achievements was the founding of *The Zoist*, a journal devoted to the investigation and promotion of the use of animal magnetism.[2] Published from 1843 to 1856, *The Zoist* provided a forum for discussing the details of hundreds of specific cases of mesmeric treatment.

Elliotson's support for the use of animal magnetism in medicine encouraged others to investigate other aspects of mesmerism. It was largely due to Elliotson that animal magnetism had much of a history in England at all. When in 1846 he delivered the Harveian Oration before the Royal College of Physicians, he succinctly summarized his stand:

> The chief phenomena are indisputable: authors of all periods record them, and we all ourselves witness them, some rarely, some every day. The point is to determine whether they may be produced artificially and subjected to our control: and it can be determined by experience only. . . . It is the imperative, the solemn, duty of the profession, anxiously and dispassionately to determine these points by experiment, each man for himself. I have done so for ten years, and fearlessly declare that the phenomena, the prevention of pain under surgical operations, the production of repose and comfort in disease, and the cure of many diseases, even after the failure of all ordinary means, are true. In the name, therefore, of the love of truth, in the name of the dignity of

2. *The Zoist* was not the first periodical published in England devoted to animal magnetism. There were two earlier journals, neither of which lasted beyond the first volume: *Magnetiser's Magazine and Annals of Animal Magnetism* (published in London in 1816 and consisting of a translation from Joseph Deleuze's *Histoire critique du magnétisme animal*) and *The Oracle of Health: A Penny Journal of Medical Instruction and . . . Amusement* (published in London in 1834 and containing an account of Mesmer's experiments).

our profession, in the name of the good of all mankind, I implore you carefully to investigate this important subject. (p. 68)[3]

Dr. Edwin Lee, a fellow of the Royal Medico-Chirurgical Society, in 1835 produced a pamphlet entitled *Animal Magnetism and Homeopathy,* which by 1838 had become a book dealing with, among other things, the report on animal magnetism of the French Academy of Medicine. Lee's evaluation of the paranormal phenomena of mesmerism as detailed there was quite skeptical. Then in 1843 Lee carried out some observations of two magnetic clairvoyants in Paris and published a report given to the Parisian Medical Society (Lee 1843). Both clairvoyants were impressive, the second, a young man called Alexis, spectacularly so. Alexis was apparently able to relay, in astonishing clarity and detail, information that he could not have known through ordinary means. Impressed though he was, in his report Lee drew cautious conclusions about mesmeric phenomena, stating that a distinction must be made between the undoubtedly genuine effects of mesmerism that may be detected in many people who are magnetized (such as a sleep-like state and catalepsy) and "psychological phenomena, as *clairvoyance,* intuition, the divining the thoughts of persons *en rapport* with somnambulists, &c.; which are powers superadded in a comparatively small number of individuals while under the magnetic influence, and in whom the lucidity is not at all times in the same degree of perfection" (p. 21). As the years passed, Lee's acceptance of magnetic clairvoyance grew less reserved, so that in 1849 he wrote, "The fact of clairvoyance, or the perception of objects by lucid somnambulists, without the assistance of the eyes, a tolerably accurate description of distant localities or persons, known to the individuals placed *en rapport,* the indication of symptoms of disease, &c., is fully proved by the concurrent testimony of innumerable witnesses, as also by the attestation of the commissioners of the Royal Academy of Medicine." In 1866 he presented data to support this position in his *Animal Magnetism and Magnetic Lucid Somnambulism.*

John Wilson, a physician at the Middlesex Hospital, undertook some tests of his own to decide conclusively whether animal magnetism was a "physical fact." He began by treating some of his own patients and found animal magnetism effective in removing or relieving their infirmities. But he did not

3. The printed edition of the oration includes a letter from Mesmer addressed to the Royal College of Physicians at London and dated 28 March 1802, in which he lamented the reception hitherto given to animal magnetism and hoped for a better reception in England, where his cause was as yet "undamaged" and where "the discovery has not yet been proclaimed" (Elliotson 1846, pp. 68–70).

consider these results conclusive and devised an approach that he thought would be foolproof:

> To the accounts which have at different times been given to the effects of animal magnetism, objections have always been raised, on the ground that they may all be resolved either into collusion between the operator and the person operated upon; or, where the character of the operator precluded all suspicion of his own integrity, into imposture on the part of the person on whom the operation was performed; and it must be admitted that these objections are not easily surmounted. It therefore appeared to me particularly desirable, to institute similar experiments on brute animals; as the results, which might be thus obtained, could not be liable to the objections above-mentioned. (Wilson 1839, pp. 7–8)

Wilson was not the first to use animal magnetism on animals. In 1784 Puységur's brother Jacques Maxime had treated an injured dog with animal magnetism, and in the same year the mesmerists at Lyon magnetized a horse (Vinchon 1936, p. 94). Wilson, however, was the first to conduct a systematic study of the effects on animals and report the findings (*Trials of Animal Magnetism on Brute Creation,* 1839). He magnetized cats, ducks, geese, fish, dogs, chickens, turkeys, horses, macaws, pigs, calves, leopards, elephants, and lions. His magnetic passes (mostly without contact) produced sleep (from which the animal could sometimes be aroused only with difficulty), catalepsy (in which the animal would hold a pose for a long period of time), convulsive movements, abnormal activity, or untypical reactions such as repeated yawning or sweating. Although these results were striking, Wilson's work was barely noted, and no one undertook to replicate his experiments.

Much more popular was a work by Chauncy Hare Townshend, *Facts in Mesmerism with Reasons for a Dispassionate Inquiry into It*; first published in 1840, it was issued in several subsequent British and American editions. Townshend was a clergyman, a scholar, and a much-traveled man of the world. He studied the historical sources and carried out many experiments, several of them in the parlors of the European aristocracy. In the book he discussed what he called "sleep-waking," a term coined by Elliotson (1840) to signify that magnetic somnambulism was a combination of the waking and sleep states (pp. 626ff). Townshend accepted the reality of the "higher phenomena" of mesmerism—a term apparently of British origin designating unusual phenomena produced by some persons in trance, including physical rapport in which the subject experiences the sensations of the mesmerizer; mental rapport, with the ability to read the mesmerizer's thoughts; clairvoyance or awareness of objects or events at a distance in space or time; and ecstasy or an elevated state of consciousness in which the subject has an awareness of spiritual things. In 1854 Townshend published a second book on the subject, *Mesmerism Proved True,* in answer to a strong attack on animal magnetism and its phenomena by William Carpenter.

Townshend's private experiments with the gentry stood in contrast to the widely attended stage demonstrations that were meant to convince all present of the power of animal magnetism. Such platform spectacles were the specialty of Charles Lafontaine (1803–1888), a Paris-born magnetizer of considerable talent who came to England in June 1841 to enthrall audiences with the more spectacular mesmeric phenomena. Lafontaine's circuit took him throughout the United Kingdom—to London, Birmingham, Manchester, Leeds, Sheffield, Nottingham, Leicester, Liverpool, Dublin, Belfast, Lisburn, Londonderry, Glasgow, and Edinburgh. Thousands of people saw demonstrations of somnambulistic trance, witnessed apparent instances of clairvoyance, and heard Lafontaine speak of the remarkable curative powers of mesmerism. Lafontaine's lectures are credited with establishing the practice of animal magnetism in Scotland. After his demonstrations in Edinburgh and Glasgow, many began to experiment on their own and, according to William Lang, "lecturers speedily sprung up, and went in every direction; and there is now no community of the slightest importance in the north which does not contain a numerous body of believers in the truths of mesmerism" (Lang 1843, p. 10).

Spenser Hall, a factory worker from Nottingham, was among those who attended Lafontaine's Sheffield demonstration. Hall began mesmerizing on his own and soon found that he was a very talented operator. A society of "leading scientific men of the town" was founded to investigate animal magnetism further, and Hall was elected an honorary member (Hall 1845, p. 3). He soon became aware of an approach called phrenomagnetism, in which the different areas of the skull (corresponding to different sections of the brain and therefore to distinct emotional and intellectual qualities) were "excited" through the application of magnetic passes. The stimulation of a particular phrenological area was believed to produce feelings and thoughts in the subject that corresponded to the quality associated with that area. Hall quickly became adept at phrenomagnetism and traveled the countryside applying his new art to the sick. In a book describing his experiences, he condemned those members of the medical profession who sought to limit the practice of magnetic healing to physicians. He claimed that the common people had the right to know about the powers of mesmerism and use them for their own benefit

(Hall 1845, pp. 19–24). Hall was in fact the first significant nonmedical practitioner of mesmerism in England.[4]

Among those Hall treated was the celebrated writer and journalist Harriet Martineau. In her account of the matter, she described how she was overcome by a debilitating illness in 1839 and for five years "never felt wholly at ease for one single hour" (Martineau 1845, p. 4). Medical treatment gave her no relief, but after a "medical friend" had attended a lecture given by Hall, Martineau invited him to apply his mesmeric powers to her illness. There was an immediate beneficial effect, and since Hall was not available for further treatments, Martineau asked her maid to imitate what she had seen Hall do. The benefit was even more pronounced, and Martineau began to feel free of pain for the first time in years. On Hall's advice, this arrangement was continued for four months. The result was the complete disappearance of all symptoms (Martineau 1845, pp. 4–6).

Hall's efforts to promote the healing use of animal magnetism for the common people were joined by George H. Barth, who in 1850 produced a practical guide, *The Mesmerist's Manual of Phenomena and Practice*, which he forthrightly declared to be "intended for domestic use and the instruction of beginners." Three years later he issued a second manual, *What Is Mesmerism?*, in which he included "useful remarks and hints for sufferers who are trying mesmerism for a cure."[5]

While Hall and Barth were battling the medical establishment, objections to animal magnetism and its phenomena were arising from quite another quarter. In 1842 the Reverend Hugh M'Neile preached a sermon in Liverpool entitled "Satanic Agency and Mesmerism" denouncing animal magnetism as the work of the devil. In 1843 George Sandby published a pamphlet called *Mesmerism, the Gift of God*, in which he took exception to this view. In the expanded version of this pamphlet, *Mesmerism and Its Opponents* (1844; 2d ed. 1848), Sandby charged that M'Neile's condemnation of mesmeric phenomena was based on secondhand information. He quotes M'Neile: "In forming a judgment of this . . . I go, of course, on what I have read. *I have seen nothing of it,* nor do I think it right to tempt God by going to see it. I have not

---

4. Cooter (1985) has pointed out that mesmerism provided an important means for the socially downtrodden to exercise power outside the established social hierarchy.

5. Some sections of Barth's manual are interesting summaries of mesmeric experience. One is his discussion of "cross mesmerism," instances in which two or more mesmerizers are treating one patient and cause distress by their dissimilar approaches. Another concerns what he considers to be the three basic mesmeric states: mesmeric sleep, mesmeric waking, and mesmeric sleep-waking.

faith to go in the name of the Lord Jesus, and to command *the Devil* to depart" (Sandby 1848, p. 65; the emphasis seems to be Sandby's). Sandby commented: "Really, any one would suppose that he were reading the ignorant ebullition of some dark monk in the middle ages, rather than the sentiments of an educated Protestant of the nineteenth century. What is this but the same spirit that called forth a papal anathema against the 'starry' Galileo?" (p. 65).

The Christian apologist William Newnham, after a brief notice of mesmerism in 1830, produced a lengthy treatise, *Human Magnetism; Its Claims to Dispassionate Inquiry* (1845). Citing the "philosophical Colquhoun," the "religious Deleuze," and the "excellent and pious Townshend," Newnham attempted to demonstrate that mesmerism held the key to a naturalistic, but not materialistic, explanation for certain paranormal phenomena. Like Sandby, he rejected the views of those who had claimed that mesmeric phenomena were due to the action of the devil, pointing out that one of the principal results of animal magnetism was the healing of disease, an activity incompatible with the notion of satanic agency. He also criticized "medical men" who had rejected animal magnetism without investigation. Newnham himself believed that the real benefit of animal magnetism was its curative power, and while accepting that somnambulistic clairvoyance and prevision might be true, he held that they should not become the focus of attention for those practicing or experiencing mesmerism.[6]

In *Zoistic Magnetism* (1849) William Scoresby denied that the devil was at work in mesmerism: "For so far from there being any ground for connecting Zoistic Magnetism with supernatural agency, or an agency, such as had been asserted, of an unhallowed nature—the close analogy of its phenomena, with the well known laws of Magnetism and Electricity, brought the subject fairly within the province of the Natural Sciences." Scoresby ascribed magnetic cures to "influences essentially belonging to the constitution of organized beings, and to an agency implanted by a beneficent Providence" (p. 4). To those who went to the other extreme and considered the phenomena of mesmerism to be the miraculous manifestation of divine power, the Reverend S. R. Maitland (1849) answered that mesmerism operates without benefit of *any* supernatural intervention, positive or negative, since it merely engages natural healing forces not as yet well understood.

Others attempted to gain a scientific understanding of these phenomena.

6. In his *Vital Magnetism: A Remedy* (1844) the Reverend Thomas Pyne said he perceived no threat in mesmerism but saw it as merely a modern manifestation of a much older art. He also rejected the notion that satanic agency could be involved, insisting that an evil spirit could not cure disease or relieve pain.

The physician John Haddock, in *Somnolism and Psycheism* (1849), described mesmerism as a young science, so young that it necessarily leaves many questions unanswered. He pointed out that cures believed to be miraculous have occurred throughout the ages, and that Mesmer was the first to provide a scientific basis for understanding them. Haddock attempted to relate mesmeric states to conditions in the brain and nervous system and investigated the psychological aspects of clairvoyance. To illustrate, he then described his experiments with Emma, a domestic of his household. He related incidents of "travelling clairvoyance" performed by this young woman and went into some detail about her mesmeric state of "ecstasy," in which she seemed to communicate with spirits of the dead and viewed the world and its affairs from a heavenly perspective.

Another advocate of the scientific study of animal magnetism was William Gregory, professor of chemistry at the University of Edinburgh. Having completed a translation of Baron Karl von Reichenbach's work on magnetism, Gregory began to practice mesmerism himself in 1842 or 1843. He set down his thoughts in *Letters to a Candid Inquirer, on Animal Magnetism* (1851), with one purpose in mind: "to draw the attention of scientific men to the existence of these remarkable phenomena. . . . My earnest desire is, that men of science should investigate Animal Magnetism, just as they would any other class of natural facts, feeling convinced that it is only in this way that they can ever be at all understood, and that if scientific men ignore their existence, and refuse to examine them, they will nevertheless continue to exist, and will be studied by others; for they cannot now be safely neglected" (p. xvi). Gregory provided a good practical basis for experimentation, describing in detail procedures and their likely effects. His discussion of the sleep-waking condition was particularly perceptive, and he showed himself to be an excellent psychological observer, pointing out, among other things, important features associated with divided consciousness.

Stout opposition to animal magnetism was offered by John Forbes, editor of the *British and Foreign Medical Review*. Starting in 1839, Forbes wrote several articles in his periodical denouncing the practice of animal magnetism and criticizing medical men for their credulity in accepting the mesmeric phenomena. He summarized his views in *Mesmerism True—Mesmerism False: A Critical Examination of the Facts, Claims and Pretensions of Animal Magnetism* (1845).

A similarly skeptical view was presented by Walter Cooper Dendy in his *Philosophy of Mystery* (1845), which included mesmerism among the doubtful "mysteries" of human experience along with spectral visions and fairy mythology. Dendy contended that the basic phenomena of animal magnetism were due to suggestion and delusion, while the "higher phenomena" were most often the result of trickery and fraud (pp. 403–29).

The most outspoken critic of the time was undoubtedly John Hughes Bennett, professor of medicine at the University of Edinburgh. In *The Mesmeric Mania of 1851, with a Physiological Explanation of the Phenomena Produced* (1851), Bennett expressed his disgust:

> During the present year, society in Edinburgh has been greatly agitated by a delusion, consisting in the supposition, that certain persons may be influenced by an external mysterious force, which is governed and directed by particular individuals. Fashionable parties have been converted into scenes of experiments on the mental functions. Noblemen, members of the learned professions, and respectable citizens, have been amusing themselves in private, whilst discourses and exhibitions to an unusual extent have been got up for the entertainment of the public. . . .
> I have been told that in some educational establishments, girls and boys throw themselves into states of trance and ecstasy, or show their fixed eyeballs and rigid limbs, for the amusement of their companions. Sensitive ladies do not object to indulge in the emotions so occasioned, and to exhibit themselves in a like way for the entertainment of evening parties. (p. 5)

As repulsive as he found all this, Bennett wrote that the reality of the facts could not be doubted: "Indeed, that a peculiar condition of the nervous system may be occasioned, in which individuals otherwise of sound mind are liable to be temporarily influenced by predominant ideas, must be admitted by all who have seen anything of the disorder. . . . It is the manner in which they may be produced and the frequency with which they are made to occur, that is new; and this certainly demands the attention of the medical and legal practitioner, connected, as it is, with human health and human testimony" (p. 5).

From this colorful preface, Bennett launched into a description of the nervous system, which probably bored and alienated many readers. This was followed, however, by a thoughtful attempt to explain the phenomena. Bennett compared mesmeric states to cases of "monomania" in which will and reason are suspended and the individual "cannot control those impressions that are most strongly fixed in the mind." He believed that about one in twenty in the general population were susceptible to being thrown into this state by gazing steadily at an object for about ten minutes. Once mesmerized not only would they act on any train of ideas suggested to them but also their "motion and sensation may be controlled in a variety of ways" (p. 10).

Although Bennett was dismayed by the public exhibition of mesmeric phenomena, he admitted that it had its medical uses. In this regard he referred to a Manchester physician, James Braid, who was, in Bennett's opinion, carrying out medical experiments with the mesmeric state that were quite promising.

James Braid and Hypnotism

Born in 1795 in Fifeshire, James Braid earned a medical degree at the University of Edinburgh and practiced among the miners in Lanarkshire and then in Dumfries and Manchester, where he lived until his death in 1860. Braid first became known to the medical world for his work with clubfoot, strabismus, curvature of the spine, and stammering. He had performed hundreds of surgical operations on clubfoot and strabismus before he began his study of mesmeric phenomena.[7]

Braid's interest in animal magnetism seems to have begun when he attended Lafontaine's first demonstration in Manchester on November 13, 1841. In his *Neurypnology* (1843) Braid recalled that he had gone to that meeting a total skeptic, believing the phenomena to be a result "of collusion or delusion, or of excited imagination, sympathy, or imitation" (p. 98).[8] That first demonstration did nothing to convince him. But at the next meeting, six nights later, he noted that the mesmerized subject could not open his eyes. He conjectured that the subject had been called upon to maintain a fixed stare, which induced a paralysis of the muscles of the eyelids. To test his theory, Braid decided to have subjects stare at fixed points, such as the top of a wine bottle or an ornament on a china piece, placed in a position high enough above their heads to cause a strain (pp. 98–99). He found that not only were his subjects' eyelids paralyzed; the state of their whole bodies and their minds was affected. Now Braid was sure he had discovered the secret of all mesmerism:

> The phenomena of mesmerism were to be accounted for on the principle of derangement of the state of the cerebro-spinal centres, and of the circulatory, and respiratory, and muscular systems, induced, as I have explained, by a fixed stare, absolute repose of body, fixed attention, and suppressed respiration, concomitant with the fixity of attention. That the whole depended on the physical and psychical condition of the patient, arising from the causes referred to, and not at all on the volition, or passes of the operator, throwing out a magnetic fluid, or exciting into activity some mystical universal fluid or medium. (pp. 101–2)

---

7. See Kravis 1988 for a complete bibliography of Braid's publications.

8. Quotations from Braid's *Neurypnology* are from the 1899 version, edited by Waite.

Braid challenged Lafontaine with his new theory. Though many remained solidly behind the mesmerist, Braid began to attract attention. Lafontaine moved on to Birmingham, where, according to his own account (1866, 1:311), he received letters requesting his return to Manchester, where Braid had now started to demonstrate and lecture on his method of inducing mesmeric phenomena. Lafontaine attended one of his demonstrations but pronounced himself unimpressed. The populace and the press, however, were intrigued by Braid's new ideas and gave them a warm reception.

Other medical men began to experiment with Braid's method with notable success, and in the March 12, 1842, issue of the *Medical Times* an anonymous correspondent gave an account of Braid's recent "excellent lecture" on animal magnetism, which, wrote the correspondent, Braid preferred to call "neuohypnology [*sic*], or the rationale of nervous sleep." This is the first printed (albeit misprinted) reference to the new terminology that Braid was evolving for mesmeric phenomena.[9] Following this account, the journal printed a letter from Braid himself in which he recounted how he used his method on Herbert Mayo, and how Mayo had tested other subjects at the time and confirmed their state.

In the same letter, Braid claimed success in using his "neurohypnology" to cure illness, leaving the animal magnetizers no space to claim healing as their exclusive domain: "I have operated successfully on the blind. . . . I may add, that last night I was called to a lady suffering the most agonizing Tic-Doloroux. In five minutes by my mode of inducing refreshing sleep, I succeeded in putting this patient into comfortable sleep, from which she did not awake till Sunday in the morning, being five and a half hours, and was then quite easy. By what other agency, I now ask, could such an effort have been induced?" Braid also stated his view that many of the extraordinary phenomena attributed to animal magnetism could be explained in a different way: "After a certain time, and frequency of being operated on in this way, the brain has an impressibility stamped on it which renders the patient subject to be acted on entirely through the *imagination,* and this is the grand source of the follies which have misled . . . the animal magnetisers" (Braid 1842a, p. 283).

In a letter published in the *Medical Times* two weeks later,[10] Braid pro-

9. There are many excellent histories of hypnotism. Especially useful for the period covered here are Foveau de Courmelles 1891, Harte 1902, Bramwell 1903, Podmore 1909, Dingwall 1967, Pattie 1967, Ellenberger 1970, Edmunston 1986, Inglis 1989, and Gauld 1992.

10. In this letter Braid corrected phrases in the earlier letter, changing "the follies which have" to "the fallacy which has," and "effort have been induced" to "effect have been produced" (Braid 1842b, p. 308). Braid was not a great writer, and his providing an errata list for a previous letter seems to show that he was aware of his shortcomings in this area. Further

vided what he considered further evidence that the higher phenomena of animal magnetism were not what they appeared to be:

> The supposed power of seeing with other parts of the body than the eyes, I consider, is a misnomer, so far as I have yet personally witnessed. It is quite certain, however, that patients can tell the shape of what is held at an inch, or an inch and a half from the skin, on the back of the neck, crown of the head, arm, or hand, or other parts of the body, but it is from *feeling* they do so; the extremely exalted sensibility of the skin enabling them to discern the shape of the object so presented, from its tendency to emit or absorb caloric. In like manner I have satisfied myself and others, that patients are drawn, or induced to obey the motions of the operator, not from any peculiar inherent magnetic power in him, but from their exalted state of feeling enabling them to discern the currents of air, which they advance to, or retire from, according to their direction. This I clearly proved to-day to be the case, and that a patient could feel and obey the motion of a glass funnel passed through the air at a distance of *fifteen feet*. (1842b, p. 308)

Braid then stated why he was so concerned to find these more ordinary explanations for what appear to be extraordinary phenomena: "I have been thus particular in noticing these points, because they may tend to remove the prejudice which must ever prevail against the introduction of this as a curative agency, whilst invested with so much mystery" (p. 308).

Braid then published a pamphlet that must be considered a foundation work in the history of hypnotism. *Satanic Agency and Mesmerism Reviewed* (1842) was a reply to the sermon in which Hugh M'Neile declared that mesmerism was the work of the devil and that lecturers on the subject were dishonest men, refusing to state the "laws of nature" by which the phenomena were produced. When he heard about the sermon, Braid sent M'Neile a newspaper account of one of his lectures explaining the nature and causes of the phenomena on the basis of physiological and psychological principles. M'Neile never responded, and when his sermon was put into print, Braid felt he had to defend himself.

In the pamphlet—quoting heavily from the *Macclesfield Courier* report of his lecture—Braid stated that there were three common views of the phenom-

---

evidence of his fussiness can be found in one of the two extant copies of his *Satanic Agency* (1842c). It includes a minor stylistic correction, apparently in Braid's own hand, which he seems to have felt compelled to ink in, even though the pamphlet was highly unlikely to be republished.

ena of mesmerism: that they are due to a system of delusion and collusion; that they are genuine but the result of imagination, sympathy, and imitation; and that they are due to the influence of a magnetic medium (the mesmerists' view). Braid then added a fourth view, his own, that the phenomena were solely attributable to a particular physiological state of the brain and spinal cord. He stated that magnetic sleep results from a rapid exhaustion of the sensory and nervous systems, which produces a feeling of "somnolency" in which the mind "slips out of gear." This special state could be used for many beneficial purposes, such as extracting teeth, relieving chronic pain, removing paralysis, and restoring hearing and sight.

Braid used the terms "neurohypnology" and "neurohypnotism" to replace "mesmerism." He spoke of the subject's being under "hypnotic influence" and undergoing "hypnotic sleep," and called the process of producing that state "hypnotizing." These words constitute the first appearance in print of the terminology that would become standard, making the pamphlet a central document in the history of animal magnetism and hypnotism.

In 1843 Braid published his magnum opus on hypnotism, *Neurypnology; or the Rationale of Nervous Sleep Considered in Relation with Animal Magnetism.* In *Neurypnology* Braid expanded on *Satanic Agency,* providing a chart of terms and a fuller explanation of his practice of hypnotism. He insisted that a new terminology was necessary to distinguish his mode of operating from "other theories and practices on the nervous system"—an obvious reference to animal magnetism and mesmerism. Braid combined the Greek words for nerve (*neuron*) and sleep (*hypnos*) to form the term "neuro-hypnotism" or "nervous sleep." As this word was too awkward, he suppressed the prefix and was left with "hypnotism." From this term he derived "hypnotic" (the state or condition of nervous sleep), "hypnotize" (to induce nervous sleep), "dehypnotize" (to restore from the state of nervous sleep), and "hypnotist" (one who practices hypnotism) (p. 94).[11]

Braid believed that hypnotism had important medical uses and attributed to it cures and ameliorations that previously had been claimed for animal magnetism. From the start he made it his mission to take over from mesmerists all claims that could be reinterpreted in terms of his new theory and expressed in his new terminology. In this way he hoped to make acceptable to the medical profession those mesmeric practices that he considered valid and based on

11. As noted in chapter 7, terms using the "hypno-" prefix had been applied to animal-magnetic states before Braid. Hénin de Cuvillers had developed several such terms. In the second edition of his *Le magnétisme animal retrouvé dans l'antiquité* (1821) he presents no fewer than 654 terms that he suggests would be useful in accurately designating various aspects of animal magnetism. Of these, 314 begin with "hypn-". See also Gravitz and Gerton 1984.

genuine phenomena. To do this Braid not only had to restructure and rein-
terpret the acceptable phenomena, he also had to reject all phenomena that he
could not fit into his theory. This task occupied Braid for the rest of his life.

Among the first things Braid noticed about the hypnotic state were certain
peculiarities of memory that were already quite familiar to the mesmerists.
Like Deleuze some thirty years before, Braid noted that the hypnotic subject
was often able to recall memories from the distant past with astounding
accuracy and with such vividness and immediacy that it might be called
"revivification." Braid also described the division of memory (often remarked
on by Puységur) that occurred between the normal and the hypnotic state,
calling it "double consciousness" (p. 32). Events that occur to the hypnotic
subject in the "second state" of consciousness or "full" hypnotic state would
be forgotten when the subject returned to the waking state (1853b, p. 26).
They would, however, be clearly recalled when the subject was rehypnotized.
Thus, there were two memories—the waking memory and the "second" or
"double conscious" memory of the hypnotic state (Braid 1844, p. 47).

As Braid continued his hypnotic experiments he came to believe that
hypnotic phenomena confirmed the findings of phrenology. Originated by
Franz Joseph Gall (1758–1828), this system of thought postulated that specific human qualities and psychological characteristics are associated with
distinct centers in the brain. Based on observation, Gall's theory was backed
up by a materialistic philosophy which maintained that character and intellect
were simply the combined functions of the organs of the brain. His system
was further elaborated by Johann Gaspar Spurzheim (1776–1832) into a
comprehensive theory of education and social philosophy which used an
examination of the shape and appearance of the subject's head to diagnose
strengths and weaknesses and prescribe procedures for remedying the defi-
ciencies.

As we have seen, Hall had used phrenological theory in conjunction with
animal magnetism, exciting various "organs" of the brain through mesmeric
passes. Braid developed his "phreno-hypnotism" in the same direction, replacing passes with pressure:

> The following is the mode of operating:—Put the patient into the hyp-
> notic condition in the usual way, extend his arms for a minute or two,
> then replace them gently on his lap, and allow him to remain perfectly
> quiet for a few minutes. Let the points of one or two fingers be now
> placed on the central point of any of his best developed [cranial] organs,
> and press it very gently; if no change of countenance or bodily move-
> ment is evinced, use gentle friction, and then in a soft voice ask what he

is thinking of, what he would like, or wish to do, or what he sees, as the function of the organ may indicate; and repeat the questions and pressure, or contact, or friction over the organ till an answer is elicited. (Braid 1899, p. 212)

In this way Braid was able, in his opinion, to elicit feelings or actions appropriate to the phrenological area of the skull being stimulated. To counter objections that he might be subtly leading the hypnotic subject to produce the desired response, he constructed experiments that he believed removed any chance of the subject's knowing what was expected (pp. 168–69).

Braid's flirtation with phrenology did not last long, however; in 1845 he dismissed it as no longer relevant to research on mental functioning. In its stead he undertook investigations into how physiological functioning could be altered by voluntary and involuntary mental efforts. At this point Braid moved on to consider the crucial role of suggestion and imagination in control of the body (Kravis 1988, p. 1195).

Now Braid began to frame his idea of the nature of hypnotism in purely psychological terms: "The condition is essentially one of mental abstraction or concentration of attention, in which the powers of the mind are engrossed, if not entirely absorbed, with a single idea or train of thought, and concurrently rendered unconscious of, or indifferently conscious to, all other ideas or impressions" (1851, p. 6). This formulation came to be called monoideism, since hypnotism was described as focusing on one idea to the exclusion of others. For Braid this concentration of attention was the factor that made the extraordinary phenomena of hypnotism possible. With regard to physiological effects, it seemed obvious to Braid that if a person's thoughts affect the body in ordinary waking conditions, how much more must mind be able to influence body during hypnotism, "when the attention is so much more concentrated, and the imagination and faith, or expectant idea in the mind of the patient, are so much more intense than in the ordinary waking condition" (1853b, pp. 3–4).

Braid attributed all peculiar phenomena of memory, all apparently supernormal perceptions, all unusual physiological effects, and all curative powers to this intense focusing of attention (1852, pp. 54ff). By studying the hypnotic state, he believed, investigators could understand more thoroughly than ever before the richness of the human mind and the basis for its power over the body. Braid admitted, however, that "as is the case in reverie or abstraction, so also is it in the hypnotic state—there are *different degrees* of mental concentration; so that, from some of them, the patient may be aroused by the slightest impression—whilst, in other stages, he can only be influenced by very powerful impressions on the organs of sense" (p. 55). In practice, he basically spoke of only two degrees or stages of hypnotism: the "sub or half-waking condition" (being partially hypnotized) and the "second conscious" or "double conscious" condition (being fully hypnotized) (1853b, pp. 5, 8, 26). Although some hypnotic phenomena occurred when a subject was only partially hypnotized, it was in the full, double-conscious state that the most striking phenomena took place.

Braid pointed out that when attention is directed in a certain way, bodily changes occur, such as milk flowing in a mother at the sight of her child, saliva produced on viewing or smelling tasty food, tears coming from grief, blushing from shame, or palpitations from fear. If mental attention could produce physical effects, what a great benefit hypnotism might be, since it focuses attention to an extraordinary degree. "Hypnotism merely enables us to control and direct the natural functions, either exciting or depressing them, as required, with more certainty and intensity than in the normal waking condition" (Braid 1844–1845, p. 297).

Recognizing the power of mind over body and ideas over physiological functioning, Braid drew attention to the presence of "dominant ideas" in the mind which, whether one is aware of them or not, powerfully affect one's psychological and physiological state. He developed a nomenclature for this, using "monoideology" to denote the doctrine of the influence of dominant ideas in controlling mental and physical action and "monoideism" to indicate the condition resulting from the mind being possessed by a dominant idea. He also used "ideo-dynamic" as a synonym for the latter term and "psycho-physiology" for the body of the phenomena involved in this condition (1855, pp. 9–10).

Braid evolved a complete framework of "hypnotic therapeutics" on the basis of these concepts:

> Since it cannot be doubted that the soul and the body can mutually act and react upon each other, it should follow, as a natural consequence, that if we can attain to any mode of intensifying the *mental* power, we should thus realise, in a corresponding degree, greater control over physical action. Now this is precisely what my processes do—they create no new faculties; but they give us greater control over the natural functions than we possess during the ordinary waking condition and particularly in intensifying mental influence, or the power of the mind of the patient over his own physical functions; and of a fixed dominant idea and physical state of the organs over the other faculties of the mind during the dominance of such fixed ideas. (1853b, p. 12)

Braid's notion of "fixed ideas" that dominate the mind and control the body led directly to a method of applying hypnotism in the amelioration or even cure of physical and mental illnesses. It involved using suggestion to create new fixed ideas of a curative nature:

> By our various modes of suggestion, through influencing the mind by audible language, spoken within the hearing of the patient, or by definite physical impressions, we fix certain ideas, strongly and involuntarily in the mind of the patient, which thereby act as stimulants, or as sedatives, according to the purport of the expectant ideas, and the direction of the current of thought in the mind of the patient, either drawing it to, or withdrawing it from, particular organs or functions; which results are effected in ordinary practice, by prescribing such medicines as experience has proved stimulate or irritate these organs. . . . The great object of all treatment is, either to excite or to depress function, or to increase or to diminish the existing state of sensibility and circulation. . . . For this purpose I feel convinced that hypnotism may be applied in the cure of some forms of disease with the same ease and certainty as our most simple and approved methods of treatment. (1853b, p. 8)

Braid described the successful treatment by hypnotism of many physical conditions, such as headache, paralysis, rheumatism, gout, epilepsy, hysteria, and spinal irritation. He pointed out that where a condition was due to an organic cause, neither hypnotism nor any other treatment could bring about cure. In all other cases, however, hypnotism was an effective remedy (1853b, pp. 15–18). Especially interesting is a case of "insanity" in which the patient was "haunted with the idea of the personal presence of a departed relative." Braid wrote: "I hypnotised her, and then excited a different idea in her mind, under the dominance of which idea I awoke her, and the unwelcome apparition never again made its appearance" (1853b, p. 19).

The use of suggestion to create constructive fixed dominant ideas or replace detrimental ones was an important contribution to the development of mental therapeutics. Although Braid's writings were quite well known in Britain in his day and became the basis for the development of a powerful stream of psychological thought in France, it would be many years before his system gained broad acceptance in his home country. William Carpenter, one of the few prominent medical men in Britain to appreciate Braid, praised his work in a *Quarterly Review* article (Carpenter 1853) and in his *Mental Physiology*.

Hypnotism in France

By their experiments with hypnotic trance as a surgical anesthetic, Eugène Azam, Paul Broca, Jean Nicolas Demarquay, and M. A. Giraud-Teulon pioneered the study of hypnotism in France (see chapter 7). The first full-length treatise in French on hypnotism was by the physician Joseph Pierre Durand (de Gros), who wrote under the pseudonym of A. J. P. Philips. Durand was particularly interested in the effects of electricity on the nervous system and how "electro-dynamism" affects health. Having spent time in America, Durand brought together certain strains of thought popular there in a book called *Electro-dynamisme vital* (1855). When he learned of Braid's work, he carried out his own experiments and then published *Cours théorique et pratique de braidisme* (1860), a work including detailed instructions about the induction and use of hypnotism in medicine, surgery, and education.

Liébeault

Ambroise August Liébeault (1823–1904) studied medicine at Strasbourg. In 1848 he read a book on animal magnetism and was immediately intrigued. He successfully mesmerized a few people but did not take the experiment further at that time. Then in 1860 he learned of Azam's experiments with hypnotism and decided to try to use that technique in his medical practice. He announced that patients who agreed to treatment by hypnotism would not be charged and soon had many volunteers. In the next few years he developed a unique approach to hypnotic therapeutics.

For Liébeault, hypnotic sleep and natural sleep were identical except that the hypnotized patient was in a state of rapport with the hypnotizer. He became a master at hypnotizing patients and used hypnotism to cure a great variety of illnesses. By the suggestion of tiredness and sleep he induced hypnotism and then made simple curative suggestions appropriate to the disease.[12] These suggestions would negate the symptoms of the disease and inculcate good habits of health maintenance.

In 1866 Liébeault published *Du sommeil et des états analogues,* in which

---

12. At one time Liébeault believed there was something in the mesmeric notion of a physical force at work in magnetic healing. In *Etude sur le zoomagnétisme* (1883) he described his success in treating infants with animal magnetism. He believed that since their psychological development was insufficient to allow for suggestion, a physical agent must have been at work. In 1891, however, he renounced the conclusion that there was a "nervous action" transmitted from one person to another by vibrations, stating that further experiments had convinced him that even infants less than two years of age had sufficient awareness to respond to suggestion.

he described his method of hypnotic induction and the various states that resulted, noting six distinct degrees of hypnotism. He speculated on the effect of hypnotism on the nervous system and the means by which healing took place. He also described hallucinations (sometimes of spirits) that could occur during hypnotism and the central role of suggestion in creating them.

Over the decades of Liébeault's active medical practice he cured many patients who had been unable to find help elsewhere (see Liébeault 1866, 1873, 1883, 1889, 1891). His reputation as a medical healer eventually brought him to the attention of Hippolyte Bernheim. The result was a professional association that would have important repercussions.

Bernheim and the Nancy School

Hippolyte Bernheim (1837–1919), like Liébeault, received his medical training at Strasbourg and developed a practice at Nancy. Bernheim had a scholarly bent and wrote well-received papers on tuberculosis, typhoid fever, and cardiovascular disorders. He first learned of Liébeault's work when Liébeault successfully treated a patient suffering from sciatica after Bernheim had failed. The cure had been accomplished by hypnotic suggestion, a method unknown to Bernheim. He visited Liébeault's clinic in 1882 and was impressed by what he saw. At once he began his own hypnotic work, and over the following three decades he presented the results in many journal articles and several longer works (Bernheim 1884, 1886, 1891, 1897a, 1911, 1920).

In *De la suggestion dans l'état hypnotique et dans l'état de veille* (1884) Bernheim strongly distinguished his views from the theory of Jean Charcot, who saw hypnotism as an appendage of hysteria (see below). Bernheim emphasized that hypnotism was a psychological state in its own right, one intimately connected with suggestion. He noted nine degrees of hypnotism, as opposed to the six of Liébeault, and described his methods of induction. He entered into a lengthy discussion of suggestion and proposed a new definition of hypnotism: the induction of a peculiar psychical condition that increases susceptibility to suggestion. Bernheim claimed that suggestion is involved in practically all human interaction and stated that paralysis, contracture, anesthesias, sensorial illusions, and hallucinations could be obtained through suggestion *without* hypnotism. He also described the phenomena associated with hypnotic sleep, including automatic movements, automatic obedience, suggested hallucinations during the hypnotic state, and posthypnotic hallucinations. There was as well an important chapter on the physiological effects of hypnotism.

Bernheim's systematic approach to the problem of hypnotism, set off clearly from Charcot's theories, earned him a strong following among clinicians and theorists. Chief among them were Henri Etienne Beaunis (1830–1921) and Jules Liégeois (1833–1908). Beaunis, a professor of physiology at Nancy, directed the first laboratory of psychophysiology at the Sorbonne. He accepted Bernheim's view of the nature of hypnotism and conducted examinations of the physiological changes that occurred under hypnosis. His findings were published in journal articles and in *Le somnambulisme provoqué: études physiologiques et psychologiques* (1886). Jules Liégeois was a professor of law at Nancy. He too accepted Bernheim's view of hypnotism and suggestion. In *De la suggestion et du somnambulisme dans leurs rapports avec la jurisprudence et la médecine légale* (1889) Liégeois wrote about the relevance of hypnotic suggestion to civil and criminal law, investigating the legal implications of suggested hallucinations, hypnotically induced amnesia, suggestions in the waking state, and other phenomena.

Bernheim and Liébeault, together with Beaunis and Liégeois, constituted what came to be called the Nancy school of hypnotism.[13] A protracted feud developed between this group and the one Charcot founded at the Salpêtrière Hospital (Barrucand 1967, pp. 166–202). Bernheim and his followers insisted that Charcot unjustifiably tied hypnotic phenomena to physiological pathology and did not realize to what degree his own experiments with hysterics were contaminated by suggestion. In fact, Bernheim argued, hysteria was not a physiological state at all but a psychogenic condition, a reaction to emotions that had not been integrated into the psyche. He showed that hypnosis and suggestion could be used to ameliorate or cure hysterics (pp. 119–28). In this way Bernheim added the effective use of hypnosis for psychological illnesses to Liébeault's mastery of the hypnotic technique for treating physical illnesses.

Over time the tenets of the Nancy school came to be widely accepted. Among those outside France who held the Nancy views were Albert Moll and Albert von Schrenck-Notzing in Germany, Vladimir Bechterev in Russia, Milne Bramwell in England, Boris Sidis and Morton Prince in the United States, Otto Wetterstrand in Sweden, Frederik Van Eeden in Holland, and Auguste Forel in Switzerland.

Charcot and the Salpêtrière School

Jean Martin Charcot (1825–1893) received his medical degree in 1853, and in 1862 was appointed to the Salpêtrière Hospital, where he established a neuro-

13. Boring (1957) has pointed out that, unlike Charcot, Bernheim was not a propagandist and therefore cannot be called the founder of a school. He argues that the school of Nancy became famous not because of effective promotion but because it was right, whereas Charcot's school of Salpêtrière was famous but wrong (p. 699).

logical clinic that would become world famous. From his experiments with hysterics, Charcot arrived at a theory of hypnotism that was the basis for the so-called Salpêtrière school.

At the Salpêtrière Charcot closely observed patients afflicted with hysteria and classified their symptoms. The results of his observations were given systematic form by Charcot's coworker Paul Richer in a work entitled *Etudes cliniques sur l'hystéroépilepsie ou grande hystérie* (1881). Richer identified four distinct phases in a typical hysterical crisis: epileptoid, clownish, passionate, and hallucinatory. He stated that once the phases had begun, they would continue inexorably to their conclusion, after which the patient would "awaken" with full memory of the crisis.

If the patient was a woman, the onset of a hysterical crisis was usually indicated by a hysterical "aura" consisting of a pain in the region of the ovaries. Many of the movements and poses of the crisis were erotic in appearance and for some patients involved overt sexual pleasure. In "major hysteria" the patient possessed what were called "hysterogenic zones" or insensitive areas on the body, usually near the breasts or ovaries in women and often in the testicles in men. Stimulation of these zones could induce or inhibit a hysterical crisis. Belts that pressed on the hysterogenic zone were sometimes worn by patients to prevent the occurrence of crises at awkward times.

In his first paper on hypnosis, delivered to the Académie des Sciences on February 13, 1882, Charcot presented his view of the nature of hypnotism: "Hypnotism, considered in its perfectly developed form—such as that frequently presented by women suffering from hystero-epilepsy with mixed crises—includes many nervous states, each distinguished by a particular symptomology. According to my observations, these nervous states are three in number, namely: (1) the cataleptic state, (2) the lethargic state, (3) the somnambulistic state" (p. 403). He proceeded to describe these states in common medical terms, speaking of "reflex movements," "muscular states," and "sensory alterations." This way of describing hypnotism was highly palatable to medical orthodoxy, and this one short speech paved the way for the broad acceptance of hypnotism among physicians. Although before this speech Charcot had already published a *Contribution* (1881) about hypnotism and hysteria, it was this paper that was historically decisive.[14]

14. The only physiologist of high repute to experiment with artificial somnambulism before Charcot's paper was Charles Richet (Richet 1875, 1880). Richet's approach, however, had mixed together the physiological and psychological and had made historical reference to the work of the mesmerizers. For these reasons it had not significantly appealed to the medical world. Charcot's reputation as a neurologist and his penchant for reducing the psychological to the physiological accomplished what had not been previously possible.

Charcot held that the three "stages" were organically determined and not the result of suggestion; hypnotism was therefore an artificially created neurosis. The symptoms of hypnotism and those of hysteria were considered to be very similar, and for that reason Charcot thought of hypnotism and hysteria as essentially identical.

Charcot's study of traumatic paralysis confirmed his ideas about hypnotism. Noting that traumatic paralysis differed from organic paralysis but seemed to be identical with the hysterical paralysis he had observed in his patients, he succeeded in using hypnotic suggestion to reproduce the symptoms of paralysis resulting from trauma. From this he concluded that the shock of the trauma produced a spontaneous hypnotic state in which the individual induced the paralytic symptom through autosuggestion.

Probably without his realizing it, Charcot's doctrine of hypnotism harked back to animal magnetism as practiced before Puységur. Like Mesmer, Charcot defined the effects of hypnotic (magnetic) induction in terms of physiological rather than psychological states and drew a direct connection between the "crisis" of hypnotism (animal magnetism) and a pathological "crisis" (hysteria) spontaneously occurring in nature.

The relationship between the Salpêtrière view of hypnotism and animal magnetism was not lost on two of Charcot's followers, Alfred Binet and Charles Féré. In 1887 they published *Le magnétisme animal,* which traced the history of animal magnetism and presented Charcot's hypnotism as its purified and modernized form:

> The history of animal magnetism has shown that if, up to late years, the existence of the nervous sleep, and of various phenomena allied with it, has been doubted, it is chiefly because the experimenters wanted method. . . . The method which led to the revival of hypnotism may be summed up in these words: the production of material symptoms, which give to some extent an anatomical demonstration of the reality of a special state of the nervous system. . . . It is to Charcot that the honor must be assigned of having been the first to enter on this course, in which he has been followed by numerous observers. (Binet and Féré 1890, pp. 85–86)

Binet and Féré also recalled the techniques of the magnetizers when they described various means that could be employed to produce hypnotic symptoms, including the application of magnets, vibrating tuning forks, electrical currents, and various metals. Charcot experimented with all these agents and early in his investigations made use of "metallotherapy" as developed by Victor Burq. Burq employed metal plates, usually copper, to heal disease. He had published his findings (Burq 1853), but it was not until he came to the notice of Charcot that his ideas gained popularity.

In 1876, at Burq's request, the Société de Biologie named a commission to study the effects of metals placed on the surface of the skin. The committee, made up of Charcot, Jules Luys, and A. Dumontpallier, noted that Burq had, with apparent success, applied metals to the skin in cases in which sensitivity had been impaired to restore normal sensitivity. They decided to limit their inquiry into metallotherapy to testing the efficacy of gold, copper, iron, and zinc. The hysterical patients of the Salpêtrière were put at the disposal of the committee, and Charcot applied the metal plates. The results confirmed Burq's claims, for anesthetic areas of some patients with even long-term insensitivity were restored to full feeling.

In the process of experimentation, the committee discovered that the application of small electrical currents could have an effect analogous to that of the metallic plates. They also discovered a phenomenon that they considered entirely new: when the sensitivity of an anesthetic area on one side of the patient's body was restored through the metal plates or an electric current, the sensitivity of the corresponding area of the opposite side of the body was lost. The committee named this phenomenon "transfer of sensibility," and for many years physiologists tended to accept this phenomenon as proven (Charcot, Luys, and Dumontpallier, 1877).[15] Bernheim rejected all claims that transfer of sensibility could be effected by the application of physical agents. For him such phenomena were due totally to suggestion, and he drew attention to defects in the methodology of transfer experiments that allowed the patient to respond to unspoken suggestions of the experimenters.

In time the main tenets of the Salpêtrière School—the physiological explanation of hysteria, the three stages of hypnotism, and the transfer of sensibility—lost favor in the medical world. At the same time the main theme of the Nancy school—the central role of suggestion—gained widespread acceptance.[16]

15. Jules Luys, a neurologist, in *Les émotions chez les sujets en état d'hypnotisme* (1887), discussed the transmission of states of emotion from one hypnotized subject to another and the effects on the nervous system of certain substances and apparatuses placed at a distance from the subject. Materials used included laudanum and glass tinted with various colors. Luys seems to have been the first to use a rotating mirror to induce the hypnotic state (see *Compte rendus et mémoires des sciences de la Société de Biologie* 107 [1888]: 449).

16. Significant contributions to the understanding of hypnotism were made by a number of investigators outside the two main schools, including A. Dumontpallier, Paul Magnin, and Edgar Bérillon of the Pitié school of hypnotism, who pursued ideas about hysteria and hypnotism based on experiments in metallotherapy (see Dumontpallier 1879, Dumontpallier and Magnin 1882a and 1882b, and Bérillon 1884). Although there were differences of opinion with Charcot in regard to the symptoms of the three stages of hypnotism, their ideas were quite close to those of the Salpêtrière school.

Part III

The Magnetic Psyche

Chapter 9

The Paranormal

From the beginning, phenomena were reported in connection with animal magnetism that were ordinarily associated with the mysterious or occult. Mesmer, for instance, claimed to be able to magnetize people from behind a wall, and he had a shocked Baron Horeczky de Horka to back up this assertion. The marquis de Puységur reported that his first somnambulist, Victor Race, responded to mental commands and could read his thoughts.

Yet Mesmer abhorred superstitious and occult practices of all kinds and viewed animal magnetism as a purely natural phenomenon, explainable in terms of matter and motion. It was Mesmer's fond hope that scientists, through experimentation, would eventually discover the laws of animal magnetism. Even those who followed Puységur, with his deemphasis of the magnetic fluid, considered all the phenomena, including thought reading and clairvoyance, to be susceptible to explanation by natural, scientific laws. Those who explained animal magnetism in terms of the Romantic notion of "sympathy" believed it to be a normal manifestation of the forces of nature and simply asked scientists to expand their limited methods of obtaining scientific knowledge. Needless to say, those who explained the phenomena of animal magnetism in terms of suggestion had no enthusiasm for the occult. Only the spiritualists—a minority among magnetizers—believed magnetic action to transcend the natural.

Those who studied these more extraordinary mental aspects of animal magnetism (clairvoyance, thought transference, and so on) kept the question of the psychology of mesmeric phenomena in the forefront. They described how the mind demonstrated new and surprising facets when in the state of magnetic sleep. They attempted to explain these novel mental manifestations in various ways. In the process they kept attention focused on human psychological functioning, both normal and paranormal.

Michael Thalbourne defines as paranormal "any phenomenon which in one or more respects exceeds the limits of what is deemed physically possible on current scientific assumptions" (1982, p. 50). This definition has the advantage of being neutral with regard to the beliefs of those who report the phenomena. Moreover, it does not rule out the possibility that the phenomena might fit into the schema of a future science. By Thalbourne's definition, the unusual phenomena described in magnetic literature can be called paranormal.[1]

The Phenomena

The paranormal phenomena of magnetism, which include thought reading and mental commands, medical clairvoyance and general clairvoyance, precognition, discerning the magnetic fluid,[2] magnetizing and healing at a distance, and ecstasy, were in some cases so widely reported in the first eighty years of animal magnetism that they were simply taken for granted as ordinary magnetic experiences. Other phenomena, although rare, were not unique.

Thought Reading and Mental Commands

In the literature of animal magnetism it is difficult to distinguish between reading thoughts and sensing mental commands. This is true of Puységur's Victor Race: "Then he became calm—imagining himself shooting a prize, dancing at a party, etc. . . . I nourished these ideas in him, and in this way I forced him to move around a lot in his chair, as if dancing to a tune. While singing it (*mentally*), I made him repeat it out loud" (Puységur 1784, p. 28).

As Puységur continued to experiment with somnambulism, he showed little interest in the paranormal aspect of his experiences, seeming to take it for granted. For instance, in testing the limits of the commands a somnambulist would follow, Puységur mentally compelled a young woman in the state of magnetic sleep to strike him with a flyswatter (1784, p. 182). In relating the story, he concentrated not on the woman's obedience to a mental command but on whether a somnambulist could be forced to do something against her will.

1. The review that follows contains both descriptions of paranormal phenomena and explanations by those who made the observations. In discussing these accounts, I shall make some reference to the possibility of intentional or unintentional deception, errors in observation, and defects of reporting. There will, however, be no attempt to evaluate the validity of the explanations offered by the reporters from a supposedly objective modern standpoint. This approach makes it possible to avoid "presentism," a common defect of historical writing which claims to make "objective" judgments about the validity of the views described, based on the assumption that modern scientific knowledge is the true knowledge and that everything that led up to it is naive.

2. Tardy de Montravel's Mlle N. was the first in a long line of somnambulists to claim that she could see the magnetic fluid. Deleuze said that magnetic subjects could commonly see and sense magnetic emanations (1813, 1:163; 1850, p. 79). Meyer wrote that most often the fluid was seen radiating from the extremities of the body—the fingertips and toetips (1839, p. 86). It was also commonly thought that somnambulists could use this ability to distinguish magnetized from unmagnetized objects (see Passavant 1837, p. 46).

Deleuze asserted that the somnambulist is so attuned to the magnetizer that she can read his thoughts while receiving no impressions through the senses. In fact, for the somnambulist it is not sensations that produce thoughts but thoughts that produce sensations (1850, p. 92). Commonly this ability to read the magnetizer's thoughts was linked to the intimate rapport between somnambulist and magnetizer (e.g., Wirth 1836, pp. 190ff; Weserman 1822, p. 50). This connection was thought to be so immediate that slight changes in the magnetizer's mood or emotion were directly felt by the somnambulist. Sometimes the somnambulist enjoyed a similar rapport with others in his presence, and although he might not be able to read their thoughts, he would strongly sense their feelings and intentions (Deleuze 1850, p. 77).

Deleuze offered an explanation for the somnambulist's ready obedience to mental commands: "It is necessary to think of somnambulists as infinitely mobile magnets. . . . We know that if you place two similar instruments next to each other and play chords on the first, the second will resonate with corresponding chords. This physical phenomenon is similar to what happens in magnetism" (1813, 1:181).

Lausanne (1819) conducted a number of experiments in which he controlled the somnambulist through what he believed to be the action of his will alone, without any possibility that the subject could perceive physically what he wanted. His mental commands were simple and carried out with ease by his subjects. In Lausanne's view, this kind of mental communication can take place only if the subject is very sensitive and there is a strong rapport between subject and magnetizer (p. 145).

Medical Clairvoyance

Probably the most frequently reported paranormal magnetic phenomenon was the somnambulist's apparent ability to diagnose a disease, predict its course, and prescribe effective remedies. Early magnetic literature is replete with reports of this kind of "medical clairvoyance" applied both to the somnambulist's own illness and to those of others.

Puységur routinely asked his patients, while they were somnambulistic, what their illness was and how it should be treated. He considered their answers reliable and followed their instructions closely. He also used somnambulists as assistants in his healing work, bringing them into the presence of the ill and asking them to determine the nature of the disease. One of his somnambulists described his experience of clairvoyant diagnosis: "It is . . . a sensation that I experience in the part directly corresponding to that where the person I touch is suffering. My hand naturally goes directly to the illness" (Puységur 1784, p. 88).

Lausanne felt himself able to diagnose disease conditions on the basis of the sensations he experienced when physically drawing near to the problem area. He thought of the body of the magnetizer as a diagnostic instrument that could sense the fluidic "currents" of the magnetic subject.

For many somnambulists, diagnosis involved "seeing" into the body. It was thought that the somnambulist could see not only the interior of his own body but that of others with whom he had been placed in rapport (e.g., Chardel 1826, pp. 249–250; Fischer 1839, 3272ff). While in the somnambulistic state Mme Lagrandré, whose mother, Mme Plantin, had had her breast removed during magnetic sleep, provided a graphic description of the condition of her mother's internal organs immediately following the surgery. Mme Plantin died two weeks after the operation (from causes unrelated to the surgery). When her surgeon, Jules Cloquet, performed an autopsy, he discovered that Mme Lagrandré's visual impressions were accurate and that her vision of certain peculiarities in the position and color of her mother's organs was correct (Brierre 1853, pp. 253–55).

Another aspect of medical clairvoyance was prediction of the course of the illness, both the somnambulist's own (e.g., Strombeck 1814, pp. 34ff) and the illnesses of others (e.g., Deleuze 1813, 1:181). This phenomenon was not considered to be as extraordinary as other aspects of medical clairvoyance since it was merely a heightened ability to discern the effect in the cause.

Somnambulists commonly prescribed medicines, herbs, and physical treatments for illness. This too was largely seen as a natural instinct heightened to a degree not available in the normal state. There were, however, cases in which the somnambulist prescribed herbs apparently unknown to him or her and used technical names for medicines ordinarily recognized only by trained physicians.

General Clairvoyance

The literature abounds in descriptions of somnambulists who were able to see objects or be aware of events that were out of the range of their senses—even at great distances. Some attributed this clairvoyance to the somnambulist's ability to perceive the magnetic fluid that emanates from all things and can radiate over distances and penetrate solid barriers. Others believed that the somnambulist was partially freed from the bonds of the body while in the magnetic state and so could explore at will any location, no matter how distant or hidden.

Chardel, who held the latter view, described a case in which the younger of two somnambulistic sisters clairvoyantly perceived the stroke of a scalpel used, out of her sight, on the foot of the elder and immediately fainted (1826, pp. 254–55). Strombeck's somnambulist described in detail the contents of her doctor's closed bag, although she could not see them, nor were they likely to be guessed (1814, p. 76).

There were frequent reports of somnambulists becoming aware of events taking place at great distances from them. Chardel related the case of the somnambulist wife of an army colonel who described in detail the suicide of a young officer known to her. She saw the young man, who lived a few miles away, point a pistol at himself, preparing to fire. She begged that someone ride to his aid immediately, but the horseman found the man already dead by his own hand (Brierre 1853, p. 265).

Sometimes the somnambulist experienced the sensation of leaving the body and being present at the location she was viewing. Kerner (1824, pp. 274ff) wrote of this kind of "traveling clairvoyance" in his young somnambulist Caroline, who accurately described events at a distance as they were happening and was totally unaware of her surroundings, except for her magnetizer.

Magnetic clairvoyance sometimes seemed to operate as an extension of the rapport between somnambulist and magnetizer (or other persons placed in rapport): "This rapport entails a transfer of sensibility from the magnetizer to the somnambulist. She sees, hears, smells, tastes and feels in the magnetizer" (Wirth 1836, p. 179). Reports of cases in which sensations experienced by the person in rapport were felt by the somnambulist are common. Kieser (*Archiv für den thierischen Magnetismus* 3:3) wrote of a magnetizer who took snuff only to hear his somnambulist complain of a tobacco itch in her nose. A somnambulist of Gmelin developed such a close rapport with her sister that she felt a sucking sensation on her breast when her sister's infant was feeding (Wesermann 1822, p. 69). In another case (*Archiv* 1:1) a magnetizer attended a flute recital at some distance from the room of his languishing patient. While he was there, the somnambulistic patient suddenly spoke of hearing beautiful music. Those in the room heard nothing and attributed it to a dream. When the magnetizer returned, they discussed the experience and discovered that the somnambulist heard the music only when the magnetizer was within earshot of the flute and ceased to hear it when the magnetizer left the recital.

Many magnetizers held that physical movement could also be conveyed from magnetizer to somnambulist. Chardel stated: "The spiritualized life [magnetic fluid] is the agent the soul uses to move the body. That of the magnetizer can sometimes, during somnambulism, give him the ability to cause movement in the members of his somnambulist—not with the same facility as in his own, but in the same manner" (1826, p. 266).

Precognition

Puységur discovered that somnambulists sometimes spoke as though they knew what was to come in the development of their illnesses and those of others. He came to trust this foreknowledge and depend on it in his healing endeavors. Mesmer, too, accepted that somnambulists could predict the future course of a disease and even other events. He did not consider this amazing, however, attributing this power to an instinct that can read future events in present causes. Strombeck's somnambulist Julie was at times so in touch with this instinct that she was a "perfect animal," garnering knowledge of her illness and the treatment it required not from reason but from the same kind of instinct animals use in healing themselves (1814, p. 133). He claimed that she was able to predict with astonishing accuracy the precise moment of her fainting spells and convulsions, and averred that she was not faking these symptoms (p. 153).

Deleuze held that somnambulists could foresee developments, but he was cautious about the matter. Taking his lead from Mesmer, he said that effects are discerned in the causes and that the prevision of somnambulists derives from natural causes. He counseled magnetizers to avoid excessive enthusiasm: "Do you want to avoid errors? Then accept only well-attested facts—that is, when a somnambulist has predicted an event close at hand that follows from known causes. Beware of believing the same about more distant and less familiar events. Above all, guard yourself from thinking that predictions are infallible. That would be dangerous to morals and absurd in the light of physics, because this supposes that all events occur from necessity, and we thereby descend into fatalism" (1813, 1:172).

Bertrand, usually cautious in his observations, said that "prevision is incontestable" (1823, p. 123). But like Deleuze, he insisted that it applied only to those cases in which the somnambulist can see the effect in the cause. He pointed out the danger that credulous people might treat the somnambulist as an oracle, and then, through "excesses of the imagination," the somnambulist might manufacture predictions to satisfy their expectations.

Magnetizing and Healing at a Distance

Mesmer claimed that animal magnetism could be applied when the magnetizer was some distance from the subject since the magnetic fluid was capable of traversing space and penetrating all objects. Barberin took this a step further, holding that distance was of no concern since the magnetic operation was basically spiritual, directed by the will of the magnetizer. In the decades that followed, the notion of magnetizing and healing at a distance remained a part of magnetic tradition, whether the explanation emphasized the action of a fluid or a spiritual power.

Wienholt, for instance, carried out experiments in which he induced magnetic sleep in a patient more than a mile away (Wesermann 1822, pp. 71–72). In England, de Mainauduc declared that for diagnosis and treatment "direction and distance are in every sense of the word immaterial, provided the Attention is properly fixed on the object" (1798, p. 102). Lausanne described in detail the procedure to be used in applying magnetic healing to an absent person (Montferrier 1819, 1:117, 146ff). These and many other examples of magnetizers using action at a distance are often mentioned in passing, as though the reality of the phenomenon need not be demonstrated. This attitude became so commonplace in England that Samuel Stearns could write satirically: "Our effluvia is not supposed to be confined to the narrow limits of one kingdom or country; no, it may be sent by the powers of imagination, in a moment, against wind and tide, and through all the different currents of air, to the remotest corners of the globe, if the Operator sets his affections upon a Person, and gives the word of command" (1791, pp. 11–12).

There were, however, magnetizers who felt the need to prove magnetic action at a distance under controlled conditions. In 1820 Jules Du Potet was invited by Dr. Henri Husson, director of the Hôtel-Dieu of Paris, to demonstrate before medical witnesses the induction of magnetic sleep in a patient isolated from all sensory clues. Du Potet, who claimed that he had successfully induced magnetic sleep at a distance many times and was convinced that the agent was the magnetic fluid, not suggestion, undertook the experiment confidently. At the time, Du Potet was using animal magnetism in the treatment of Catherine Samson, who had proved to be a particularly good somnambulistic subject. She was brought into the usual place of treatment in the hospital. Husson was in attendance, but Du Potet was concealed in a specially constructed oaken closet a few feet from her. Samson wondered why Du Potet had not arrived, and Husson intimated that he must have been detained. At a previously agreed-upon signal from Husson, Du Potet began magnetizing from his concealed position. Within three minutes the young woman was in a state of magnetic sleep. The experiment was repeated the following day, with equal success. Two days later, Samson was again made somnambulistic by the secret magnetic action of Du Potet, this time in the presence of the renowned surgeon Joseph Claude Récamier, who gave the signal to commence. To the astonishment of Du Potet, Husson, and other witnesses of this experiment, Récamier, although he himself had determined the signal, later stated that Du Potet and Samson must have been in collusion.

Ecstasy

Jung-Stilling believed that in magnetic somnambulism the soul was temporarily separated from the body: "The human soul, by an artificial stroking, or
magnetizing, can be detached from the nervous system in various degrees,
and even become a free agent if the degree of detachment is sufficient. . . .
When the soul is in a state of detachment from its sensitive organs, while still
in the body, consciousness of the visible world ceases as long as that detachment lasts. The soul, however, lives and acts in the sphere of its knowledge,
and eventually, by frequent repetition of this state, enters into connection with
the world of spirits" (1854, pp. 48, 49). This out-of-the-body state was commonly called "ecstasy."

As Bertrand pointed out (1823, p. 330), the striking similarity between the
states attributed to religious ecstatics over the centuries and those of magnetic
somnambulists was sufficient to place them in the same psychological category. Both involve impressions of separation from the body, and both entail
some kind of communication with a higher spiritual world. While in the state
of magnetic ecstasy, somnambulists might find themselves communicating
with angels, demons, saints, or souls of the departed. Experiences of magnetic ecstasy and combination with the discarnate world initiated animal
magnetism into a new phase, which might be called magnetic spiritism (see
chapter 10).

The Critics

From the first reports of paranormal phenomena in connection with animal
magnetism there were those who expressed skepticism. Some critics doubted
some of the phenomena and accepted others, some denied the reality of any of
the phenomena, and some believed in the reality of the phenomena but attributed them to the devil. Skepticism could be found both within the magnetic
tradition and among those who rejected the reality of animal magnetism.

Within the Magnetic Tradition

From very early there were magnetic practitioners who urged caution in
accepting the more startling phenomena, particularly those associated with
magnetic somnambulism. Würtz, for instance, a strong supporter of Mesmer,
admitted the reality of magnetic somnambulism but decried as excessive
certain claims made by somnambulists, such as the ability to see the internal
organs of the body. He laments: "Is it not humiliating that our century should
fall back into the ancient time of ignorance and superstition and give such a
blind faith to everything these people say in their state of crisis? One con-

cludes that the human spirit has become decadent when one sees enlightened men consulting as oracles people who speak in a dream, in order to get information from them that derives from beyond their external or internal senses, and which they believe magnetic sleep inspires in them" (1787, pp. 52–53).

Though Würtz and a few others in the magnetic tradition dismissed the paranormal phenomena of somnambulism, most magnetic practitioners who criticized the phenomena believed that some of them were genuine and valuable. Their objections were usually leveled against what they considered to be credulity and lack of care in observing the phenomena. Thus Deleuze, while accepting the reality of mental communication, medical clairvoyance, general clairvoyance, and precognition, urged the utmost caution in dealing with somnambulists while they are engaged in these activities and a certain skepticism regarding their pronouncements, particularly their predictions of the future (1813, 1:171).

Hénin de Cuvillers rejected all magnetic phenomena that could not be explained in terms of suggestion. He called claims of thought reading, clairvoyance, and precognition absurdities and held that magnetizers who accepted them were superstitious and naive. Bertrand also asserted that insufficient attention had been given to unconscious suggestion on the part of experimenters. He doubted that somnambulists could see the magnetic fluid or the interior of physical organisms. Yet he held that the state of somnambulistic ecstasy was a genuine psychological phenomenon in which individuals could sometimes perform paranormal feats, such as thought reading (1825, pp. i–xxiv, 431ff, 472ff).

Chardel, a great supporter of animal magnetism, was also one of its most capable critics. He pointed out the possibilities of self-deception in supposed magnetization at a distance and cautioned that apparitions of departed spirits were often nothing more than the action of a pathological imagination; yet he did not rule out the possibility of communication with spirits and even warned imprudent experimenters against possession by an "enemy will" (1826, pp. 263–74, 287 ff).

Outside the Magnetic Tradition

Critics of animal magnetism who did not practice or support it, like its adherents, pointed to errors in observation due to excessive zeal or naivete. They, too, referred to the power of implicit or explicit expectations over the "heightened imagination" of the somnambulist. But they also cautioned against deliberate deception or fraud—most often perpetrated by the somnambulist—and claimed that some phenomena were mere cheap tricks. Others held that the phenomena were genuine but the product of devils and the prince of devils— Satan himself.

Faulty Observation, Suggestion, and Fraud

Among the ablest critics of the phenomena of animal magnetism was Julien Joseph Virey (1775–1846). His article in the *Dictionnaire des sciences médicales* (also published as a monograph; see Virey 1818) exemplifies thoughtful analysis of an extremely complex issue.

Virey began by dismissing the testimony of witnesses, for their claims to have seen miracles "can be the effect of illusions or marvels, of the seduction of enthusiasm, of credulity, or of haste in judgment or other favorable prejudices which too often lead to believing what one wants to believe." To balance their reports, which he allows are made in good faith by intelligent people, Virey pleads "to oppose confidence with skepticism" (1818, pp. 54–55).

But why, Virey asked, had animal magnetism been able to persevere despite the opposition of savants, and despite widespread sarcasm and ridicule? Why did able physicians in Germany and elsewhere support it, and why, if it is mainly the product of charlatanism and cupidity, did so many generous people, interested only in helping others, give so freely of their time and energy to practice it? And why were there so many incontestable cures resulting from animal magnetism—often where conventional medicine had completely failed?

Virey accepted the magnetizers' assertion that if skeptics want to make a judgment on the phenomena of animal magnetism they should put aside their prejudices, try the operations themselves, and judge from their own experiences. Virey undertook to do just that, showing a wide knowledge of the literature of animal magnetism and a generous empathy for the magnetizer's viewpoint. At one point he wrote: "I say to our adversaries: When it is a matter of transmission of disease [contagion], you acknowledge the reality of miasma, but when it is a matter of the transmission of health you say it is imagination. If it is possible in one case, why not in the other?" (1818, p. 62). Since action at a distance is widely accepted by scientists in regard to electricity, for example, why, he asked, should it be denied when it comes to animal magnetism?

But for all his empathy, Virey did not accept the theory of animal magnetism. He argued that the somnambulists' purported capacities to discern microscopic objects, see in the dark, trace scents, recognize by taste substances that are medicinally beneficial, and foresee crises and other bodily modifications they will experience in the course of an illness are simply manifestations of heightened sensibility to the inner workings of the organism, and therefore explainable in terms other than the action of a magnetic fluid. He said that we can detect in animals a sense of what needs to be done to heal themselves—a kind of "medical instinct"—which applies to human beings in the somnambulistic state as well. Virey also claimed that it was not a large step to assume that the somnambulist can so identify with another suffering person that he or she can sense what would be medicinally beneficial to that person.

Virey believed that somnambulism could produce such an inner concentration that any one of the senses could be greatly magnified. A somnambulist might distinguish a glass or other object touched by the magnetizer through a heightened sense of smell, discerning his or her characteristic scent on the object. Thus one does not need to resort to a theory of magnetic fluid to account for such a phenomenon.

Virey stated that internal voices, communication with spirits, and possession can all be explained in terms of an internal instinct that is experienced as alien to the individual and therefore assumed to be another entity. The operation of this internal instinct, a spontaneous principle of action, is obscured in the state of illness and in dreams by the distractions of waking life. In sleep or somnambulism, however, it encroaches on one's awareness. Nevertheless, even in these states the "chain of reasoning" that has led to the conclusions reached is lost. All the person perceives is a striking communication, prescription, or command which he believes comes from "somewhere else" (1818, pp. 74–75).

Communications that derive from this source, whether nocturnal or somnambulistic, may be quite accurate. For instance, a person may have a strong presentiment about a friend that proves to be true. This is not, however, because of some communication at a distance, but rather, Virey asserts, because the strong sympathy and identification that occurs in friendship works within the dreamer or somnambulist, enabling her to reach certain conclusions about experiences the friend is likely to have—conclusions which are then communicated, fully formed, to the conscious mind. Thus for Virey, precognition of future events in the life of another is always the result of a combination of sympathy and unconscious, instinctive reasoning. This also accounts for the apparent ability of the magnetic somnambulist to read the thoughts of the magnetizer: "A somnambulist is subject to the will of her magnetizer; she submits and accommodates her will to his movement and influence. He thereby directs her chain of thoughts, and she is subordinated or even abandoned to the action of this 'mover.' This communication is so intimate that even before the magnetizer has spoken, the somnambulist has, so to speak, divined it" (1818, p. 76).

The inner concentration characteristic of somnambulism also accounts for other marvels in magnetic literature, declared Virey. He cites examples such as Victor Race, who showed signs of an intellectual brilliance uncharacteristic of his ordinary state of consciousness. In other cases a peasant known to speak only his native patois was able to discourse in good French, or a person who spoke only one language showed the ability to speak in another. Numerous instances are described in magnetic literature in which the somnambulist, while in a state of ecstasy, held forth on profound philosophical or religious subjects about which he had little to say in the waking state. Virey attributes these phenomena to a concentration and "cerebral tension" that gathers together fleeting bits of information in the brain and organizes them into surprising intellectual products. The process, however, is not voluntary: "What is needed is an unstable disposition" (p. 78).

As for the somnambulists' supposed ability to see magnetic fluid emanating from the head and fingers of their magnetists, this according to Virey is just the result of the expectations imbued in them by the theory of animal magnetism itself. For him the extraordinary effects of animal magnetism follow from a faith shared by magnetizer and magnetized in its reality. This faith leads to the powerful intervention of the imagination and, combined with a profound sympathy that works below conscious awareness, produces all known effects. It was Virey's belief that a kind of psychological submission was always involved in producing the phenomena: "Concerning those who operate as magnetizers: they only act on individuals inferior to them, either in terms of physical or moral qualities. It would be impossible to act without this superiority. . . . So I must seek out my inferiors in spirit and character. Then, vigorously taking the ascendancy over these people, who see me as endowed with a powerful energy, I strike them with a blow of the imagination" (1818, pp. 85–86). Whether it is officers magnetizing their soldiers or men magnetizing women, the cause is the same: a psychological submission that heightens faith in the magnetic process. While affirming that most magnetizers use this power in good conscience, Virey stated that a small number are charlatans deliberately abusing the imaginations of their dupes. What would it take, then, to convince Virey of the existence of magnetic fluid and its transformative power independent of the imagination? "When they have brought a woman somnambulist who can read a closed book placed on her stomach in the presence of the Academy of Sciences—then they will deserve to be believed. Until then, we will be permitted to attribute their cures and other genuine results to nervous communications and to those routes that marvels and illusions have continually been imposed on minds through the ages" (1818, p. 93).

Virey lamented that many magnetizers were not capable of viewing animal magnetism dispassionately. He held that once a believer in animal magnetism, always a believer: "Once the rivet of this firm belief has been clinched, one persists, and dies with the sign of the beast inscribed on the forehead" (1818, p. 81). Whereas Virey used this biblical phrase figuratively, there were those who would have meant it most literally: theological writers who considered animal magnetism to be an antireligious movement. For they held that, far from being the products of imagination and bad observation, the paranormal phenomena of animal magnetism were the work of Satan himself.

The Work of the Devil

Probably the most vocal and systematic critic of magnetic phenomena as works of the devil was the abbé Wendel-Würtz. In *Superstitions et prestiges des philosophes, ou les démonolatres du siècle des lumières* (1817) he argued that Satan no longer found his chief supporters among gross and ignorant practitioners of superstition. Rather, he had now found the secret of bringing the great philosophers to his service, influencing them to declare that he was merely an imaginary being. Those who would combat Satan must now contend with men and theories held in respect by society. And, according to Wendel-Würtz, "among the workers of satanic prodigies of our time, the magnetizers occupy the first rank. . . . The discovery of this agent [animal magnetism] is usually attributed to a French physician named Mesmer. . . . But actually one should give him the ghastly credit of having discovered a diabolic secret that has existed in all ages and has been rediscovered a thousand times in a thousand different forms" (pp. 44, 46).

Focusing on Deleuze's recently published *Histoire critique du magnétisme animal,* Wendel-Würtz saw in the phenomena of animal magnetism described there the signs of diabolical intervention. The ability of the magnetizer to suspend normal sensation in his somnambulistic subject and the somnambulist's ability to reveal the name, seat, nature, and cause of an illness and prescribe effective remedies; to read the magnetizer's thoughts, without benefit of any external sign; to predict the future course of a disease; to recall detailed memories of things long forgotten in the waking state; the startling change in personality and intelligence in the somnambulistic state; the sure control the magnetizer has over his somnambulist, by which he can command her to do things of which she would ordinarily be incapable—all of these phenomena were, to Wendel-Würtz, alarming signs that the devil had insinuated himself into the daily fabric of contemporary life, working his evil magic under the guise of a naturalistic healing practice.

Wendel-Würtz made a strong case for the reality of the paranormal phenomena of animal magnetism. Their widespread and frequent occurrence, their attestation by thousands of witnesses, and their acceptance by persons of intelligence and substance convinced him that he was not dealing with fraud or illusion. Given the reality of the phenomena, he could only conclude that Satan was at work.

Some years later, Honoré Tissot attacked animal magnetism on the same ground:

> What is animal magnetism? It is an operation by which a person is rendered possessed by a demon by means of certain gestures, by a look, or even by the will alone. The evil spirit takes hold of the magnetized person, puts her in a state of somnambulism or ecstasy and, speaking by mouth from this state, reveals hidden things, reads from closed books with closed eyes, tells about things happening in nearby rooms, houses, and distant places, responds to thoughts, etc. All of these signs are, according to the ritual of the Church, indications of demonic possession. The agents of these magnetic phenomena are the same as in the witches or oracles of the pagans, in the inspirations of the Shakers, in the illuminati of Swedenborg or the Martinists, in the Jansenist convulsionaries, in cataleptics, epileptics, lethargics, and frenetics, and all who are affected by diabolical ecstasy. (1841, pp. 84–85)

Tissot admitted that the Catholic church had not explicitly denounced animal magnetism. But it had condemned black magic, of which, Tissot averred, animal magnetism was a form. These arguments would recur in one shape or another throughout the rest of the century. As will be seen, when American spiritualism came on the scene, animal magnetism and spiritualism were often tarred with the same theological brush, not only by Catholic writers but also—and often more vehemently—by Protestants.

The New French Commissions

In the first decades of the nineteenth century, claims and counterclaims for the genuineness of the paranormal phenomena of animal magnetism grew. In the early 1820s the controversy focused in Paris and eventually produced new investigatory commissions and a different outcome from that of the commissions of 1784.

In 1825 Louis Rostan wrote an article for the *Dictionnaire de médecine* that appeared simultaneously in pamphlet form. Rostan, who had been experimenting with animal magnetism for some time, found that it did produce "a

modification of the nervous system" and in this way could yield results beneficial to the health of the magnetic subject. He had serious doubts, however,
about the reality of the paranormal phenomena attributed to animal magnetism. Rostan pointed out that the will of the magnetizer is involved in producing the magnetic state and that the magnetizer's wishes or expectations could
prejudice his observations.

Nevertheless, Rostan's basically positive review fueled a movement that,
the supporters of animal magnetism hoped, might finally lead to the official
approbation of the medical establishment. The proximate trigger for the establishment of the new investigatory commissions came principally from the
experiments conducted at the Hôtel-Dieu in Paris by Jules Du Potet in 1820
(see chapter 7). The director of the hospital, Dr. Husson, was so impressed by
Du Potet's work with the patient Catherine Samson that he suggested that
more definitive experimentation be carried out to determine the reality of
animal magnetism. Husson described Du Potet's experiments to his colleague
Pierre Foissac, who at the time was extremely skeptical about animal magnetism. After conducting some of his own magnetic experiments, Foissac in
1825 presented a *mémoire* to the Académie Royale de Médecine arguing for
an investigation of animal magnetism and its possible uses in medicine. After
heated debate, the academy set up a five-member commission (including
Husson) to look into the matter (Foissac 1833, pp. 6–11).

The Commission of 1826

The commission's report, drafted by Husson and issued on December 13,
1825, was extremely favorable to animal magnetism. It began by pointing out
that in the decades since the commissions of 1784 many reputable persons had
been practicing animal magnetism. Mesmer's theory of magnetic fluid had
been superseded by a theory of a nonspecific fluid directed by the will. The
baquet had fallen out of use and public, communal treatment had been replaced by private, individual treatment using less elaborate gestures and little
or no equipment. Effects obtained through animal magnetism were strikingly
different, convulsions having been supplanted by somnambulism as the dominant type of magnetic crisis (Foissac 1833, pp. 12–28).

The phenomena of magnetic somnambulism emerge as the central theme
of the report. Husson listed those phenomena for which he found unanimous
acceptance among contemporary authors: the exterior senses are shut down
and interior senses are used that are not available to the somnambulist in the
waking state; he perceives only those individuals with whom he is in rapport
and those objects to which his attention is directed; he submits to the will of his

magnetizer in matters that are not against his conscience, and senses that will directly; he sees the fluid, can perceive the interior of his own body and those of others, and is able to describe the nature of an illness he may perceive there and prescribe a remedy; he remembers things that have perished from his memory in the waking state; he has visions and presentiments of things to come, although these can easily be erroneous; he expresses himself with force and sometimes with vanity; returning to the natural state, he forgets all the sensations and ideas of the somnambulistic state, so that in the two states he is like "two different beings"; and he can fall into a state of insensibility so deep that he hardly breathes and is immune to pain (pp. 28–31).

The report ended by concluding that the academy should carry on its own examinations and not be influenced by the conclusions of the commissions of 1784; that the method of examination employed by those earlier commissions was faulty; that animal magnetism then was different in theory, practice, and effects from animal magnetism at present; that French physicians should not lag behind German physicians in the investigation of the phenomena; and that the academy ought to study and teach the proper use of animal magnetism so as to prevent abuses.

After a long and often argumentative discussion of the report, Husson presented to the academy a paper replying to the objections raised. A secret ballot approved the commission's conclusions by a vote of 35 to 25. The academy then appointed a permanent commission to study animal magnetism. This commission, constituted in 1826, continued its work for a number of years and submitted a report in 1831, with Husson once again the reporter. Its conclusions were a complete vindication of animal magnetism and the paranormal phenomena of somnambulism (see Foissac 1831, pp. 115–208).

The commission of 1826 attempted to delineate the genuine effects of animal magnetism, including agitation in some subjects and calming in others, acceleration of breathing and heartbeat, convulsive movements, numbness, drowsiness, and, in some cases, somnambulism. Although questions remained about why some people become somnambulistic, the reality of the state could not be doubted, said the commission, and in that state certain striking phenomena occur: the subject develops new faculties—clairvoyance, intuition, and prevision. Also, somnambulism produces certain profound physiological changes, such as insensibility and a sudden increase of vigor.

The commission noted that somnambulism could be simulated and that therefore charlatans could use this phenomenon to their own advantage. Also, it advised investigators of somnambulism to take the greatest precaution in conducting their experiments to prevent errors due to illusion. The commissioners stated that they had observed cases in which magnetic sleep was produced at a distance and without the subject's awareness that he was being magnetized. They also pointed out that once individuals had been made somnambulistic, a mere look or act of the magnetizer's will could return them to that state.

The commissioners described examples of what they considered to be genuine clairvoyance: somnambulists who could identify the number and color of playing cards without the use of their eyes; some who could read from a book with their eyes completely covered; examples of prevision in which the onset of epilepsy, the accomplishment of a cure, and other biological events were predicted with accuracy. The commissioners regretted that they had not observed enough cases of magnetic healing to make important observations on the therapeutic potential of animal magnetism, but they had seen enough to say that animal magnetism should be counted as a medical tool and that its use should be supervised by the medical profession. Their final recommendation was that "the academy should encourage research on magnetism as a very curious branch of psychology and natural history" (Foissac 1831, p. 206).

With this report, the partisans of animal magnetism could breathe a sigh of relief. They had overcome the stigma imposed by the commissions of 1784 and had apparently gained for magnetic practice acceptance and respect from the medical establishment. Four years after publication of the report and related material, however, the controversy arose anew in the halls of the academy.

The Commissions of 1837

In 1835 Charles Hammard conducted an experiment to determine whether a subject's ailing tooth could be extracted without pain using animal magnetism as an anesthetic. He invited a physician named Oudet, a member of the Paris Academy of Medicine, to observe. The procedure was so strikingly success-ful that Oudet published a note on it in the *Journal de médecine et de chirurgie pratiques* in 1837. This resulted in a strong rebuke delivered by the physician Capuron at a meeting of the academy. Capuron dubbed the experiment a fraud and called for an end to the "imposture" of animal magnetism.

At this point, Didier Jules Berna, a member of the academy who had written a treatise in support of animal magnetism (Berna 1835), offered to bring before that body certain experiences of his own with magnetic somnam-bulism to show the reality of the phenomenon. The academy's president, Renauldin, hoping to discredit animal magnetism, responded to this offer by setting up a nine-man commission. Three members—doctors Roux, F. Du-bois, and J. B. Bouillaud—had already declared themselves enemies of animal magnetism, and many felt that the commission was so heavily weighted with critics that its investigations could not be fair.

The final report of the commission contained seven conclusions: (1) magnetic somnambulism did not exist; (2) insensibility to pain was not proven; (3) claims that a magnetizer could, by a mere act of the will, restore sensibility to a somnambulist must, following the first two conclusions, be denied; (4) claims with regard to the magnetizer's ability to control the movements of the somnambulist must similarly be denied; (5) claims that the somnambulist would obey mental commands were without foundation; (6) claims that a somnambulist can see without the eyes were not valid; and (7) clairvoyance, or vision through opaque objects, was an illusion.

Berna was so upset by both the tone and the substance of this report that he published a lengthy refutation (1838) of what he considered to be the easy assumption of many members of the academy that all magnetic phenomena were the products of charlatanism. He described how the commission was set up and how its members had resisted following recommendations made by Berna that would enhance the possibility of good observations. Husson also strongly opposed the later report, pointing out that paranormal phenomena are very fragile and that attempts to induce them often meet with failure. His main objection to the report was that the commissioners had overreached their mandate from the academy and exceeded what they could legitimately conclude from what they had observed. All the commission had been asked to do was study and pronounce on experiences made available by Berna. To reach general conclusions was both illegitimate and absurd (Burdin and Dubois 1841, pp. 547–58).

The report was put to a voice vote on September 5, 1837, and was accepted by a large majority. This was not, however, the end of the matter. A friend of Dubois's, Charles Burdin, announced at the meeting that he would offer a prize of three thousand francs to any somnambulist who could demonstrate the phenomenon of eyeless vision to the satisfaction of a commission of three persons from the Academy of Medicine and three members of the Academy of Science. On September 12 the Academy of Medicine accepted Burdin's offer and, instead of the judging committee he had proposed, appointed a commission of seven of its own members. A two-year limit was placed on the offer of the prize.

Both Dubois and Husson were elected to the new commission. Husson proposed nine precautions to be taken by the commission in its examination of candidates for the prize, intended both to insure the fairness of the commission and to preclude attempts at deceit by candidates. They were not adopted.

The commission received a letter from a Dr. Jules Pigeaire describing his

experiments with his own eleven-year-old daughter, Léonide. In the somnambulistic state, she seemed to be able to read when blindfolded and describe
objects concealed in boxes. At the commissioners' invitation, Pigeaire and
Léonide came to Paris in May 1838. The commissioners and Pigeaire could
not, however, agree on the final form of the trials, and the experiments never
took place. Pigeaire wrote *Puissance de l'électricité animale, ou du magnétisme vital* (1837), describing the feats of his daughter and criticizing the
conduct of the commissioners and other opponents of animal magnetism.[3]
    After unsuccessful attempts by other clairvoyants to capture the prize, the
academy adopted a resolution that after October 1, 1840, when Burdin's offer
was to expire, it would no longer respond to requests to examine the claims of
animal magnetists (Burdin and Dubois 1841, pp. 630–31).

---

3. His friend and associate Dr. N. N. Frapart later wrote his own account (1839) of the
Pigeaire affair.

Chapter 10

Magnetic Magic and Magnetic Spiritism

With so much interest in paranormal phenomena arising in magnetic sleep, animal magnetism was in danger of being seen as a framework for the revival of old superstitions. This danger, feared by Mesmer, was heightened by two streams of thought that developed within mesmerism very early: magnetic magic and magnetic spiritism.

Nevertheless, investigations by proponents of these approaches further advanced the understanding of the psychological dimensions of mesmeric phenomena. They even served to bring certain aspects of the alternate-consciousness paradigm into sharper focus. Concentrating on the part played by the mind of the magnetized subject in producing healing and clairvoyant effects, they forced the conclusion that some kind of mental activity occurred that did not involve the participation of the conscious mind of the subject. In the process they posed a question that would sooner or later have to be answered: If the phenomena showed evidence of intelligent thinking outside the awareness of the subject, what was its source? Spiritists answered that the hidden mental activity emanated from angelic or discarnate human spirits. Those who could not accept this conclusion had to look elsewhere. The only other possibility was that unconscious thinking took place within the psyche of the subject. Indirectly, the proponents of magnetic magic and magnetic spiritism brought this issue further into the light.

Magnetic Magic

The theory and practice of animal magnetism from the beginning put observers in mind of natural magic, a feature of all the ancient western traditions and systematically developed by the philosophical and medical writers of the Renaissance. The earliest critics of animal magnetism believed that Mesmer took his ideas from writers of the fifteenth to seventeenth centuries, such as Paracelsus, Fludd, von Helmont, Maxwell, and Kircher, who wrote of the magical power of a universal magnetism and whose philosophies of "world souls" and "astral influences" draped primitive magical thinking in scientific robes (Thouret 1784). It is not surprising, then, that a group of animal-magnetic theorists came to see animal magnetism as the continuation of ancient magical traditions. For them, "magic" involved mobilizing the innate, invisible forces of nature. The first truly methodical spokesmen for this viewpoint were magnetizers steeped in the German Romantic movement.

Eschenmayer

The prevalence of Romantic ideals in early nineteenth-century Germany had assured the early establishment of animal magnetism in that country. Many of the first German magnetizers had come under their sway. Hufeland's *Über Sympathie* (1811) called for a broadening of the notion of science so that all of nature would be seen as interconnected. Following a similar line of thinking, Passavant (1825) asserted that by means of "life magnetism"—a synonym for animal magnetism—the absolute spirit of nature operates through the spirit of the mesmerizer to heal the sick. The magical action of *Lebensmagnetismus*, he stated, makes clairvoyance possible and accomplishes other marvelous but natural feats. Later Carl Gustav Carus (1857), discussing the role of sympathy in the cure of illness, stated that the application of *Lebensmagnetismus* mobilized the unfailing healing power of the unconscious present throughout nature.

The physician Carl Adolph von Eschenmayer (1770–1852) elaborated these ideas into a fully articulated system. Early in his professional career Eschenmayer developed an interest in the occult and was particularly influenced by the nature philosophy of Schelling. He saw animal magnetism as a practical demonstration of the truth of ancient magical beliefs. In 1817 he became coeditor, with Kieser, Nasse, and Nees von Esenbeck, of the newly established *Archiv für den thierischen Magnetismus*.

Eschenmayer (1816) believed that with the emergence of animal magnetism, magical practices that in the past had been tinged with superstition could at last be understood for what they were. Although he took pains to deny that animal magnetism was the product of some fevered mystical imagination, he did believe that it dealt with the same phenomena as the older magical tradition and that, extraordinary as those phenomena might seem, they were nonetheless real. Direct experience of the phenomena in the everyday practice of animal magnetism, he said, had made their existence undeniable. This reacceptance of phenomena that in the past were considered magical necessitated a reexamination of what could be regarded as possible and impossible. The phenomena indicated that human beings were not as limited by space and time as had been thought, that they were able in some cases to perceive things at a distance and look into the future (p. 19).

Eschenmayer believed that nothing supernatural was involved in all this. Rather, the phenomena were the result of the awakening of faculties that followed natural laws. "Those who are in the forefront of animal-magnetic practice cannot be accused of mysticism. The phenomenon should not be thought of as a pet system, or the result of the passionate imagination of the Rosicrucians, or of a Campanella, a Fludd, or a Maxwell. Rather it has come before us spontaneously and, from my knowledge of the writings and accounts, has found moderate and open-minded observers" (p. 9). Animal magnetism was an extraordinary opportunity to observe how those laws work, "an experiment we conduct with the soul to seek out its inner nature, just as the natural scientist experiments to discover the inner properties of corporeal things. . . . The inner nature of the soul seldom reveals itself to us and chooses this or that individual (often unbeknown to himself) as the means of revelation; but if it happens, we are astounded to look into the hidden depths" (pp. 26–27). Examining the depths of the soul in this way, one discovers remarkable powers, such as the clairvoyant ability to see things despite distance or obstacles, or to look into the future. But these powers are natural, not supernatural: "[Precognition] is not divination; it is a time-intuition of the imagination. Each organism is altered according to laws, but laws for which we do not yet have any comparison" (p. 47).

Whereas the astronomer makes calculations according to laws of motion to pinpoint the future position of planets, in somnambulism an organism is at work, and changing factors—such as receptivity to energy—must be taken into account. The faculty through which the information comes is the imagination. A kind of unconscious inner calculation takes place according to the tendencies of observed phenomena, and a conclusion is reached about what is likely to occur. This removes any quality of the miraculous from the phenomena of precognition and avoids the undesirable conclusion that some spirit dwells in the subject and provides information about the future (pp. 47–48).

But for Eschenmayer there is yet another dimension to the phenomena: "For me the individual soul is also a universal soul, in much the same way as Paracelsus spoke of an 'Adamish' body and a sidereal body. The individual soul works with the Adamish body and concentrates on earning a living . . . while the universal soul in the sidereal body gives itself up to ideas and speculation. The former is preoccupied with a finite cycle, while the latter breaks out and strives for the unending and eternal" (p. 66). In the highest grade of functioning, the level of the universal soul, time and space are overcome, and the distant and the future are both present. The only limitation on this clairvoyant ability is that acts dependent on human free will cannot be determined in any absolute way, so that error in predicting the future is still possible (pp. 67–68).

Ennemoser

Joseph Ennemoser (1787–1854), a professor of medicine with a penchant for anthropology, wrote prolifically on the history of magic and religious traditions. Ennemoser was a dedicated practitioner of animal magnetism, and his *Anleitung zur mesmerischen Praxis,* published in 1852, was one of the best German manuals on the practice of mesmerism. He also had a burning interest in the psychological dimensions of magnetic theory and the relationship of occult traditions to mesmeric phenomena.

In his first book, *Der Magnetismus nach der allseitiger Beziehung seines Wesens* (Magnetism according to the universal relation of its being; 1819), Ennemoser described animal magnetism as the heir to magical phenomena and thought. In the preface to the second edition he wrote: "Magnetism is able to give the meaning of the symbolic enigmas of ancient mysteries, which were considered quite insoluble. . . . In the same manner the manifold declarations of ecstatic seers and mystic philosophers . . . will now become more intelligible by means of magnetism" (1854, 1:viii). It was Ennemoser's conviction that the natural laws involved in the phenomena of animal magnetism could be formulated, that the task was "to render the mystical scientific, rather than science mystical" (1:xv–xvi).

Whereas Eschenmayer generally rejected the notion that spirits intervene in the lives of men, Ennemoser accepted the possibility. Describing the operation of clairvoyant powers in the seer, he stated that since these powers are limited by the degree of inner purity, sometimes the seer needs to call upon an external spiritual agent to assist. "Thus it becomes clear why, either under natural or induced circumstances, foreign spiritual agency is generally present" (1819, 1:13).

In *Der Magnetismus im Verhältnis zur Natur und Religion* (Magnetism in relation to nature and religion; 1842), Ennemoser argued that the main phenomena of animal magnetism and magnetic sleep have all occurred in the principal religious and magical traditions of the past. The only difference is the deeper awareness brought to the contemporary investigation of magnetic phenomena. Now for the first time, the natural sources of the phenomena and the laws governing their manifestation could be uncovered. Ennemoser believed that through exploration of the inner life of the spirit in magnetic somnambulism, people discover their connections with an objectively existing spirit world, with its good and bad spirits, and that communication with that realm in magnetic trance is attested in all ages and cultures. Although in some cases apparitions and visions of the spiritual world may be illusory, genuine experiences of this kind do occur. The state of "magnetic ecstasy" in which spiritual visions take place is one in which the somnambulist is most deeply in touch with his own spiritual nature and thereby gains contact with other spirits and a dimension beyond space and time (1842, pp. 115–20, 227–29).

Du Potet

Baron Jules Du Potet de Sennevoy (1796–1881) from early in his life sensed that he had an unusual ability to influence people. Taught the art of animal magnetism by the abbé Faria and Deleuze, he believed he had found the vehicle for putting this natural ability into action. Du Potet came to be considered one of the most notable magnetizers of the nineteenth century.

As we have seen (chapter 8), on the invitation of John Elliotson of University College Hospital in London, Du Potet reintroduced animal magnetism to the British. He lived in London for some time, demonstrating his art and teaching his technique. In France, Du Potet taught animal-magnetic technique and published his lessons in various editions. *Cours de magnétisme animal* (1834) was the first; the third edition, called *Traité complêt de magnétisme animal* (1856), had twelve lessons and included a history of animal magnetism, along with discussions of theory and practice. Du Potet also edited two journals of animal magnetism: the short-lived *Le propagateur du magnétisme animal* and the important *Journal du magnétisme,* which went to twenty volumes between 1845 and 1860.

Over the years Du Potet left the theories of Puységur and Deleuze behind and evolved an approach closer to Mesmer's. He rejected the notion that will or belief was primary in the production of magnetic phenomena: "I see that to produce sleep there is no need to desire it or will it; [the magnetic] agent produces this state like opium. . . . A lack of belief on the part of the magnetic subject is no hindrance—merely a mark of stupidity" (1852, p. 54). Du Potet held that the magnetic medium or fluid—which he simply termed "magnetism"—was an agent of nature that caused the phenomena automatically. This magnetism had been used by the ancients in their magical practices, although they were not in a position to understand it or investigate it scientifically. Du Potet saw himself as a modern-day magician and did not hesitate to use the title. In *La magie dévoilée* (1852), he claimed to have grasped both the real meaning of animal magnetism and the true inner nature of ancient magical practices.

In Du Potet's opinion, animal magnetism had been given too much importance. Although Mesmer's discovery was significant, it was merely one step in the development of human powers. Du Potet wanted to push beyond the boundaries magnetizers had artificially established and to seek properties in magnetism "even more mysterious and more important to recognize in connection with moral science, and consequently more capable of revealing to us the divine faculties of the human soul. To do this it is necessary to isolate oneself from the magnetic world, from those who have eyes but do not see, and who always retrace the same circle without discovering anything new" (1852, pp. 61–62).

According to Du Potet, the "new" phenomena of magnetism, somnambulism, trance, and thought projection were of the same source as the magical enchantments of witches and sorcerers in ancient times. Only the form of the phenomena has changed, not their substance; "the agent is the same in all the phenomena" (1852, p. 64). And what is this agent? Du Potet said it is *thought before it has become action* (1852, p. 65). Thoughts can be planted in material objects like seeds and then, at the time intended by the thinker, can make their power felt by persons who come near the objects. This, he asserts, was the principle by which amulets and blessed tokens operated, and why talismans, altars, and holy places can influence those who come into their presence (p. 219).

Du Potet carried out magical demonstrations of this power of thought. For example, he drew lines and symbols with chalk on a wooden floor and used the power of his mind to invest them with a specific meaning. Then he had volunteers placed within the chalked area and found that they reacted to the lines and symbols exactly as he had intended. Their reactions were so strong and unambiguous that he considered them undeniable proof of his contention (1852, pp. 67–84).

But the power of thought in the sense intended by Du Potet did not derive from mere cogitation. It was necessary to produce a concentration of personal force to empower the thought with the energy to affect another person. Du Potet saw thought in this sense as a mental creation that was invested with a "semi-material" envelope, a creation conceived in the brain and sent out via the fingers into the person or object touched. This, he said, is how magnetizers always operate, although they do not realize what they are doing. He saw magnetism as the link between the soul and matter, the agent that is invested by the soul with the power to affect matter. That is why to be successful in magical (or magnetical) operations it is necessary for the operator to be vitally energetic and alive, fully alert in every sense—not preoccupied or benumbed by excessive eating or drinking.

Du Potet dismissed Puységur's view that magnetic action takes its direction from the goodwill of the magnetizer. He believed that to be effective the will had to gather a force and energy far beyond that imagined by the marquis. The human will in its normal state was "stagnant," needing to be awakened:

> Something has to happen to your mental intention. A bell will not make a sound or produce a vibration unless it is struck. A sponge does not release the water it holds in its cells until it is squeezed. Observe what happens with concentrated passion, when a man is about to have a passionate outburst, although not yet letting it go. He burns, he freezes, or he trembles. In his body all is in tumult. The force within reaches his skin, while his heart beats harder and harder, like a drum. When it reaches a pitch, his volcano erupts over the human landscape in an outpouring of lava and a whirlwind of sulphur. Behold! This is how you must use your desires which are like a fire that glows and shines in you unseen. It is exactly like the basic act of reproduction, except that here the emitting organ is the brain. (p. 234)

Du Potet makes this bold comparison of magical (and magnetical) action to the sex act even more explicit: "It is necessary . . . that a fire run through you, that a kind of erection (which is not erotic) happens that allows an emission of the brain to depart from your being. Your hand must conduct this animated essence, this living magnet, to the chosen surface, and it must immediately establish the spiritual rapport and attraction proper to it. It is not the female organ that receives this emission . . . but purer and more active elements which the senses cannot see" (1852, p. 253).

Du Potet believed that magnetizers had no idea of the power of the forces they were dealing with. They were like children playing with adult tools, not knowing how to use them, ignorant of "the substance behind the form . . . the key that can open all the locks of nature's laboratories" (p. 254). With this thought Du Potet formulated what is perhaps the most direct equation of animal magnetism and ancient magic to be found in the magnetic literature.

Magnetic Spiritism

In the first decades of the nineteenth century, a trend developed that connected magnetic somnambulism with communication with spirits. Those spirits might be the souls of departed human beings or angelic spirits, good or bad. The roots of magnetic spiritism can be found in the practices of the Swedenborgian Stockholm Exegetic and Philanthropic Society, whose *Lettre sur la seule explication satisfaisante des phénomènes du magnétisme animal et du somnambulisme* (1788) taught that evil spirits and negative supernatural forces were involved in the production of disease, and that cure was dependent on their removal. The society claimed that human beings have an "inward" sense of sight and hearing that enables them to communicate with angels and spirits. "If this be so, which we do not at all doubt, then Somnambulism may become to the Magnetizer and those present, who will rightly improve it, an adumbration, though feeble, of the first immediate correspondence with the invisible world" (translation from Bush 1847, p. 265).

The assumption was that an evil spirit in the ailing person produced disease. The successful induction of magnetic somnambulism was considered an indication that a good spirit was overcoming the power of the evil spirit in the patient's body (Bush 1847, pp. 268–69). This document contains the seeds of what would shortly become a powerful spiritistic movement within the magnetic tradition.

Jung-Stilling

Johann Jung-Stilling, as we have seen, became the principal spokesman for the Swedenborgian magnetic tradition in the first years of the nineteenth century. He believed that, through magnetizing, the human soul can be separated from the body by degrees and that, once detached from its organs of sense—once "ecstatic"—it experiences everything differently. When the soul has ceased to perceive the physical world, it begins to operate in what Jung-Stilling called the "world of knowledge." Here the somnambulist sees things with a faculty of perception not hindered by physical obstacles. When the state of detachment from the body has been experienced many times, it is possible for the somnambulist to enter into communication with spirits, seeing them with an inner vision and listening to them with an inner faculty of hearing. But since in the ecstatic state the imagination is particularly active, one must be cautious about such visions. Jung-Stilling therefore suggested that wisely devised criteria be applied to discern whether information coming from such ecstatic apparitions truly originates in the world of spirit.

But it is not enough, said Jung-Stilling, to be confident that one is in touch with the world of spirits. One must also find out the state and motivation of the spirit. For a somnambulist may just as easily encounter a mischievous or destructive spirit as a good, angelic one. Then, too, the somnambulist might come across the spirit of a departed human being, and that spirit will have all the qualities, good and bad, that it had when in the body (Jung-Stilling 1854, pp. 25–62).

Because human beings are so easily misled in this matter, Jung-Stilling generally discouraged somnambulists from seeking ecstatic visions: "When a man dies, the soul gradually divests itself of the body, and awakes in Hades. . . . We can learn nothing from spirits that are still in Hades, for they know nothing more than we do, except that they see farther into futurity. . . . Besides this, they may err and willfully deceive. We ought, therefore, by all means to avoid intercourse with them. Spirits in a state of perfect bliss, or such as are really damned, never appear. Every man has one or more guardian spirits about him: these are good angels, and perhaps also the departed souls of pious men. . . . Good spirits have power over evil spirits;—but the will of man is free: if it incline to evil, the good cannot help him. We ought not to seek intercourse with guardian spirits, for we are nowhere referred to them" (pp. 231, 233).

As it turned out, Jung-Stilling's admonitions paled beside his fascinating stories of spirit intercourse with the world of men, so his treatise did not instill the caution about spirit communication he intended. On the contrary, it became an inspiration and guidebook for those who wanted to follow spiritistic pursuits in the practice of animal magnetism.

Kerner

The work of Justinus Kerner (1786–1862) further strengthened the connection between magnetic magic and magnetic spiritism. In his *Geschichte zweier Somnambulen* (History of two somnambulists; 1824), Kerner described how he first practiced "magical healing" on Christiana Kapplinger of Weinsberg. In 1822 Christiana began falling into spontaneous somnambulistic states. While in trance she was led by a "guide" into a region of "indescribable beauty and sweetness." Unfortunately, at the same time she developed painful symptoms, including weakness of the limbs, stomach pains, inhibition of menstruation, and disturbances of sleep. No one was able to help the young woman until her father applied animal-magnetic passes. That brought some relief but also caused the father to experience sympathetic pains. Eventually he called upon Kerner for help. Kerner brought in a physician from Heilbronn named Seiffer, who showed him how to apply the magnetic technique "more as a spiritual than a corporeal power, using belief and will to magnetize the patient" (Kerner 1824, pp. 3–11).

The second case history Kerner presents is of his work with the seventeen-year-old Caroline S., who in 1822 was overtaken by spontaneous somnambulism while visiting her mother's grave and thereafter frequently entered sleep-walking and sleep-talking states. From the experiences and pronouncements of these two somnambulists, Kerner began to formulate a theory of magnetic somnambulism in conjunction with a notion of communication with the spirit world. He came to believe that spontaneous magnetic states are not as uncommon as had been thought and that at the time of death everyone enters a state of magnetic somnambulism in which they communicate with departed loved ones. Some, however, such as Christiana and Caroline, experience this magnetic communication with spirits in their everyday lives, aided by spiritual guides who lead them through spiritual realms and teach them about spiritual matters. These guides, commonly beloved family members, can take on any form they choose. When someone becomes somnambulistic, he or she returns to a state of paradisiacal innocence and possesses a kind of primitive purity and freedom that was the common state of mankind before the Fall. In this special state, the somnambulist is incapable of falsity or deception. Through communication with spirits, the somnambulist learns much about animal magnetism and the healing arts. The magical healing action of plants and foods, for instance, are understood and can be conveyed to the ill. The power of these remedies is best evoked when used in conjunction with animal magnetism. This "magical-magnetic" treatment makes use of the healing forces of nature to bring about cures never before believed possible (pp. 347–81).

The paranormal psychic phenomena that accompanied the magnetic state were also of great interest to Kerner. Perhaps nowhere were these abilities demonstrated more clearly than with one of his somnambulists who eventually became world-famous: Friederike Hauffe, the seeress of Prevorst.

Born in 1801 in a village near Löwenstein, Friederike was raised in a poor and very pious family. Early she showed a strong religious inclination and was conscious of what she thought was the presence of spirits. While still a young girl she often experienced apparitions, sometimes calmly and sometimes with alarm. These occurrences were still taking place in her late teens when, after a generally happy life, she sank into an extreme depression. While in this state she became engaged to be married. On the day of her marriage she attended the funeral of a local dignitary whom she greatly respected and had a profound spiritual experience, accompanied by a feeling of great calm and inner peace, which left her entirely indifferent to what happened in the world. From that point on she became more and more involved in her inner life; marriage became a mere shell in which what she considered her real life took place. On February 13, 1822, she had a dream in which she saw herself suffering from a great illness and sitting next to the dead body of the dignitary at whose funeral she had been so affected. In the dream she felt comforted by being close to the corpse and was convinced that this would cure her. The next morning Friederike was attacked by a fever that lasted two weeks. When the fever left she suffered repeated muscular spasms and fell unconscious. Massage and bleeding by local doctors did not help her, but when a physician from another town placed his hand on her forehead she became calm and showed all the signs of a subject in the state of magnetic sleep. She remained very ill, however. Nevertheless, she had become pregnant, and she gave birth to a child in February 1823. The child died after a few months, and Friederike again became extremely depressed. Her physical symptoms redoubled and she neared death. Although she revived somewhat, from then on she spent the rest of her life—some seven years—almost constantly in a dream-like somnambulistic state (1890, pp. 48–59).

She then began to perceive spirits almost continually. Her dead grandmother appeared to her and magnetized her, and Friederike believed that this treatment was the only one that could help her. Friederike also saw the spirits of other departed persons known to her when living. Those who took care of her noted what they considered to be paranormal physical effects produced by spirits, describing objects near Friederike slowly moving through the air. Friederike also related prophetic dreams, gave divinations, had prophetic visions in tumblers and mirrors, and had apparitions that foretold the impending death of people she knew (pp. 60–64).

Again Friederike became pregnant and again her condition worsened, although in this instance the child was born healthy. Magnetic passes intended to relieve her condition simply made her more ill. Her symptoms became more and more bizarre, so that those close to her speculated that she was possessed and brought in an exorcist. At one point Friederike ceased using the local dialect and spoke only in high German; she also developed a peculiar language of her own (1890, pp. 64–70).

Kerner was then called in to give his opinion. Hoping to draw the woman out of her almost continual magnetic state, he suggested that all magnetization cease and that she be treated with ordinary medical means. Still she got worse, suffering now from dysentery and bleeding gums that led to the loss of all her teeth. Her family next brought her to Weinsberg so that Kerner could give her continual medical attention. By now Friederike was a mere skeleton.

Kerner first attempted to treat her exclusively with homeopathy, refusing to encourage her somnambulistic state. When this produced no positive results, he began to consult with her while she was in magnetic sleep about what should be done, and he allowed some magnetic passes to be made. She immediately improved with this treatment. The extreme degeneration of her state when he began treatment, however, coupled with the traumatic death of her father, made it impossible to save her life. She died on August 5, 1829.

In the course of treating Friederike, Kerner came to view her somnambulistic state as her true waking state, since in this state she was extremely alert, intelligent, and perceptive. She was also, however, physically very weak, and anyone who stood near her for any period of time seemed to feel a loss of energy and sometimes experienced muscular contractions or tremors. Those who encountered her spoke of an unusual brightness in her eyes and a sense that she existed more in the world of spirit than the physical world. Friederike often believed herself to be out of the body and sometimes thought she saw her own double. Kerner also wrote of peculiarities of weight associated with her: he claimed that she always floated in her bath and that when he placed his fingers against hers, there was so much magnetic attraction that he could lift her from the ground by his fingertips (pp. 74–80).

Friederike also exhibited certain clairvoyant peculiarities. When she looked into a person's right eye she saw, behind her own reflected image, another figure that she believed to be the person's inner self. If she looked into someone's left eye, she perceived whatever internal disease the person suffered from and could describe its location and prescribe remedies. Kerner carried out experiments in which he placed folded pieces of paper with various written messages on the woman's stomach. Her reaction to the different messages convinced him that she was able to absorb their meaning directly into her body. Also, like so many somnambulists before her, Friederike appeared to be able to see the internal organs of the body as well as accurately trace nerve patterns.

From his experience with Friederike and other magnetic subjects, Kerner developed a theory of magnetic spiritism. In his view, the magnetic somnambulist in the state of ecstasy becomes familiar with a world of spirit, in which there is a constant war between good and evil spirits; the somnambulist becomes a participant in the battle (1824, pp. 392–93). Good spirits are associated with healing and kindness; bad spirits with disease and harm. From this schema Kerner developed a vocabulary to describe the magnetic states associated with each. The state associated with good spirits and healing was termed "agato-magnetic" while that associated with bad spirits and disease was called "demonic-magnetic." Just as good spirits are present to aid in agato-magnetic sleep, so also bad spirits, who fight against God and produce certain kinds of illness, are present to cause trouble in demonic-magnetic sleep. The artificially induced state used to treat disease was dubbed "magic-magnetic."

Kerner believed that physicians sometimes overlooked the presence of demonic spirits in their patients—especially in cases diagnosed as mania and epilepsy—and so could not clear up the disease. What was needed, he said, was a treatment that induced the magic-magnetic state so that the possession could be removed and the illness cured. "In this magic-magnetic healing the psychic power of believing must be used in connection with the organic power [of medicine]. Often the healer has the first but not the second, and vice-versa" (p. 4).

Thus Kerner tried to incorporate in the tradition of animal magnetism the very element that Mesmer, in his criticism of Gassner, had sought to obliterate—the notion of disease as possession and cure as exorcism. Kerner stated that he himself had at first been astounded to encounter possession in treating illness. After he had recognized it, he studied the New Testament, exorcistic writings, and especially "the book of Nature" to learn how to deal with this condition. He claimed that often in the process of removing the evil possessing spirits, the demons would speak—sometimes with more than one voice, revealing that a number of entities are present. He also held that there are two principal kinds of possession: "that with closed eyes and subsequent amnesia and that with open eyes and conscious awareness of all that takes place" (1836, pp. 10–14).

Magnetic treatment of possession took a lot out of the magnetizer. Sometimes the process involved calling in good spirits to remove the evil ones, but usually the magnetizer had to do most of the work. Kerner stated that it required much more energy to magnetize a possessed person than someone who was not possessed. Repeated magnetization, he said, tired out the demon. As the demon's hold on the victim weakened, it would begin the process of leaving the body from the bottom upward. This was often accompanied by choking and convulsions on the victim's part, with a total collapse of the body when the evil spirit finally left (pp. 17–25).

Kerner described many cases successfully treated through his method. One was that of an eight-year-old girl who in 1835 was suddenly taken ill with severe pains in her arms and feet. She suffered greatly and continually cried out in pain. Then she underwent convulsions and was confined to her bed, where she remained for eight weeks. Suddenly a rough male voice began to speak from her mouth. At the mere mention of God or prayer, the girl's stomach would swell up like a drum and "one could feel blows on it." Every kind of conventional medical treatment was attempted, but to no avail. But when Kerner used his magic-magnetic approach, the girl was fully healed in just a few days (pp. 38–40).

Kerner represents the ultimate in the spiritistic magnetic tradition. His view of certain illnesses as possession-induced and his readiness to use magic-magnetic exorcistic procedures frequently put him at odds with those magnetic practitioners who continued the traditions of Mesmer or Puységur. Kerner was often criticized for his naivete in accepting so readily the paranormal and spiritistic phenomena described by his magnetic somnambulists. He did not hesitate to use their subjective experiences as the basis for a magical-mystical philosophy of life which in certain features resembled those of Eschenmayer and Ennemoser but was much more spiritistic in emphasis.

As questionable as Kerner's judgment may have been, he was nonetheless a man of intelligence. His works were read by a broad spectrum of the population, and his influence in Germany and beyond continued for many decades.

Billot

Although spiritistic magnetism was most frequently associated with the Stockholm Society and the German writers, there was a fringe tradition of spirit communication connected with proponents of animal magnetism in France from the late 1780s (see Darnton 1968, pp. 127–59). It was not until the 1830s, however, that French spiritistic magnetism gained respectability. Its first influential proponent was G. P. Billot (b. 1768).

Billot's vision of animal magnetism derived from an unlikely combination of influences. He was first of all a devout Catholic, and his magnetic writings show how highly he valued that theological tradition. But he was also affected by the spiritistic tradition of animal magnetism as embodied in the Stockholm Society's Swedenborgian views and in Jung-Stilling's writings. He was highly respectful of the Puységuran approach to animal magnetism and especially admired the writings of Joseph Deleuze.

Billot claimed that magnetizing merely disposes the subject to the influence of spirits. All healing action and all ability to see at a distance or peer into the future are produced by the good guardian spirits or guides that are assigned to every human being. All inhibition of healing, all distortion of thinking, all mistaken impressions come from evil spirits. In this framework, Billot virtually annihilated the role of a magnetic fluid in the magnetic process and drastically downplayed the importance of imagination in the production of magnetic somnambulistic phenomena.

In the introduction to his two-volume *Recherches psychologiques sur la cause des phénomènes extraordinaire* . . . (1839), a collection of letters between Billot and Deleuze and brief accounts of magnetic phenomena observed by Billot, he describes how he first became acquainted with animal magnetism. A friend of Billot's family, a scholar and practitioner of animal magnetism, organized a local society of magnetizers. Billot, a physician, was curious about the effects of animal magnetism and attended some meetings. As it turned out, his friend, a Roman Catholic, claimed that all somnambulistic gifts were due to spirits who mediate with God and that the phenomena of somnambulism proved the existence of a spiritual soul and its survival of death; in this and many other ways, he believed, they supported the basic tenets of the Catholic creed.

Billot became convinced that spiritual intelligences assist in animal magnetism and formulated his theory in five propositions:

(1) That the influence that one man exercises over another by magnetic action takes place by means of an assistant, either unknown or little known, whose presence provides the only explanation for magnetic phenomena. (2) That magnetic sleep (so-called) and its effects must be attributed to this assistant. (3) That in magnetic sleep, a person is dominated by this assistant, and that all that the person does or says is instigated by this assistant. (4) That this assistant may be either a friend or an enemy to the person, considered as an intelligent being subject to the laws of God, and that as an enemy it causes illusory visions, false promises, deceptive premonitions, and, in a word, all the errors to which somnambulists are subject, errors which disclose the undeniable dangers of magnetism. And (5), consequently, that magnetic phenomena are definitely not produced by an internal sixth sense proper to human beings, still less by a heightened or deranged imagination. Rather they are based on the fact that human beings are intelligences united to matter who can be put in rapport with intelligences not united to matter, proving thereby who is the supreme intelligence—God. (1839, 1:xii–xiii)

In an attempt to persuade Joseph Deleuze, the most important and influential magnetizer living at the time, of these ideas, Billot initiated a correspondence that would last from March 1829 until Deleuze's death in 1835. It is clear that Deleuze respected Billot and paid serious attention to his ideas. He was not easy to convince, however, and despite Billot's belief that Deleuze was coming around to his views (1:x), Deleuze's final letters show that he remained skeptical to the end with regard to the action of spiritual beings in the phenomena of animal magnetism.

When he began the correspondence, Billot was aware that Deleuze had consistently opposed those mesmerists who considered magnetization to be purely the action of the spirit or mind and allowed no role for the magnetic fluid. He was also aware that Deleuze had little use for those who spoke of spirit communications during magnetic somnambulism. Nevertheless, from the start he boldly stated his spiritistic persuasion: "I cannot agree with the French school that does not admit any influence of a spiritual power distinct and separate from that which animates the body of somnambulists. . . . But you too think that the notion that spirits intervene in magnetic operations is not based on solid ground" (1:24). Deleuze took this position, Billot believed, because he considered animal magnetism to be a natural phenomenon and feared that if it were a purely spiritual operation it could not be studied by natural science. Billot countered that "invisible" does not mean "not natural," and insisted that when science rules out the invisible it has become too narrow.

This argument touched Deleuze in a vulnerable spot, for he considered himself a loyal Catholic and rejected the position that only the material is real. Having gotten Deleuze's attention, Billot laid out his panorama of magnetic somnambulism. He postulated "(1) That immaterial beings exist who, under the rule of the divinity, exercise a palpable influence on the acts, both physical and moral, that men perform in their lives. (2) That the religious belief (adhered to in all ages and by all peoples, ancient and modern) in spiritual guides attached to man during his terrestrial life should not be rejected by thinking physicians who want to promote the progress of physiology, since it alone can satisfactorily explain many of the phenomena of life and resolve the great problem of what causes the extraordinary effects observed in magnetic somnambulists" (1:29).

To prove his position, Billot described the case of Marie Thérèse Mathieu, who was severely injured at age twelve when a set of stairs collapsed under her. Her right knee had been so damaged that it atrophied and she lost the ability to walk. Also, she had been menstruating when the accident occurred, and after the fall menstruation ceased and did not return. In 1819, when Marie was thirty-two, her parents called in Billot to treat her with electricity because they had heard about the success of electrical treatment in similar cases. This treatment was of little value to the woman, but six or seven years later, after becoming acquainted with animal magnetism, Billot remembered Marie and wondered if she could be helped through magnetization. He reestablished contact with her and in 1825 began this new treatment.

Billot used a variety of techniques in an attempt to place the woman, now thirty-nine, in a state of somnambulism, but could never put her in trance. Nevertheless, her treatment turned out to be successful beyond his expectations. In April 1825 he induced an unaccustomed movement in Marie's atrophied knee. When he placed his right finger on her kneecap and held his left hand under her knee, an undulation began in the joint, which had been immobile for years. He then discovered that he could initiate and stop the movement by verbal commands. To find out if the agent of the movement was separate from Marie, Billot asked her to resist his command with all her might: "I commanded. She resisted. But the movement took place in spite of her. From this moment a singular dialogue was established between myself and the atrophied member" (1:48). Believing now that the mover was independent of Marie, Billot set up a simple yes-no code for communication between himself and the agent. Then, to rule out the possibility that this was an evil spirit, he asked the mover whether it was from God, to which it responded by tracing a cross. From further questioning he received the answer that it was Marie's guardian angel or spirit guide.

As Billot proceeded with this dialogue through a number of sessions, the answering signals switched to head movements. Soon the spirit was forming words with Marie's vocal cords and communicating verbally, although in an altered tone of voice. This method of communication, however, seemed to leave Marie feeling miserable. Upon questioning, she revealed that she could hear the "little voice" of her guide within her. To avoid the discomfort of direct possession, the dialogue was continued between Billot and the guide simply by having Marie relay the words of the "little voice." Billot asked Marie how the guide acted on her, causing the involuntary movements. She replied: "Spirit acts on spirit; that is, on the soul which is me; and the 'me' obeys the impulse coming from him and causes my organs to move as you observe. If I resist, the spirit (my angel) acts forcefully on my organs—if God permits it" (1:60).

Billot insisted to Deleuze that Marie was not somnambulistic in these sessions. She was fully conscious and remembered everything afterwards. He said that one of the main tasks of the guide was to tell Marie what she needed to do to improve her health. These prescriptions were not Marie's own remedies but those of a wise spirit adviser. Billot also claimed that at times Marie was clairvoyant; this too was not due to any faculty of her own but was simply the result of relaying information given her by her guide. The voice, he argued, was not a part of Marie but a separate agent. He attempted to prove this by emphasizing that there were clearly two different voices and two different wills. It would be absurd, he argued, to suppose that the same person could both will and not will the same thing at the same time (1:64–65). He claimed that there was much theological and historical evidence to support his position, citing biblical references to small voices that reveal the will or God and quoting Plato's description of the small voice of the "demon" that guided and admonished Socrates. Billot claimed that there was nothing remarkable or new in the notion that people could communicate with their guardian angels. The only novel element was that now communication could be brought about at will by using magnetic techniques that dispose the subject to perceive the inner voice of the guide.

Billot also described occasions in which Marie's guide produced waking hallucinations, intended, he said, to remind her in a vivid way of procedures that she was supposed to carry out as part of her health regime. For example, if Marie forgot to fumigate her knee with balsam smoke at the prescribed hour she would suddenly see a smoking incense burner in front of her and, on smelling the odor, realize her memory lapse. According to Billot, Marie's spirit guide could also affect her body in a more striking way. If Marie's vein was opened and blood was allowed to flow, it could be stopped and started again at Billot's command. Billot believed that this control of the emission of blood was one of the strongest proofs that an exterior agent was at work (1:93–94).

Deleuze treated these reports with interest and respect, but he was not convinced that the incidents Billot described proved the presence of spirits. It is just as reasonable, he argued, to suppose that animal magnetism triggers some as yet unknown natural human powers, which then manifest under the trappings of voices and visions. In other words, countered Deleuze, to demonstrate the existence of occult human faculties is not the same thing as to demonstrate the intervention of spirits (1:138–39).

Billot's reply was that since the phenomena of animal magnetism are so extraordinary and, by Deleuze's own admission, require an explanation outside the known laws of nature, why not accept the explanation already accepted throughout human history: that spirits do these things? Besides, we should be willing to listen to somnambulists themselves in this matter. In the past magnetizers have looked to somnambulists for confirmation of their theories, not bothering to find out the somnambulists' subjective experience of the phenomena. When somnambulists were given the chance to talk about how they experience magnetic sessions, they spoke of apparitions and spirit healers (1:118–20, 151).

It was Billot's contention that any theory of animal-magnetic fluid was questionable on philosophical grounds. He chided Deleuze for his notion of a partly material, partly spiritual fluid that functions as the agent for magnetic activity. Billot claimed that only spirit can act on spirit. Therefore, if one person affects another, it is only by his spirit affecting the spirit of the other. If any physical effects occur in the body, it can only be through the mediation of the spirit, not in any direct way (1:164–66). Somnambulism is one thing and contact with the spirit world is another, as demonstrated by Marie Thérèse Mathieu, who was never somnambulistic. Somnambulism was of no particular interest to Billot; it was only significant as the state of consciousness through which the subject was put in touch with spirits.

Although most apparitions and communications described by Billot were considered to emanate from angelic spirits, he also wrote of apparitions of departed human spirits. He presented one case in which an elderly woman consulted a somnambulist who had a vision of a young girl named Lucile. Through the somnambulist, Lucile told the woman to stop mourning for her, since she was perfectly happy and content in heaven. The woman had in fact been grieving for her granddaughter, Lucile, who had died at age six or seven (pp. 280–82).

Billot also related instances of objects materializing out of nowhere. In one case, a woman consulted one of Billot's somnambulists about a condition of partial blindness. The somnambulist prescribed the application of a certain flowering plant. Since it was winter, Billot asked where such a plant could be procured; the somnambulist said that it would be provided. The patient then said that she had found such a plant on the table in her room, although she did not know where it had come from. When Billot went to her room and examined the plant, he discovered that it was a species of thyme native to Crete. Neither he nor anyone else could explain how it got there or how it could be in flower in winter. In another case an envelope, after being seen in a vision by three somnambulists simultaneously, suddenly appeared at the feet of one of the participants; the envelope was found to contain the relics of three saints (2:5–9).

In his response to Billot's communication of these experiences, Deleuze said that he respected Billot's religious sentiments, particularly since it was Deleuze's own experience with magnetism that had led to his return to the church (as was also true of Puységur). Further, he admitted that the phenomena of animal magnetism demonstrate the spirituality of the soul and the possibility of communicating with intelligences separated from matter. But he had a hard time believing that purely spiritual beings could affect material objects—although he admitted that if he had himself witnessed phenomena such as those described by Billot, he might feel differently. In any case, Deleuze stated that at this point he held that magnetism proves the spirituality of the soul and, as a consequence, its immortality; that somnambulists can acquire knowledge beyond their senses; and that souls separated from the body can, in certain circumstances, get in touch with other living beings and directly communicate their thoughts and feelings. Deleuze believed that all the phenomena—from the lowest degree of somnambulism with its imperfect clairvoyance to the highest degree of ecstasy with its precognition and thought reading—all ought to be explained by a single principle based on latent faculties of the soul, not the intervention of spirits. If somnambulists—or others—do in fact communicate with spirits, that should be considered a separate issue, not the cause of the phenomena of animal magnetism (2:14–29).

Billot perceived what he thought to be serious omissions in Deleuze's reply, for although objecting to Billot's explanation of magnetic phenomena, he failed to offer an alternative. Billot also felt that Deleuze's objection to the notion that spirits could affect matter was philosophically and theologically unsound. Billot stated his belief that magnetic passes and the resulting effects served merely to dispose the subject to communicate with spirits, creating a state similar to that enjoyed by Adam and Eve before the Fall. All the phenomena that followed were due to the action of spirits. Also, Billot capitalized on Deleuze's own argument that the lowest and highest of magnetic phenomena must be explained by a single principle, insisting that the most convincing single explanation of all phenomena was the intervention of spirits (2:56–57, 65, 101). In response, Deleuze acknowledged that the phenomena of somnambulism proved the spirituality and immortality of the soul, but he steadfastly maintained that "it does not follow that spirits, angels, or demons are the agents of magnetism. Its phenomena vary considerably depending on the opinions of the magnetizer, the relationships that occur among those attending the magnetic sessions, and what is willed by those who magnetize" (2:73).

Deleuze was suggesting, in a roundabout way, that if the magnetizer or the group involved with magnetizing believe that spirits communicate during the sessions and cause all magnetic phenomena, then that is what the somnambulist will experience. She will see and hear spirits, and she will be convinced that all occurs because of them. While admitting that spirits may communicate with somnambulists, Deleuze refused to take that final step and agree with Billot that spirits are always involved in magnetic phenomena and that all such phenomena are caused by their action. This remained Deleuze's position to the end of the correspondence in 1833 and, to all appearances, to his death in 1835.

Although Billot did not succeed in converting animal magnetism's most respected proponent to his spiritistic position, it seemed that magnetic spiritism was an idea whose time had come. For from this point on, it became a force to be reckoned with in the history of animal magnetism.

Cahagnet

Louis Alphonse Cahagnet (1809–1885) was an enthusiastic experimenter with spiritistic magnetism. He had a powerful interest in exploring the connections he believed existed between this world and the world of spirits. As a young man, he encountered magnetic somnambulists whose ecstatic experiences enthralled him. But it was not enough for him to hear about these things—he had a burning desire to experience them himself. He tried being magnetized but could not attain an ecstatic state. He tried physiological techniques, such as compression of the carotid arteries, hoping to enter trance by depriving his brain of blood, but to no avail. The use of galvanic machines proved equally fruitless.

Then he turned to drugs. He first ingested opium, but this seemed to act more as a poison than a narcotic; when smoking it also failed he burned incenses, took belladonna and ether, and even tried magical invocation of spirits—all to no avail. Finally, in 1848, Cahagnet drank coffee laced with hashish and entered into an altered state of consciousness in which he experienced very unusual perceptions and seemed to be able to see into the interior of his body. In his ecstatic condition he felt a mystical oneness with the world and experienced himself as a microcosm of the universe. He also felt that he had gained the power to read thoughts and the clairvoyant ability to see any object he desired simply by thinking about it (Cahagnet 1850b, pp. 94–117).

While carrying out these personal experiments, Cahagnet also investigated the powers of a number of magnetic somnambulists, which he described in *Arcanes de la vie future dévoilés* (1848–1854). In their ecstatic states, these somnambulists believed they were regularly in touch with angelic spirits and the spirits of deceased human beings. Cahagnet first came across this phenomenon in the somnambulist Bruno Binet, a young man of twenty-seven, "of mild disposition, and very limited in intelligence in point of spiritualism, having read and heard but little of magnetism" (Cahagnet 1850c, p. 1). Cahagnet was conducting experiments in medical clairvoyance with Bruno, who was an exceptionally good magnetic subject, when suddenly the somnambulist was thrown off his seat and filled with fear. When Cahagnet asked what was happening, Bruno said he had heard a voice on his right telling him that the disease of the person he was treating was incurable. When he saw no one standing in the direction from which the voice came, he was catapulted into a state of terror. Cahagnet told him to ask who it was who spoke to him. Bruno got the reply that it was his guide, Gabriel. It took Bruno some time to get used to the voice, but after a while he accepted it as his source of information in medical consultations (Cahagnet 1848, pp. 1–2).

Taking advantage of this opportunity to gain inside information about animal magnetism and healing, Cahagnet began to ask questions of Gabriel. Soon Cahagnet found that Gabriel could also provide information about angels and departed spirits. Gabriel informed him that Bruno would eventually be able to see his deceased father, who was "alive and happy." He also said that people have bodies in the afterlife and that friends and family are reunited there.

In the process of talking to Gabriel through Bruno, Cahagnet was given a disturbing revelation:

> G: "I should be much more lucid were you not possessed by an evil spirit." C: "What do you mean by that?" G: "I mean that a person with whom you no longer have friendly relations is the cause of your being beset by an evil spirit." C: "What power can this spirit have over me? I in no wise dread him; I have never experienced any effect from him . . ."
> G: "You say that you have never felt the influence of the one I speak of;

do you not recollect that about three years ago, you experienced a painful oppression in your slumbers; you had dreadful visions, and, if you would confess it, you were raised up in your bed, and even now you must experience much difficulty in going to sleep." (Cahagnet 1850c, p. 7)

Cahagnet admitted he had experienced the symptoms Gabriel mentioned but had attributed them to nervous irritation. But now he discovered he had indeed exchanged disturbing letters with a friend at the very time his problem began. Gabriel said that was because at the very time Cahagnet was desirous of seeing a spirit of some kind, the "bad choleric fluid" conveyed by this man's letters had made him vulnerable to the intrusion of an evil spirit. It was only through the intervention of Cahagnet's protective "angel of light" that he had not been "thrown out of his head" and fully possessed (Cahagnet 1848, pp. 11–14). This incident greatly impressed Cahagnet, and he began to take Gabriel much more seriously.

Eventually Bruno began to regularly experience ecstatic states in which he went to "heaven" to see his father and observe how he was faring. He described heaven as a place which has no horizon and is illumined by a superb light. He saw what he thought to be God, gray-bearded, wearing a brilliant cap and sitting on a throne. "I was raised into the air; I beheld the earth under my feet, and all those insignificant beings, men, so vain, so proud, that they appeared to me paltry and mean compared with the divine beings before me! How dirty and dark, too, appears to me this room compared with the places I beheld" (Cahagnet 1850c, p. 12).

Another of Cahagnet's "ecstatics" was Adèle Maginot, a "somnambulist by birth" whom he had known for a long time. Like Bruno she specialized in medical clairvoyance, but she also had visions of departed persons and answered questions about the nature of the afterlife with great alacrity. She successfully identified deceased individuals unknown to her and supplied information that was corroborated by their relatives. In one such case a curious cleric asked if she could see his departed father. She accurately described the man's appearance and dress and gave details of his life that no one could have known but the abbé (Cahagnet 1848, p. 122).

Through his own experience and questions put to the spirits, Cahagnet evolved a methodical approach to conducting ecstatic magnetic sessions. He said that the magnetizer should recognize and utilize the special gift of the clairvoyant he is dealing with. He should also carefully formulate his questions to spirits, realizing that spirits too have their specialties and can answer some questions better than others. Cahagnet pointed out ways to circumvent the evil spirits who sometimes try to interfere with these processes. He believed that ecstatic sessions should not be prolonged beyond ten minutes and that too long a session could lead to death (Cahagnet 1848, pp. 291–96).

The Revival of the Intrusion Paradigm

With Puységur's discovery of magnetic sleep and the emergence of the alternate-consciousness paradigm, unusual phenomena of consciousness could be explained in purely natural terms, for the psyche was seen as sufficiently complex to harbor operations that were hidden from consciousness. Magnetic spiritism bypassed all that and brought the intrusion paradigm back into consideration.

Jung-Stilling, Kerner, and Cahagnet all believed that the psyche is an arena in which good and evil spirits battle each other, with the good healing and guiding and the bad undermining and causing illness. People do not have complete control over their own fortunes; they are intruded upon by both combatants in the cosmic battle and subjected to intentions and designs that originate outside themselves. In this framework, magnetic sleep was considered to be important in two ways: as a tool for providing experiential knowledge of how body and soul interact, and as a weapon in the battle with evil spirits, putting the somnambulist in touch with helpful spirit guides and giving the magnetizer the means to remove harmful intruders.

Although working from within the intrusion paradigm, the proponents of magnetic spiritism actually furthered the development of the alternate-consciousness paradigm, for they unwittingly forced the question of unconscious mental activity. Spiritists emphasized that certain of the phenomena of mesmerism could only be explained in terms of the activity of some intelligence outside the conscious awareness of the magnetic subject. While they believed that this intelligence was a spirit of some kind, others could not accept that explanation. Nonspiritists were forced to look elsewhere, and, as we shall see, they eventually concluded that the elusive intelligent agent was concealed in the human psyche itself. There it carried on a mental life unconscious to, but concurrent with, ordinary conscious thought.

In the meantime, the revival of the intrusion paradigm in mesmerism resulted in a rather unstable amalgam. Magnetic spiritism was only a modest force in the magnetic tradition of Europe in the first half of the nineteenth century. Beginning around 1850, however, its influence would be strongly augmented by ideas emanating from across the Atlantic. There a new religio-philosophical movement was stirring and rapidly gaining acceptance. American spiritualism was about to become a significant factor in the fortunes of animal magnetism.

Chapter 11

The Rise of Spiritualism

In the late 1840s a new religio-philosophical movement arose in the United States that would soon spread to England and Europe. Spiritualism, as it came to be called, began in 1848 with the "spirit rappings" heard by two sisters in a poor farmhouse near Hydesville, New York. Despite these humble beginnings, within a few years it captured the imaginations of millions of people and launched a controversy that would continue for many decades.

Spiritualism could only have enjoyed such a sudden and widespread popularity because the way had been prepared by various developments,[1] among them the emergence of animal magnetism in the United States in the 1830s.

The Beginnings of Mesmerism in America

The history of mesmerism in the United States falls into two distinct periods. The first began at the same time as mesmerism was reaching its height in France. Mesmer made a concerted attempt to bring animal magnetism to the attention of the most influential leaders of the newly formed American nation. His first overtures were to Benjamin Franklin, then ambassador to France. Franklin had known about Mesmer since at least 1778, and he met him for the first time in 1779. The occasion was a visit to Mesmer's quarters to discuss the glass harmonica, the instrument Mesmer used to provide background music for his magnetic treatments. Franklin had been involved in the invention of that unusual musical device, and he hoped to hear Mesmer, an accomplished performer, play it. It seems, however, that while Franklin tried to keep the focus on music, Mesmer was more interested in discussing animal magnetism. In December 1779 Mesmer invited Franklin to lunch to talk further about animal magnetism, but it is not known whether that meeting ever took place (Lopez 1966, p. 170).

1. For a discussion of the socio-religious factors favoring the rise of spiritualism, see Lawton 1932, Judah 1967, Nelson 1969, Kerr 1972, Isaacs 1975, Moore 1977, Berry 1985, Oppenheim 1985, Barrow 1986, and A. Braude 1989. Among the more useful histories of spiritualism are Spicer 1853a and 1853b, Capron 1855, Britten 1870 and 1884, Podmore 1902 and 1910, Isaacs 1975, Moore 1977, and Brandon 1983.

George Washington too became the object of some attention from Mesmer. Mesmer made contact with the American hero through the marquis de Lafayette, Washington's friend, who was an enthusiastic member of the Paris Society of Harmony. Lafayette seems to have seen himself as something of a missionary for mesmerism. He initiated Benjamin Franklin's grandson, Temple, into the Society of Harmony and wrote to Franklin when he was heading the commission, praising Mesmer and protesting the choice of D'Eslon as the magnetic practitioner who would demonstrate animal magnetism to the commissioners (Gottschalk 1950, p. 78). On May 14, 1784, he wrote Washington about Mesmer and animal magnetism, which he termed "a grand philosophical discovery." He promised to get Mesmer's permission to reveal the "secret" of his doctrine to Washington. Lafayette's letter was followed up by one from Mesmer himself, who wrote to Washington on June 16, 1784:

> The Marquis La Fayette proposes to make known in the territory of the United States a discovery of much importance to mankind. Being the Author of the discovery, to make it as diffusive as possible, I have formed a Society, whose only business it will be to derive from it all the expected advantages. It has been the desire of the Society, as well as mine, that the Marquis should communicate it to you. It appeared to us that the man who merited most of his fellow men should be interested in the fate of every revolution which had for its object the good of humanity. I am, with the admiration and respect that your virtues have ever inspired me with, Sir, Your Obedient Servant, Mesmer. (*George Washington Papers,* Library of Congress, S. 4, P. 5.)

Washington wrote a diplomatic but noncommittal reply on November 25, 1784:

> Sir: The Marqs. de la Fayette did me the honor of presenting to me your favor of the 16th. of June; and of entering into some explanation of the powers of Magnetism, the discovery of which, if it should prove as extensively beneficial as it is said, must be fortunate indeed for Mankind, and redound very highly to the honor of that genius to whom it owes its birth. For the confidence reposed in me by the Society which you have formed for the purpose of diffusing and deriving from it, all the advantages expected; and for your favourable sentiments of me, I pray you to receive my gratitude, and the assurances of the respect and esteem with which I have the honor, etc. (Fitzpatrick 1938, p. 498)

Nothing further came of this exchange, and Washington seems to have shown no interest in the matter.

Thomas Jefferson, however, was not as indifferent to the doctrine of animal magnetism. This political thinker and statesman, so highly revered by reform-minded Frenchmen, seems to have considered Mesmer's teaching to be dangerous. In 1785 Jefferson sent antimesmerist pamphlets from France (he was then the American representative in Versailles) to influential friends whose curiosity about animal magnetism had been aroused, among other sources, by Lafayette.

Lafayette's enthusiasm for animal magnetism in America was noted even by the king of France, who is reported to have said to the marquis, "What will Washington think when he learns that you have become Mesmer's first apothecary apprentice?" (Hirsch 1943, p. 12). This comment was supposed to have been delivered in the summer of 1784, when Lafayette was about to leave on a trip to the United States. After his arrival Lafayette delivered a lecture on animal magnetism before the American Philosophical Society—this at a time when that society's founder, Benjamin Franklin, was in the midst of investigations that would reject that doctrine. During this sojourn, in September 1784, Lafayette also visited the mother house of the Shakers, located near Albany, New York. He was fascinated with this religious sect that healed through the laying on of hands, and while there he magnetized one of the Shakers, Abijah Wooster. This man spoke "under the strong operations of the Spirit," and Lafayette seemed interested in both observing his unusual state of consciousness and intervening with magnetic operations of his own (Hirsch 1943, p. 13). Lafayette later wrote to the prince of Poix, "If you had the smallest grain of faith, my dear prince, I should talk to you of a new sect of shakers who make contortions and miracles; all this is connected with the great principles of magnetism" (Hirsch 1943, p. 14).

It seems that Lafayette's attempts to promote animal magnetism in America bore no fruit. In fact, animal magnetism did not gain a foothold in the United States until some decades later. In the closing years of the eighteenth century, a superficially similar healing technique called "perkinism" did briefly enjoy some recognition. Its inventor, Elijah Perkins (1741–1799), a physician of Plainfield, Connecticut, with no knowledge of animal magnetism, developed a set of two metallic tractors, constructed from dissimilar metals, for curing illnesses. The tractors, about three inches in length, were held, one in each hand, and stroked toward the heart across the affected area of the patient. The similarity in technique to that of magnetic practitioners was not overlooked by Perkins's medical colleagues, and in 1796 he was censured for promoting a healing approach that had been "gleaned up from the miserable remains of animal magnetism" (Quoted in Carlson and Simpson 1970, p. 16). Although perkinism enjoyed some acceptance in America, England, and even the Continent, its success was short-lived, dying out in the first decade of the nineteenth century.

After the period of perkinism, there was an interval in which animal magnetism seems to have more or less slipped out of sight. Then in 1828 Grant Powers published his *Essays on the Influence of the Imagination on the Nervous System, Contributing to a False Hope in Religion.* Powers, pastor of the Congregational church in Haverhill, New Hampshire, emphasized the influence of the imagination on the human nervous system, where it could produce phenomena that had formerly been attributed to spiritual sources. In the past, he wrote, sudden remarkable healings would automatically be seen as the work of supernatural forces. But enlightened people should learn from the phenomena of animal magnetism that the imagination is sufficient to account for such happenings (pp. 15–23).

Powers compared the powers of animal magnetically stimulated imagination to those demonstrated by Elisha Perkins and his "metallic tractors." Perkins had enjoyed success in healing people by passing his tractors over the diseased or inflamed area. As with Mesmer, thousands were ready to certify their cures, attributing the cause—mistakenly, said Powers—to an external force rather than recognizing the extraordinary power of the imagination over the body (pp. 24–25).

It was Powers's belief that religious phenomena had to be reinterpreted in terms of this power. To illustrate this he proposed that his readers consider the following fantasy:

> Let us suppose that Mesmer and Deslon [*sic*] had been ecclesiastics; that they had inculcated the idea on this class of person, that religion, in a high degree, produced similar effects on the human body; and that without religion they must be damned;—suppose they had endeavoured by all possible means to excite their apprehensions, to raise their animal feelings, and by hurried, boisterous, and long addresses, they had kept their minds strained intensely for hours in succession, yea, whole days and nights;—and have we not reason to believe, that similar effects would have followed? and when one had exhibited these symptoms, another, and another, would do the same? Such a result would be natural, as in the case of animal magnetism; especially, if when one arose from the paroxysm, he was taught by those whom he considered his superiors, to believe that he emerged from a state of endless condemnation to a state of justification, life and peace. (pp. 23–24)

To emphasize his point, Powers recalled the Kentucky revival of 1800–1803, when people gathered for meetings lasting three to five days, praying incessantly, singing, shouting, laughing, crying, and gesticulating. Under the stimulation of these actions, some would experience violent muscular convulsions, sometimes accompanied by laughing or barking, sometimes taking the form of a trance-like stillness. Those overtaken by the spasms would sometimes have visions and begin to prophesy, believing they were speaking for God. Once started, the experience would spread throughout the meeting, much like what the French commissioners investigating animal magnetism had witnessed. Powers insisted that these phenomena must be shunned because of the great evils that result; the phenomena bring about a mockery of God, promote infidelity, and create false hope (pp. 36–43, 102–9).

Amariah Brigham took a similar view of animal magnetism in his *Observations on the Influence of Religion upon the Health and Physical Welfare of Mankind* (1835), in which he compared phenomena traditionally attributed to supernatural causes to those now seen to be the result of disturbances of the nervous system and the power of the imagination, and noted their similarity to the phenomena of animal magnetism.[2] He believed that revival meetings, far from exhibiting outpourings of the holy spirit, induced disorders of the nervous system, leading to an increase in disorders of the mind such as insanity, apoplexy, epilepsy, and tic douloureux (pp. 267ff).

Animal magnetism was first promoted and taught as a technique by Joseph Du Commun, a Frenchman who had emigrated to the United States and took a teaching position at the U.S. Military Academy at West Point. In 1829 Du Commun, who had practiced animal magnetism for some years in Paris, delivered three lectures dealing respectively with the history of animal magnetism, its applications and effects, and its theory. Published as *Three Lectures on Animal Magnetism* (1829), they became the first treatise on animal magnetism published in the United States.

One of those who read Du Commun's book was Samuel Underhill, an Ohio physician and professor of medicine. He felt that animal magnetism offered an important technique for dealing with disease and set about to learn all he could about it. In 1838 he began to publish a monthly periodical, *Annals of Animal*

2. Brigham was influenced in his view of the relationship of religious phenomena to animal magnetism by Dugald Stewart's *Elements of the Philosophy of the Human Mind* (1827), which examined the "principle of imitation" at work in the contagious nature of convulsions and hysterical disorders. Stewart recommended a thorough study of the details connected, both with the use of Perkins's tractors, and with the practice of animal magnetism to extend "our knowledge of the laws which regulate the connection between the human mind, and our bodily organization." He believed that their doctrines elucidated the role of the imagination and must "entitle Mesmer and Perkins to the gratitude of those who cultivate the Philosophy of the Mind; whatever the motives may have been which suggested the experiments of these practitioners, or whatever the occasional mischiefs of which they may have been the authors" (3:135).

Magnetism, and in 1839 he introduced the subject to the State Medical Convention of Ohio. In that year he also wrote to O. S. Fowler about experiments he was conducting using animal magnetism as a means of exciting the "mental organs" of phrenology. This seems to establish Underhill as the originator of "phrenomagnetism" (see below). Underhill practiced and lectured on animal magnetism for more than thirty-five years. His lectures were eventually published in 1868.[3]

The most important early promoter of animal magnetism in the United States was without question Charles Poyen (d. 1844). Born in France, Poyen had learned the techniques of animal magnetism when, as a medical student in Paris in 1832, he was searching for a way to cure an illness from which he suffered at the time. He came to the United States a few years later and began to spread the word on animal magnetism. Poyen was a good lecturer and used that ability to arouse the interest of a public that knew little of the subject. His first publication was an English translation of Husson's report of 1831 (see Husson 1836), with an introduction by Poyen on the history and theory of animal magnetism. In 1836 Poyen toured the northern Atlantic seaboard demonstrating animal magnetism. He published a description of his tour and the response he received under the title *Progress of Animal Magnetism in New England* (1837).[4]

Trained in the tradition of Puységur and Deleuze, Poyen emphasized the importance of magnetic somnambulism and accepted the reality of magnetic clairvoyance. Largely because of the great curiosity aroused by Poyen's lectures, people began experimenting with magnetic somnambulism and magnetic healing. Soon the sight of itinerant magnetizers and their clairvoyant somnambulistic partners became common in America. A pamphlet written in 1843 estimated that there were at that time between twenty and thirty men lecturing on animal magnetism in New England and more than two hundred magnetizers practicing in the Boston area (Fuller 1982, p. 30). These teams roamed the countryside and offered their services to the ill. The pair would give public demonstrations of magnetic clairvoyance to draw the attention of the local populace. Then, for a fee, the magnetizer would make his somnambulist's powers available to diagnose health problems clairvoyantly and prescribe treatment.

3. In 1835 the first American periodical devoted to animal magnetism appeared. Published in Hanover, New Hampshire, and titled *Magnet,* it lasted only a year. A general treatise on animal magnetism was published anonymously in Philadelphia in 1837. Called *The Philosophy of Animal Magnetism Together with the System of Manipulating Adopted to Produce Ecstasy and Somnambulism,* this work has been attributed to Edgar Allan Poe.

4. For further information about Poyen's role in bringing animal magnetism to the attention of the American public, see Carlson 1960.

The popularity of animal magnetism was spurred by the publication in 1837 of a number of American translations of French magnetic literature. A complete translation of the Franklin commission report, along with "An Historical Outline of the 'Science,'" an abstract of the committee report of the Royal Academy of Medicine of 1831, and other comments was published in Philadelphia in that year (Franklin 1837). This little treatise concluded with the caution: "We warn females from submitting themselves to the action of magnetism; *so gross have been the indecencies committed, that the arm of the law has more than once interposed to put a stop to its proceedings*" (Franklin 1837, p. 58). Despite his reservations, the anonymous author of the remark reluctantly conceded the reality of animal magnetism and called for serious experimentation with its phenomena.

In 1837 a work compiled by John King, a so-called professor of animal magnetism, was published in New York. It contained a review of the same 1831 report, a summary of views antagonistic to animal magnetism that had appeared in the *Journal of Commerce,* and King's own theory of why animal magnetism was an effective physical and mental agent. King's book is significant in that it contains an English translation of a small section of Puységur's *Magnétisme animal* (1807). This seems to be the only English translation of any part of Puységur's work ever published.

The publication of an American translation of Deleuze's *Instruction pratique* gave those who were interested in practicing animal magnetism much more than the sketchy outline offered by Poyen's lectures. The translator, Thomas C. Hartshorn, included an appendix describing local cases of mesmeric practice known to him. Instances of medical clairvoyance were described, such as a case in which a magnetic somnambulist insisted that a very ill patient had a disease of the spleen although her doctor had not made that diagnosis. The patient died a few days later, and an autopsy revealed that an infected spleen was in fact the cause of death.

Although Deleuze's manual emphasized animal magnetism as a healing technique, in America the emphasis was on the more spectacular aspects of somnambulism, particularly clairvoyance (see Fuller 1982). In 1837 William Leete Stone published a *Letter to Doctor A[mariah] Brigham, on Animal Magnetism: Being an Account of a Remarkable Interview between the Author and Miss Loraina Brackett while in a State of Somnambulism.* Brackett had received a severe blow on the top of the head and had for some time been in a state of impaired speech and blindness as a result. When she was treated with animal magnetism, her speech became normal and her eyesight improved slightly. But the most striking result of the treatment was her somnambulistic clairvoyance. Stone described her as being able, without eyesight, to recognize specific colored flowers and cloths and to read the contents of sealed letters.[5]

Stone's enthusiastic pamphlet was countered by Charles Ferson Durant, who considered Stone to be naively uncritical in his evaluation of Brackett. In *Exposition, or a New Theory of Animal Magnetism with a Key to the Mysteries* (1837), Durant claimed that there was no reason to accept the reality of somnambulistic phenomena and clairvoyance. He had conducted experiments of his own with animal magnetism and concluded that its apparent effects were simply the result of suggestibility and self-delusion.

At this point Poyen, who had himself worked extensively with a clairvoyant (a Miss Gleason), came to Stone's defense. In his *Letter to Col. Wm. L. Stone* (1837) Poyen said there was nothing in Durant's objections that had not been used for decades by others who sought to discredit animal magnetism. Poyen described tricks used by skeptics to get the suggestible somnambulist to give wrong information and drew upon material from his own classes on animal magnetism to support the claim for the genuineness of magnetic phenomena.

While many of the medical profession took a wait-and-see attitude in regard to the validity of the phenomena of animal magnetism,[6] others joined Durant with scathing criticism. One of those was David Meredith Reese, who placed animal magnetism on his list of popular delusions in *Humbugs of New York* (1838). "Humbug" to Reese meant "any system of science, philosophy, or religion which seems to be what it is not" (p. vi). After presenting a cursory history of animal magnetism in Europe and its introduction into the United States, he described the work of Poyen and his followers in most unflattering terms:

> He commenced his public lectures during the winter of 1836–7; and having found a girl who was simpleton enough to favour his designs, by becoming his somnambulist, he visited Boston and other places, and to the present hour is itinerating with one or more "sleeping beauties" who

5. Contemporary cases of somnambulistic clairvoyance were not limited to magnetic somnambulism. In 1834 an instance of spontaneous somnambulism accompanied by clairvoyance and double memory was described in what is probably the first full-length published account of this phenomena in the United States (Belden 1834). The clairvoyant, Jane C. Rider, was apparently able to see in the dark, carry out complex physical operations with her eyes closed, and reveal the contents of sealed boxes. The author of the account was reluctant to attribute her state to animal magnetism, however, stating that one must possess "more than an ordinary share of credulity" to admit the claims of mesmerism (pp. 98–99).

6. See, e.g., *Boston Medical and Surgical Journal* 13 (1836), pp. 418–19; 14 (1836), pp. 8–12, 36, 83, 322, 349–51; 15 (1836), p. 409; 17 (1837), pp. 113–14.

are *trained for the purpose,* and by whom multitudes have been gulled into a belief in the "new science." . . . A number of third rate doctors, merchants and mechanics, having failed in their appropriate employments to realize either fame or emolument, have become converts to the opinion that "the world is a great goose, and every man a fool who does not pluck its quills." Accordingly, providing themselves each with a factory girl, who would *rather sleep than work,* they have scattered themselves abroad in the villages, towns, and cities of the land. . . . These itinerating mountebanks and their misses, have been most successful in Schenectady, New-York. (pp. 35–36)

Reese also made dark references to what he perceived as the moral dangers of animal magnetism, citing a case of a German professor practicing magnetism who, accused by a somnambulist's father of having taken advantage of her, was brought to court. The magnetizer was acquitted and retained his faculty position, yet Reese implied that the mere fact that an accusation was made should be sufficient to condemn the practice (pp. 53–55).

Looking into the future, Reese foresaw dire consequences if animal magnetism were not quashed. Magnetic treatment would supplant reputable hospitals, medical colleges, and physicians, and "the infatuation and crimes of the days of Salem witchcraft [would] be re-enacted in our times." He trusted that his and others' testimonies would expose animal magnetism to the educated public as a "humbug," and that the "ignorant and depraved" would give it up once it ceased to turn a profit for them (pp. 60–62).

Mesmerism Comes of Age in America

Despite Reese's warning, in the early 1840s animal magnetism began to come into its own as a popular movement in the United States. In 1842 Charles Caldwell, a Philadelphia physician of some note, spoke publicly in support of animal magnetism. Not one to avoid controversy (he had publicly debated the issue of the contagious character of yellow fever with Benjamin Rush in 1793 and in 1820 had taken a public position in favor of phrenology), Caldwell believed that the debate about the usefulness of animal magnetism would be a short one since the supportive facts seemed to him so easy to demonstrate. He was not worried about the dangers enumerated by Reese but considered the closed minds of the "sombre-souled and ill-boding antimesmerists" the greater danger, since such rigidity stood as a barrier to the progress of the human spirit (Caldwell 1842, pp. xxiii–xxv). Caldwell saw Elliotson mesmerize in England, spoke with magnetizers in Paris, and conferred with M. J. Cloquet about his removal of the breast of Madame Plantin while she was in mesmeric trance. In his short book *Facts in Mesmerism and Thoughts on Its Causes and Uses* (1842) Caldwell presented a history of the rise of animal magnetism, an outline of its chief phenomena, and an explanation of its cause. He accepted the validity of all the phenomena of animal magnetism, including clairvoyance and precognition. In discussing the theory he provided no new ideas but simply echoed the views of Mesmer, Puységur, and other mainstream mesmerists.

Caldwell's book was followed by a rash of treatises on animal magnetism. In 1843 alone, five notable works on the subject were published, by Samuel Gregory, K. D. D. Dickerson, John Bovee Dods, Robert Collyer, and LaRoy Sunderland.

Dickerson, himself a "practical magnetiser," put together a history of the progress of animal magnetism in America from Poyen to Sunderland. His *Philosophy of Mesmerism* is historically valuable, including as it does the names of the best-known mesmeric practitioners and somnambulists of the day. Samuel Gregory's *Mesmerism, or Animal Magnetism and Its Uses* also contained a brief but useful historical sketch of American mesmerism.

The popularity of the lectures of John Bovee Dods (1795–1872), delivered to an audience of two thousand on six consecutive nights, is indicative of the great interest animal magnetism was now generating. Although Dods appreciated Mesmer's discovery, he rejected his terminology, preferring "mental electricity" or even "spiritualism" to "animal magnetism." Dods, a minister of the Universalist church, called mesmerism a "power of God" and compared its healing effects to those produced by Jesus.

By 1850 Dods had developed an elaborate new framework for understanding the phenomena of animal magnetism, explaining them in terms of "electrical psychology."[7] He presented his formulation in a series of lectures delivered to the U.S. Senate at the invitation of seven members, including Daniel Webster, Sam Houston, and Henry Clay, and later published. Dods stated:

> One human being can, through a certain nervous influence, obtain and exercise a power over another, so as to perfectly control his voluntary motions and muscular force; and also produce various impressions on his mind, however extravagant, ludicrous, or wild—and that too while he is in a perfectly wakeful state. I have stated that it is one of the most

7. Dods lamented that certain persons had stolen his theory and changed the name to "electro-biology." He said that electro-biologists claimed that their theory differed from that of electrical psychology, but he had discovered from close investigation that it was identical (p. 201).

powerful remedial agents to alleviate the pains of the suffering, and to cure those diseases that set the power of medicine, and the skill of the ablest practitioner, at defiance. (1850, p. 29)

After describing the cure of a young woman from Virginia who had been unable to walk for eighteen years, Dods took pains to distinguish his electrical psychology from mesmerism. He said first of all that mesmerism involved an intimate rapport, or sympathy, whereby the mesmerized subject senses what the operator senses and does what the operator wills. In electrical psychology, however, the senses of the subject remain entirely independent of those of the operator. And whereas in mesmerism the subject has no memory of what occurs while mesmerized, in electrical psychology the subject "is a witness of his own actions, and knows all that transpired" (p. 31).

In this way Dods identified mesmerism with somnambulism and attributed to electrical psychology those phenomena associated with animal magnetism that occur without somnambulism. He held that both were accomplished through the action of the same "nervous fluid," but nonetheless must be distinguished from each other (pp. 31–32). Dods saw this nervous fluid as basically electrical and argued that diseases "begin in the electric force of the nerves, and not in the blood" (p. 76). According to his "doctrine of impressions," disturbances of this electro-nervous fluid may be produced by destructive "mental impressions" (for example, grief or unwanted passion) or "physical impressions" (for example, exposure to damp air or overeating) that intrude from the outside. These impressions throw off the balance of the electro-nervous system and cause diseases of all kinds.

Cure is effected through the patient's being "*electrically* and *psychologically* controlled," so that the operator can directly affect his or her body and mind. Positing that people have a kind of charged aura around their bodies, Dods said that this "electric or magnetic circle" surrounding the patient meets that of the operator at two points, and a communication is established. The operator, through various suggestions, induces what Dods called the electrical state (wherein the operator controls the patient's muscular movement) and the psychological state (wherein the operator can create any sort of mental hallucination in the patient). Dods believed that one in twenty-five people is naturally in this electro-psychological state and subject to powerful waking suggestions. Others had to be coaxed into it by artificial means. In any case, the resulting waking suggestibility allowed the operator to make healing suggestions and effect cures (pp. 207–32).

Among those generally recognized at the time as making original contributions to animal magnetism in the United States were Robert Hanham Collyer and LaRoy Sunderland. Collyer was an English physician and pupil of John Elliotson who toured the eastern United States in the late 1830s lecturing on phrenology. When he began to include mesmerism in his lectures, he drew large crowds. Encouraged by his success, Collyer took up residence in America, joined the Massachusetts Medical Society, and became editor of the *Mesmeric Magazine*.

Collyer performed experiments with what he called "phreno-magnetism" in 1841 and developed a theory and technique of using animal magnetism directly to excite the faculties of the mind described in phenological language by Franz Joseph Gall and Johann Gaspar Spurzheim. Soon, however, he changed his view of the phenomena he had observed and denied that specific areas of the brain—identified as sources of such human qualities as benevolence and destructiveness—could be physically affected. In *Psychography* (1843) Collyer insisted that what appeared to be a magnetic excitation of parts of the brain producing corresponding sentiments or attitudes was in reality a process by which the magnetizer imposed the expectations of his own mind on the mind of the magnetic subject. The resulting "phrenological" sentiment in the subject was therefore produced by his or her unwitting response to the mental impressions conveyed by the magnetizer, not by stimulation of the subject's brain by magnetic fluid (pp. 8–16).

Collyer did not deny the reality and power of animal magnetism, but he gave it a new twist and offered a new set of terms. To Collyer, the importance of animal magnetism lay in its ability to affect the brain in such a way that clairvoyance could occur. In that state "the faculties seem to have hardly a limit of action; time and space are annihilated; the secrets of the past, present and future are brought within the immediate range of THOUGHT" (p. 26).

The mechanism behind magnetic clairvoyance involved the action of "nervous fluid" that obeyed the same laws as electricity, light, and heat. Collyer believed that thoughts were conveyed from one mind to another by the transmission of this fluid, and he described this process as one brain painting an image on another brain—an operation that he called "psychography": "I might adduce a hundred such instances [of clairvoyance], showing the *embodiment* of ideas which by a *concentrated* and *undivided* effort of the will, may be depicted on a recipient brain. This mental photographic process depends on the resident principle of the brain, being subject to the same laws as that of electricity, light, heat, &c. The brain, in all its operations, uses this vital electricity" (p. 31).

In the same year that Collyer published his book on psychography, LaRoy Sunderland (1804–1885) published a work dealing with animal magnetism under a different name. In *Pathetism* Sunderland rejected the notion of a magnetic or nervous fluid, separating himself from both Mesmer and his compatriot Collyer.

Sunderland was born in Rhode Island in 1804 and became a Methodist minister. As a young man he developed an interest in the effects of the mind on the body and read books on mesmerism and phrenology in an effort to understand the relationship between the two. As editor of the religious periodical *Zion's Watchman,* in 1841 he introduced articles on animal magnetism into the magazine. When they encountered a less than enthusiastic reception, he left *Zion's Watchman* and in 1842 founded *The Magnet,* a journal devoted to the study of phrenology and animal magnetism. Like many others at that time (see Sizer 1884), Sunderland used the term "phrenomagnetism" to designate the connection between the phenomena of animal magnetism and the "organs" or regions of the brain that were supposed to control specific mental-emotional faculties.

Sunderland formulated his particular explanation of the relationship of mind to body in his theory of "pathetism":

> I use this term to signify not only the AGENCY by which one person by manipulation, is enabled to produce *emotion, feeling, passion,* or any physical or mental effects, in the system of another, but also that SUS-CEPTIBILITY of *emotion* or *feeling,* of any kind, from manipulation, in the subject operated upon, by the use of which these effects are produced; as also the *laws* by which the agency is governed. I mean it as a substitute for the terms heretofore in use, in connection with the subject, and I respectfully submit it to all concerned, whether this be not a far better term for the *thing signified,* than either Magnetism or Mesmerism. (p. 3; see also Jervey 1976)

Sunderland carried out a far-reaching revision of nomenclature. To mesmerize was now to "pathetise." Taking up the term "somnium" (already used to designate somnambulism by Samuel L. Mitchill in *Devotional Somnium,* 1815), Sunderland called the somnambulist a "somnist," artificially induced somnambulism "somnipathy," and the person placed in the state of artificially induced somnambulism a "somnipathist." Declaring that pathetism could be used to develop the faculties affected through phrenology, Sunderland named this process "phrenopathy" (1843, pp. 3–4). (By 1845, however, he had ceased to incorporate phrenomagnetic theory into his teachings.)

Sunderland emphasized the effects of pathetism on the consciousness of the subject: "My control over the consciousness of patients, is just in proportion to the susceptibility. The functions of all the mental organs may be increased to insanity, or subdued into a state of perfect repose, where the patient seems lost to this world, as really as though he had ceased to live. And from this state of unconsciousness, he may be waked up, as it were, into another world, where all his feelings, views and perceptions, differ, toto caelo, from those peculiar to him in his normal condition" (p. 83).

This double-consciousness experience, involving a second state very different from the normal state, had by now become a familiar subject in the literature of animal magnetism. Sunderland connected double consciousness with clairvoyance, which he believed operated preeminently in diagnosis of disease. Although not ruling out other manifestations of clairvoyant powers, he believed they occurred when the knowledge obtained was already known by the pathetizer, with whom the subject was in sympathetic rapport.

Sunderland rejected the notion that a physical fluid passed from pathetizer to pathetized. Rather he held that the action of pathetism was based on vital forces possessed by all living things, which constitute a "sympathetic system" that makes them subject to the laws of sympathy and antipathy: "A peculiar *connection* between two entities, organs, or substances, which differ in certain *qualities* or *functions*, produces a *positive relation* or the law of sympathy. . . . By establishing a *positive relation* between two persons, the mind of one may thereby control the *susceptibility* of the other; or by applying the hand of one to any part of the other, different mental and physical changes, may thus be produced" (p. 101).

These profound changes could be induced in a subject who was awake or somnambulistic, and they could be brought about by suggestion or even the mere power of the pathetizer's will. Sunderland attributed the resulting mental and healing phenomena to the action of the subject himself. To Sunderland, the effects of pathetism are self-induced; it is due to the action of the subject that "the mind withdraws itself from the consciousness of pain; it cures diseases; it induces the so-called 'change of heart' in 'revivals of religion;' it brings on the Trance, and often induces other changes, which have been attributed to God or to the Devil" (pp. 7–8).[8]

Among those with whom Collyer and Sunderland felt most in competition was Joseph Rodes Buchanan (1814–1899), dean of the faculty of the Eclectic Medical Institute in Covington, Kentucky. Buchanan claimed that he, not Sunderland, was the discoverer of "phrenomagnetism." He also coined the word "psychometry" to refer to the alleged ability of certain gifted individuals

8. Like Collyer, Sunderland seemed very sensitive to any criticism. When, in 1845, an anonymous work appeared entitled *The Confessions of a Magnetiser; Being an Exposé of Animal Magnetism,* Sunderland reacted quickly to its criticism of his way of practicing pathetism. Within months he published *"Confessions of a Magnetiser" Exposed!*

to sense emanations given off by objects and thereby obtain detailed information about the history of those objects.

According to Buchanan, the phenomena of animal magnetism fall into six categories: (1) attraction, whereby the magnetic subject is drawn to the magnetizer and obeys his will; (2) sympathy, through which the subject feels the sensations, emotions, and thoughts of the magnetizer; (3) intuition, which includes clairvoyance and precognition; (4) volition, or the power of the magnetizer to control the actions of the subject; (5) sleep-waking, which is analogous to the natural sleep-walking state; and (6) therapeutic benefits, or physical healing (pp. 252–57).

Buchanan's view of how therapeutic benefits accrue implies certain dangers for the operator. The subject, ailing and in pain, is liable to exert a "malign" or "morbidic" influence on the magnetizer, just as the "sympathetic relations" between the two persons allows a healthy influence to pass the other way and cure the sick (pp. 257, 260).

Buchanan claimed that one cannot practice animal-magnetic healing effectively without a thorough knowledge of the principles of phrenology. The magnetizer must know the locations of the various phrenological regions on the skull, corresponding to the various brain "organs," and use that knowledge in the positioning and direction of his magnetic passes. For instance, he suggested that upward passes should be made from the region of Disease (at the cheek bones) toward that of Health (at the crown of the head). This will be felt by the subject as a "rousing, bracing, refreshing operation" (p. 261).

Buchanan lectured widely on his elaborate system, to which he gave the overall name of "neurology." Speaking before groups such as the Phreno-magnetic Society of Cincinnati, he impressed many who were already favorably disposed toward phrenology.[9]

In 1845 James Stanley Grimes (1807–1903), having earlier made public his theory of phrenology (1839), published *Etherology; or the Philosophy of Mesmerism and Phrenology: Including a New Philosophy of Sleep and Consciousness, with a Review of the Pretensions of Neurology and Phreno-magnetism* (1845). Under this rather flamboyant title, Grimes, a lawyer and professor of medical jurisprudence at Castleton Medical College, attempted to introduce a new theory and yet another nomenclature for the phenomena of mesmerism and phrenology.

Grimes's term "etherology" was derived from "etherium," his name for the material substance that connects all things in the universe and provides a

9. For more on Sunderland, Collyer, and Buchanan in the phrenological movement, see Davies 1955.

means of mutual influence. Knowledge of the etherium in all its manifestations was called "etherology," and "etheropathy" was the term reserved for the phenomena known as mesmerism, animal magnetism, neurology, pathetism, hypnotism, catalepsy, somnambulism, and clairvoyance (Grimes 1850, pp. 18–27). Etheropathy results from an abnormal condition of the human constitution, said Grimes, a degenerated or morbid state incompatible with health. This makes the mesmeric state "a departure from, and violation of, the ordinary laws of man and the designs of the Creator" (pp. 27–28). This abnormal condition may be generated spontaneously, as in the case of Jane C. Rider of Springfield, Massachusetts, who manifested the state of somnambulistic clairvoyance without benefit of mesmerism or any operator. It may also be produced artificially through "Mesmerising, Magnetizing, Willing, Charming, &c.," a process Grimes labeled "inducting" (pp. 29–30).

To induce the etheropathic state the operator should be of "sound and vigorous body and mind," but no particular concentration of the mind or effort of the will is required. Once the subject is in an etheropathic state, experiments may be performed that involve sympathy (communication of sensation between operator and subject), obedience to the silent will of the operator, production of normal sleep, somnambulism, mesmeric sleep, paralysis, trance, clairvoyance, sympathetic clairvoyance (mental communication), and transfer of power from one operator to another (pp. 35–37). Other unusual phenomena may occur in the state, such as an exhibition of extraordinary physical strength, the ability to read a person's character, and the illusion of communication with spirits. Grimes believed that some mesmerized persons who thought they were in touch with spirits were merely conforming to the expectations of the operator, and others were the victims of their own wishes. In a later work, *The Mysteries of Human Nature Explained by a New System of Nervous Physiology: To Which Is Added A Review of the Errors of Spiritualism, and Instructions for Developing or Resisting the Influence by Which Subjects and Mediums are Made* (1857), Grimes examined the views of the supporters of a spiritualistic philosophy. He attributed all physical phenomena of mediumship, without exception, to fraud. Even mental phenomena were seen as illusory. And whereas in 1845 Grimes admitted that clairvoyant mesmeric phenomena occurred frequently, he now believed them to be rare if not nonexistent (pp. 399–410).

In 1846 another work on animal magnetism offered yet another set of terms to denote its phenomena. *Animal Magnetism; or Psychodunamy,* by Theodore Leger, was a history of animal magnetism and by far the most complete to appear to date. Leger, a pupil of Deleuze and member of the Medical Faculty of Paris, brought to the American scene a thorough grasp of the historical events surrounding Mesmer and animal magnetism in France. He was particularly well versed in the later French commissions. His choice of the term "psychodunamy," meaning "power of the soul," seems more cosmetic than anything else, for he adds no new ideas to the already seething pot of theories current in the United States.[10]

The Swedenborgians

While Sunderland, Dods, and other ministers of religion were moving to separate mesmerism from the spiritual, one philosophical tradition in the United States sought to do the opposite. By the 1830s Swedenborgianism was establishing a following among American philosophers and theologians. The most direct effect of mesmerism on American Swedenborgianism may be seen in *Mesmer and Swedenborg; or, the Relation of the Developments of Mesmerism to the Doctrines and Disclosures of Swedenborg*, written by George Bush and published in 1847.

Sixty years earlier, the followers of Swedenborg in Stockholm were among the first to recognize the potentials of magnetic somnambulism. The Stockholm Exegetical Society explained the phenomena of mesmerism as the product of action by spirits. Bush was more sophisticated, however, viewing the phenomena as concrete, physical proof of the pronouncements of Swedenborg. He intended to show that Swedenborg's teachings, substantiated by the facts associated with animal magnetism, made a significant contribution to both physiology and psychology (p. 13).

Swedenborg often had visions in which he claimed to communicate with spiritual beings. Bush noted the similarity of mesmeric trance to Swedenborg's spiritual ecstasy and believed that this likeness lent support to the genuineness of his visions. Bush pointed out that the two states were not, however, identical and that Swedenborg himself had insisted that one must not confuse "the lower with the higher manifestations" (p. 29).

Bush affirmed the reality of mesmeric psychological phenomena such as thought and memory transfer, suggested hallucinations (here called "phan-

---

10. Other works promoting animal magnetism during this period include *The Animal Magnetizer* 1841, Morley 1841, Douglas 1842, Drake 1844, Johnson 1844, Hall 1845, Smith 1845, Wilson 1847, and *A Key to the Science of Electrical Psychology* 1850. John B. Newman (1848) wrote a work on "fascination" that equated mesmeric phenomena with those of animal fascination and charming. Works opposed to animal magnetism include Brown 1843, Jones 1846, and Blakeman 1849. A humorous parody also appeared entitled *Lecture on Mysterious Knockings, Mesmerism, &c., with a Brief History of the Old Stone Mill, and a Prediction of Its Fall, Delivered Before the A N ti Quarian Society of Pappagassett . . . by Benjamin Franklin Macy D.F., D.D.F., A.S.S., Professor of Hyperflutinated Philosophy* (1851).

tasy"), sensitivity to auras (called "spheres"),[11] magnetic vision (seeing with the eyes closed), magnetic hearing (heightened sensitivity to the magnetizer's voice and insensitivity to all other sounds), clairvoyance, repugnance at naming objects,[12] and invariable truthfulness in answering questions. Bush believed that these were precisely the kinds of psychological phenomena one would expect to find in the spiritual world if the teachings of Swedenborg were true. And, "if *Mesmerism is true, Swedenborg is true,* and if Swedenborg is true, the spiritual world is laid open, and a new and sublime era has dawned upon the earth. . . . The divine hand itself has, in the teachings of this illuminated seer, lifted the veil interposed for ages between the world of matter and the world of mind" (p. 161). In fact, Bush saw the coming of mesmerism as divinely timed: "I would [ask] whether there be not something more than a merely *plausible* basis for the position, that the ultimate design, in Providence, of the development of Mesmerism at the present era, is in fact nothing less than to pave the way for the universal admission of Swedenborg's claims" (p. 168).

As an appendix to his book, Bush included a section on the visions of a "young man whose educational advantages have been of the most limited character"—Andrew Jackson Davis of Poughkeepsie, New York. Davis was claiming at the time that some of his revelations, attained in the mesmeric state, emanated directly from Swedenborg, although he had never read any of Swedenborg's writings or heard about his doctrine in any detail. While advising caution in accepting Davis's claim, Bush himself concluded, on the basis of his examination of Davis's teachings and his own contact with the man, that there was good reason to think the revelations did originate with Swedenborg:

> It is clear, we think, from the evidence afforded, that the Mesmeric state, in its more sublimated manifestations, does enable one human

11. In his chapter on spheres, Bush develops an interesting discussion of magnetic rapport as it relates to the apparent sensitivity of the magnetized subject to the aura around an individual: "Everyone is surrounded by an invisible aura or atmosphere, which is constantly exhaling from his person and spreading to some distance on every side, and bearing to him somewhat the same relation that the aerial atmosphere does to the earth. The first effect of the Magnetic condition is to produce a blending or congenial inter-relation of the respective spheres of the operator and the subject, and the more perfect is the moral affinity of the parties, the more complete is the amalgamation in this respect; for the sphere is not merely the efflux of the corporeal system, but it emanates also from the interior spirit, the seat of sentiments and intellectual sympathies. A very decided opposition or antipathy of internal spheres is extremely unfavorable to the Mesmeric influence, and often avails to counteract it altogether" (p. 71).

12. "It is often a perplexing problem why the subject, when in that [magnetic] state, and attempting the description of the commonest material objects, should not at once designate them by their appropriate name, instead of describing them by their qualities or uses, which he is almost invariably prompted to do" (p. 144).

spirit, while sojourning in the body to come into actual converse with another human spirit similarly conditioned. If so, why may not a like intercourse occur between an embodied and a disembodied spirit. . . . There is nothing wonderful in the fact of Mr. D.'s conversing with Swedenborg than of Swedenborg's conversing with the departed spirits of other men. But the other and lower phenomena of Mr. D.'s transic state, go directly to prove, as we have shown, the truth of Swedenborg's intercourse with the spiritual world. This again, when established, reflects back a powerful evidence of the truth of Mr. D.'s intercourse with himself [Swedenborg] or some adequate representative. (pp. 212–13)

But if Bush hoped that Davis would become an apostle of Swedenborg and the forerunner of a new Swedenborgian movement, he was to be disappointed. Davis did come to be recognized as the forerunner of a new spiritual movement: spiritualism.

The Birth and Spread of Spiritualism
Andrew Jackson Davis

The 1830s and 1840s saw the rise of itinerant mesmeric teams that gave clairvoyant demonstrations and offered clairvoyant medical diagnoses. Among those who discovered their clairvoyant powers in this way was the young Andrew Jackson Davis (1826–1910). Davis was the only son of a shoemaker who was both abusive and a poor provider. He received very little formal education—the rudiments of reading and writing—and at the age of sixteen became an apprentice shoe clerk in Poughkeepsie. In 1843 the magnetizer J. Stanley Grimes came to town to demonstrate the wonders of mesmerism and phrenology. A local tailor named Levingston picked up the technique from the demonstration and began magnetizing the inhabitants. He found Davis to be an excellent subject and an unusual clairvoyant. The Levingston-Davis team enjoyed such immediate popularity that they decided to go on the road. According to contemporary reports, Davis was able to point out the location of a disease, accurately describe the patient's symptoms, and give technical information about the functioning of the organs involved. He also gave demonstrations of traveling clairvoyance, in which he mentally visited distant places unknown to him and described strange homes in vivid detail.

Soon after he began to be mesmerized, Davis had mystical experiences that convinced him he had an important role to play in educating the public about spiritual matters. He changed magnetizers and moved to New York City, hoping there to publish the contents of his mystical revelations. He asked William Fishbough, a Universalist minister, to take his dictation and in 1845, at the age of nineteen, began his career as a prolific writer of spiritual books. His first book, *The Divine Revelations,* ranged in content over the broadest of issues, including mythology, biblical archaeology, and the progress of civilization. This work and the books that followed enjoyed tremendous popularity. The public was hungry for his message, which included a belief in reason and human progress, the notion that spirit is the cause of all things, and the conviction that after death everyone enters a spiritual world that can be contacted even now while in a state of magnetic ecstasy.

One of the mysteries of Davis's writings is where he obtained the vast knowledge of history, geology, languages, mythology, and theology they displayed. He had had a mere five months of formal education, and there is no evidence that he had read any books other than a few religious works or that he had plagiarized from any. Whatever their source, Davis's works struck a note that was right for the time. His emphasis on clairvoyance and the accessibility of the spirit world received a sympathetic reception from a large readership. And the publication of *The Divine Revelations* just two years before the Hydesville knockings aided the rise and rapid spread of spiritualism in the United States.[13]

The Fox Sisters and the Spirits

In December 1847 John Fox, a blacksmith by trade and a Methodist by religion, moved with his wife and two daughters into a small rented house in Hydesville, just east of Rochester, New York. Three months later the family had barely gotten settled when the household was disturbed by some rather bizarre events. The older daughter, Maggie, was fifteen and a half, and her sister, Kate, was twelve when they became the center of phenomena that the young girls interpreted as the action of spirits.

Over the last week of March 1848 noises began to be heard in the house as if furniture were being moved, and pounding noises seemed to come from the walls. The family searched the house and found nothing to explain the disturbances. On March 31 they all went to bed in the same room, for mutual security. The noises began as soon as the four had retired. But this time the young girls treated them playfully, and Kate, snapping her fingers, asked the sounds to imitate her. The snapping was reproduced by the knocking in the walls. Maggie tried the same test by clapping her hands. These sounds too were echoed by the raps. Then Mrs. Fox asked that ten raps be given. When

13. For more information on the career of Andrew Jackson Davis, see Podmore 1902, Brown 1972, and Moore 1977.

this was immediately done, she requested that the sounds tap out the ages of her children, and this too was done.

The astonished family immediately sent for neighbors to witness what was happening. A code was quickly worked out by which questions could be answered yes or no. In this way it was ascertained that the raps were being produced by the spirit of a deceased person who had been buried in the basement of the house. Within a few days crowds of people descended on the Fox home, listening to the raps, searching for their source, and speculating about the body supposedly buried in the cellar. A neighbor, E. F. Lewis, obtained written statements from witnesses describing how the knockings went on for hours at a time and, in response to questions, gave answers in accurate numbers regarding personal family matters.

The owner of the forty-year-old house became upset by the actions of self-appointed investigators as they tore up floorboards and excavated the basement in an attempt to solve the mystery. The Fox family was asked to leave and moved into the home of an older son, David, but the raps followed them there. It was then realized that the knocking occurred only in the presence of the two young daughters. It was decided to separate the girls, and Kate was sent to stay with an older sister, Leah, in Rochester. The knockings continued at David's house, however, and also happened in Leah's.

Now the types of phenomena taking place around Kate began to expand. Objects were seen flying through the air and beds shook at night. Books and other articles were shifted from one place to another, combs of several ladies were drawn from their hair and placed in the hair of others, musical instruments played as they floated through the air above onlookers, and chairs and tables were tipped, turned over, or moved about.

Rochester was the home of Isaac and Amy Post, a Quaker couple devoted to the abolition of slavery. They were acquainted with the Fox family, and when Kate arrived from Hydesville they investigated the phenomena taking place around her and became convinced that spirits were behind them. Although Mrs. Fox had used the alphabet to obtain the name of the communicating spirit in the first days of the manifestations, that practice had not been continued. Isaac Post suggested a method of spelling out messages by reciting the alphabet and recording when a knock would occur at a particular letter. This greatly facilitated the process of communication (Capron 1855, pp. 52–53, 63–65). Because of current interest in a relatively new invention, the telegraph, this came to be called "telegraphing" the message, and eventually communication through raps was called "spiritual telegraphy."

Before the involvement of Isaac Post, the only spirit to make himself known was the ghostly tenant of the Fox house. Now, at the Post house, the first message to be telegraphed was one that would become the cornerstone of spiritualism: "We are all your dear friends and relatives." From that moment the possibility of communicating with deceased loved ones inflamed the imaginations and piqued the curiosity of the public.

With the use of the alphabet, the random physical manifestations were converted to intelligible messages, and the table became the principal locus of this spiritual communication. Raps, which had previously occurred in wall, floors, and sundry furniture, now took up their main residence in the parlor or dining room table.

It took spiritualism only a few years to spread throughout the United States. From the beginning the phenomena attracted the attention of men and women of note. Horace Greeley, William Cullen Bryant, James Fenimore Cooper, George Ripley, and Harriet Beecher Stowe were among its earliest supporters. Spiritualism often found a warm welcome in the home of abolitionists. It also gained important adherents in Congress and eventually found its way into the White House, when Mary Todd Lincoln brought in mediums to contact her dead son, Willie.

All three Fox sisters became public channels for the manifestations, but they were soon replaced by other, more engaging mediums. These newcomers developed other ways to communicate with the dead, employing automatic writing, slate writing, and direct use of the medium's own voice to get messages across.[14]

But perhaps the most significant aspect of the early practice of spiritualism was its use as a private family experience. It seemed that the spirits of the dead did not need spectacular public mediums to reach the living. Households all over the country found among their own members those who had the gift of mediumship, and family spirit circles sprang up in thousands of homes. Most of these circles included only family members and some close friends, and

14. The full list of spiritualistic phenomena is long indeed. Adin Ballou (1853) provided quite a comprehensive one. He included noises that indicate an intelligent source, such as "knockings, rappings, jarrings, creakings, tickings, imitation of many sounds known in the different vicissitudes of human life, musical intonations, and in rare instances, articulate speech"; the movement of objects such as "tables, sofas, light stands, chairs, and various other articles"; shaking, tipping, sliding, or levitating these objects; moving people through the air; spirit writing with pens or pencils or chalk; the production of "catalepsy, trance, clairvoyance, and various involuntary muscular, nervous, and mental activity in mediums, independent of any *will* or *conscious* psychological influence by men in the flesh"; speaking, writing, and preaching through mediums; the production of spirit limbs or the whole human form; and, through any of these means, the communication of "affectionate and intelligent assurances of an immortal existence" (pp. 1–3).

there was no attempt to impress the world at large. The phenomenon of private family séances made spiritualism a unique religious practice. There was no dependency on a trained clergy and no need for overall organization of any kind. Spiritualism was in its earliest years a movement without structure, and its members strongly resisted organizing.[15]

The predominance of home circles had other implications for the development of spiritualism. When public skepticism arose regarding the genuineness of the "spiritual manifestations," those whose involvement was principally in family groups were largely unmoved. Their belief in the reality of the phenomena did not depend on public demonstrations by paid mediums who might be suspected of deceit. Their intrafamily experience was personal and direct and could not be called into question on such grounds.

Spiritualism was unique in another very important way. Although there were noted male mediums, women dominated the leadership of spiritualism. It has been pointed out (Braude 1989) that spiritualism made it possible for women for the first time to assert their rights to political influence in American society. A number of charismatic and articulate mediums went on lecture tours to propound their views to the general public. Their message was not always about spiritual matters. They took the opportunity to state their positions on such topics as woman suffrage, dress reform, marriage reform, free love, and holistic health care. For these speakers the need to free women from the domination of men was at the heart of the matter. Often speaking from trance, they could claim that they were mouthpieces for the spirits and that their messages partook of the truth of the eternal realms. They kept up their campaign for women's rights for several decades and were instrumental in radically altering the country's view of relations between the sexes.

15. There are many excellent historical studies of spiritualism, including Capron 1855, Britten 1870 and 1884, Maskelyne 1875, Podmore 1902 and 1910, Doyle 1924, Nelson 1969, Brown 1972, Stemman 1972, Moore 1977, Brandon 1983, Oppenheim 1985, Barrow 1986, and A. Braude 1989.

Chapter 12

Table Turning: Speculations about Unconscious Mental Activity

With the spread of spiritualism throughout the United States, its sometimes startling phenomena became topics of interest for the public at large. Rappings in walls and on tables and movements of objects without the apparent use of physical force enthralled the popular imagination and soon became the subject of articles in newspapers and periodicals and of scientific treatises that attempted to explain their occurrence.

Phenomena so extraordinary could not long be confined to the United States, and soon came to be known in Europe and Britain. The crossing of the Atlantic was accomplished principally by one particular physical manifestation: the moving and rapping table.

The Table-Moving Fad

In April 1853 German newspapers began publishing accounts of bizarre phenomena associated with common parlor tables. A merchant in New York City sent his brother in Bremen instructions about how to reproduce physical manifestations that were becoming commonplace among spiritualists in America. The Bremen resident set up an experiment and was astounded by the results.

The man gathered eight people, three men and five women, around a table in the center of a room, instructing them to touch neither each other nor the legs of the table except in the prescribed way. They formed a chain (reminiscent of the magnetic chain) by resting their hands lightly on the surface of the table, palms down and fingers spread. Each "sitter" placed the little finger of his or her right hand over the left little finger of the sitter on the right to produce a circle of hands on the table surface. After about thirty minutes the table began to rotate, first from right to left, then left to right. Then, according to the account of observers, the whole table began to change its position and the chairs of the sitters were withdrawn while they continued to maintain the chain. The tabletop began to move about and rotate so rapidly that the sitters could hardly keep up with it. The movement continued for about four minutes. Later that evening the experiment was repeated with the same results.

This account, first published in the *Gazette d'Augsbourg* on April 18, 1853, was immediately reprinted in other newspapers in Germany and France (see Guillard 1853), and the "table-moving" fad began its sweep through Europe.[1] Within five years hundreds of books, pamphlets, and articles had been written describing table-moving experiences and attempting to explain how such things could happen.[2] One English observer wrote:

> It would be difficult to single out any scientific subject which has with such rapidity, taken so extensive a hold of the popular mind. If we travel by railway carriage, steamboat, or omnibus, this is the universal topic of conversation. From the aristocratic saloons of Belgravia to the "Parlours" of Whitechapel—the Green Park to the Cat and Mutton Fields, "table moving" is all the rage. From the Royal Institution, where the secretary pokes his head through a forest of electrical apparatus, to inform the audience that the *facts* are established, down to the humblest Mechanics' Institute, all are full of it. . . . Every evening party must of course have its experiments; accordingly, gentlemen come provided with very elegant *chapeaux* for the occasion, and many an innocent flirtation occurs consequent on the proper arrangement of the little fingers of some of the fair operators. (Nicholls 1853, p. 1)

As in America, the purely physical aspect of the experiments was soon expanded to include the mental, so that the motion of the table became a means of telegraphing messages to the participants. In some cases the mes-

1. Various names were given to the phenomena. In English they were termed "table moving," "table turning," "table rapping," "table tipping," "rotating tables," "responding tables," and "talking tables." In French, "tables tournantes," "tables parlantes," "tables répondantes," and "la danse des tables." In German, "Tischrücken," "Tischklopfen," "Tischsprechen," "Tischdrehen," "die Selbstbewegun der Tische," "das Tanzen der Tische," and "die somnambulen Tische." Chains were formed to cause the rotation of other objects, such as hats and human bodies, but table phenomena retained their place of preeminence.

2. In 1853–1854 alone there were dozens of serious treatises on the subject, for example: *Avis aux chrétiens* 1853, Ballou 1853, Bautain 1853, Beecher 1853, Birt 1853, Braid 1853b, Carpenter 1853, Close 1853, Cohnfeld 1853, *La danse* 1853, Dibdin 1871 [1853], *Examen raisonné* 1853, Faraday 1853a, 1853b, Frisz 1853, Gillson 1853, 185?, Godfrey 1853a, 1853b, Guillard 1853, Harvey 1853, Kerner 1853, Koch 1853, Nicholls 1853, *Practical Instructions* 1853, Prichard 1853a, 1853b, Richemont 1853, Rogers 1853a, 1853b, Roubaud 1853, *Satanic Agency* 1853, Silas 1853, Snow 1853, Spicer 1853a, 1853b, *Table Moving by Animal Magnetism* 1853, *Table Moving, Its Causes* 1853, *Table qui danse* 1853, *Table Turning* 1853, Townshend 1853, Trismetiste 1853, Almignana 1854, Bellanger 1854, Chevreul 1854, Dods 1854, Gentil 1854a, 1854b, MacWalter 1854, Mattison 1854, Morgan 1854, Morin 1854, and Townshend 1854.

sages were alphabetically spelled out through raps occurring spontaneously somewhere in the material of the table itself, very much as they had in the home of Isaac Post. Henry Spicer provided an entertaining account of this kind of sitting in London, when a medium from Boston was invited to form a magnetic circle at the home of one of Spicer's friends. The participants sat around a large (inspected) table. In a few minutes raps began to sound in the table:

> Those now celebrated sounds—be their origin what it may—are certainly of a novel and most peculiar character. Nothing that bears the slightest affinity to them—as mere sounds—has ever yet visited *my* ears. It is stated, and I had afterwards opportunities of observing, that the rappings are not always of a precisely similar kind . . . ; but the prevailing rap is of one especial kind, and can be, perhaps, described in no better way than by requesting the reader to fancy a bird, say, a pheasant—of considerable power of bill, confined in a strong wooden box, and pecking vigorously to get out. The working of the needles of the electric telegraph will, perhaps, supply the next approximate sound. . . . [There were heard] five raps for the alphabet. This had been placed ready on the table, and, as the querist passed her pencil slowly along the line of letters, each, as it was needed, was indicated by a clear, distinct rap. (1853a, pp. 210–11, 213)

An anonymous author of 1859 described the table feats:

> I have repeatedly seen the table incline forward to an angle of 45 degrees or more; the candle-lamp, water-bottle, inkstand, pencils, &c., remaining on the table as if they were a part of it. At other times, I have seen the table rise *perpendicularly* from the floor, our hands all resting on the *top* of the table. . . . I have seen the table-cover drawn from under our fingers, and thrown upon the floor. Once, as the table was moving, one person only lightly resting his fingers on it, I jumped on the top, and, by this novel mode of locomotion, was carried around the room. I have seen a table, with the medium (a delicate female) lightly touching it with the tips of her fingers, rise off the floor, and answers telegraphed by its movements, notwithstanding the utmost efforts of two strong men to hold it down. (*Confessions* 1859, pp. 90–91)

The means of telegraphing the message in this and many similar cases was the tapping of a table leg on the floor. Typically, one leg of the table would rise off the floor and lower itself again to produce the tap. The message would then be deciphered by tapping through the alphabet and ending at the desired letter (a laborious process), by having one of the sitters recite the alphabet until a tap came, by running a pointer over the alphabet printed on a board until a rap was heard, or by some other means devised ad hoc. It was through the attempt to make alphabetical communication easier that the "psychograph" (eventually named the "planchette," and later still the "ouija board") was invented by the composer Richard Wagner. It consisted of a sliding plate upon which the participants would place their hands. The plate had a pointer which, under the guidance of the hands, moved over a paper with the alphabet and the numbers 0 to 9 printed on it. It was claimed that the messages thus spelled out were produced by the same force that spoke through the tables. And it was further claimed that the messages were produced equally effectively with the participants' eyes opened or closed (Cohnfeld 1853, pp. 115–25).

Explanations

Along with reports of table-moving phenomena came attempts to explain their cause. Five explanations of the phenomena predominated: they were the productions of departed spirits; they were the workings of the devil; they were simply due to fraud; they arose from delusion, hallucination, and self-deception; or they were produced by the force of a "fluid" such as electricity, the "odyle" of Reichenbach, or animal-magnetic fluid.

Spirits

The spiritualists held that the spirits of deceased human beings were anxious to communicate with living family and loved ones. Spiritualists believed that spirits of deceased relatives and friends of the sitters would be drawn to the circle by the opportunity to communicate through the "talking tables."

It seemed self-evident to spiritualists that since living human beings could not move tables without physical force or make rapping or tapping sounds without mechanical means, these phenomena had to be produced by spirits. And since the resulting messages often contained detailed information about the sitters that only the dead could know, these spirits must be who they said they were—discarnate loved ones.

To those who proposed that the sitters themselves might be producing the physical effects through some kind of force (such as electricity, magnetism, or mesmerism) emanating from the human body, the spiritualists offered what they thought was an irrefutable argument: the physical effects are not random but intelligible and meaningful. As the American spiritualist Herman Snow put it, they convey "the manifestations of *mind*. . . . Connected messages do come through these phenomena" (1853, p. 45).

The notion that the dead communicate with the living was attractive to some religious ministers and priests. The abbé Almignana, "doctor of canon law, theologian, magnetizer, and medium," had little difficulty incorporating the concept of discarnate communication into the Catholic doctrine of the communion of saints—the continued fraternity of the just departed with the living. In a book on the phenomena of somnambulism (1854) Almignana took pains to prove that they were not due to the intervention of evil spirits. He reasoned that if the tables were moved by the devil, prayer should dispel the effect. Describing his many experiences with moving tables, undertaken with "pious laity," he wrote: "Desiring to find out, in the interests of religion and our souls, whether the demon was the agent who brought movement and language to the tables, we used every means (exorcism excepted) that the Church has provided to chase away the demon, and yet accomplished nothing. For neither prayer, nor the names of God or Jesus, nor the sign of the cross made on the tables, nor the crucifix, the rosary, the gospels, or the Imitation of Jesus Christ placed on the tables, nor blessed water could stop the turning, the rapping, and the responses" (p. 11).

The abbé apparently succeeded in remaining in the fold of his church while carrying out his spiritualistic experiments. By contrast, C. H. Harvey, a Methodist minister of Pennsylvania, was removed from his post as professor in the Wyoming Seminar in Pennsylvania and expelled from the ministry after embracing the tenets of spiritualism. Harvey (1853) described what he believed to be a systematic persecution of spiritualist beliefs by ecclesiastics:

> These . . . already have . . . sounded forth their threats, as the Jews did against believers in the messiahship of Jesus. . . . One who has taken his pen against them, thinks all believers and countenancers of spiritualism should be excluded from the church; another (and both are ministers of the Gospel) that mediums, and those who countenance them should be confined in a penitentiary. . . . I have been called *crazy,* and *wicked,* and *stupid,* and *silly,* and a thousand and one other things equally creditable to those from whom they came, and containing argument of equal weight and validity—but none of these things move me. I am still a believer in Spiritual Manifestations. (pp. 3–4)

For Harvey, communications coming through the tables and other means were from "glorified spirits"; they were not from the devil, nor were they due to a "fluid" or some other purely natural agent. One of his chief arguments against such naturalistic positions was that they could with equal validity be applied to supernatural manifestations of all kinds—including those described in the bible (p. 107n).

The Devil

The kind of thinking Harvey had to contend with was probably best represented by the Anglican clergyman Nathaniel Steadman Godfrey, who firmly believed in the reality of table moving, but held it to be the machination of Satan. In a pamphlet based on his own experiences, Godfrey (1853a) "proved" that this was the case. He then undertook a further series of experiments to expose how and why the devil was using the talking tables in this fashion.

Gathering a group of pious Christians around the offending piece of furniture, Godfrey prepared to ask the table a series of questions that would put the issue beyond doubt. The group soon got the table to turn and then raise a leg off the floor so that it might tap out responses. Godfrey used the method of pointing to letters printed on a board and asking that the table tap a leg when it wanted the letter to which he was pointing at the moment. Asking questions as a theologian convinced that he was querying the devil, Godfrey received answers perfectly in keeping with his belief: the respondent was indeed an evil spirit; he was a seducing spirit spoken of by Saint Paul; he was the spirit of a dead person suffering the pains of hell; and he would go back to hell when he left the table. God compelled him to answer questions but it was Satan who forced him to come to the table; table turning is of the devil. Godfrey also discovered from the spirit that madmen are possessed by devils, that epilepsy is possession, and that hell is fire and brimstone (pp. 26–32).

R. C. Morgan went even further, warning that the world was now in the "last days," and that table moving was the devil's preparation for the coming of the Antichrist (Morgan 1854). This theme was taken up by Charles Cowan (1861), who used the occasion to strike a verbal blow at "Romanism," which he called "Satan's masterpiece." But many Roman Catholics agreed with Cowan's evaluation of table moving. For example, Eugène Panon des Bassayans, comte de Richemont (1853), agreed that the facts of table turning were genuine but were the works of Satan and should be shunned by all Catholics.[3]

The most effective Catholic spokesmen for this point of view were Jules Eudes, marquis de Mirville, and Henri Roger Gougenot des Mousseaux. Mirville (1853) believed that table manifestations were genuinely supernor-

---

3. This view was also embraced by the authors of *Examen raisonné des prodiges récents d'Europe et d'Amerique, notament des tables tournantes et répondantes. Par un philosophe* (1853) and *Avis aux chrétiens sur les tables tournantes et parlantes, par un ecclésiastique* (1853), and by the Comte de Résie (1854, 1857), who added animal magnetism to the list of diabolically produced phenomena.

mal and could not be explained by natural causes. He also held that the phenomena were produced by means of a "fluid," such as that described by the practitioners of animal magnetism. He insisted, however, that the beings who manipulated the fluid and moved the tables were evil spirits. These evil spirits were also at work in the phenomena associated with animal magnetism, such as clairvoyance and thought reading.

Gougenot des Mousseaux (1860) also believed that spirits, such as angels or the just souls of the departed, could make themselves known to the living, communicating with them in their minds. But in the case of the talking tables and in the clairvoyance generally manifested in animal magnetism, he held that the participants are being manipulated by evil spirits for wicked purposes. Manifestations such as these, in his view, are merely the modern equivalent of ancient magic and must be condemned as such.

A less radical rejection of the "spiritual manifestations" in the name of religion was given by Charles Beecher, pastor of the First Congregational Church in Newark, New Jersey. In *A Review of the "Spiritual Manifestations"* (1853) Beecher states the issue simply: "Omitting as outgrown the theory of collusion, two hypotheses remain:—I. PNEUMATIC; Natural Law with Spirits. II. APNEUMATIC; Natural Law without Spirits. . . . While, then, the pneumatic hypothesis accounts for all the facts adduced by the other theories, as well as they, it also accounts naturally for other facts by which they are embarrassed. It is, therefore, probably the true hypothesis" (pp. 9, 35).

Since in his view the origin of the phenomena is "pneumatic," in spirits, he must ask, "What kind of spirits?" His answer was that they are "powers unseen, powers aerial, under the masterly guidance of some one mind of fathomless ability, and fathomless guile" (p. 74). But if the spirits are evil, Beecher did not opt for a condemnation of spiritualism. He could not believe that Christianity was so fragile that it could be destroyed by this movement: "How shall the movement be met? Obviously with kindly courtesy. Whatever be the character of the powers communicating, there is no objection to hear all they have to say. If they can logically destroy the authority of the Word of God and the truth of evangelical doctrine, let them do it" (p. 74).

Fraud

While many skeptics claimed to find collusion or deceit in the widespread table-moving phenomena, a mass of phenomena seemed to escape explanation by this means. One of the problems faced by those who sought to assign the cause of all "spiritual manifestations" to deceit was that at this stage, most phenomena were not being produced by public or paid mediums but were occurring in the parlors of thousands of homes, through the mediumship of a family member or close friend. To attribute them to deliberate deception would imply an epidemic of moral depravity, with relatives and friends tricking each other for unfathomable motives.

There were, however, individuals who took the public stage and did not hesitate to defraud audiences or sitters for money or adulation. These tricksters were denounced by a number of authors who wrote on the table-moving phenomena. S. B. Emmons (1859), for example, cited many instances from history in which people were duped by clever tricksters into believing the impossible, and he counseled those who experiment with table-moving and related matters to be very cautious. Yet he acknowledged that in some instances surprising phenomena were produced through natural agents, such as "electrobiology" (pp. 118–65, 224–68).

Only one author in the 1850s asserted that fraud could account for *all* table moving and table rapping. In *Psychomancy: Spirit-Rappings and Table-Tippings Exposed* (1853), Charles Page wrote:

> When we are told that a table is moved by the mere effort of the will, that it moves about when it is not touched, we deny the statement *flatly* at once and challenge the reproduction of the miracle, and when we are told that spirits rap upon tables, floors, doors, walls, or any thing else, we deny the statement, and challenge the production of any kind of rap or sound in these cases, which is not clearly traceable to human agency. . . . The juggler with his legerdemain far outstrips any thing that has ever been accomplished by rappers and tippers, but then he tells you that he performs by sleight of hand, and that unless your eyes are quicker than his hands, you will be deceived. (pp. 29–30, 35)

Studying the magicians' handbooks of his day and performing his own experiments, Page concluded that hidden pressure and misdirection could account for all table-moving phenomena. He warned the reader that as time went on and those who were now successfully deceiving the public were exposed, new and more amazing feats would begin to be done, since jugglers are constantly perfecting their repertoire:

> Tricks *must improve,* in order to sustain their pecuniary value or bolster reputation. . . . Divest yourselves of all ideas of the supernatural, or any new fluid, or new law, or property whatever, and, regarding the performance either as a trick or case of illusion, scrutinize sharply every movement and circumstance in connection, and you will find that either the table does not move, or, if it does move, you will see what actuates it. Remember! there are controlling and controllable agents that *can*

raise a table from the floor; but the action of the will, or the mere superposition of hands, NEVER. (pp. 95–96)

Delusion

Although deliberate fraud as an explanation satisfied a few investigators of table moving, delusion seemed to be more palatable to many. It seemed perfectly possible that those involved in the table-moving experience were physically producing the effect without any conscious awareness of doing so.

The famed British physicist Michael Faraday, for example, was dismayed that there were so many inadequate explanations being proffered for the new fad of table turning. He was especially offended by those who believed that the phenomena were produced by a diabolical or supernatural agency, affirming that for the "natural philosopher" such an explanation was "too much connected with credulity or superstition to require any attention on his part." From the first, Faraday believed the cause to be "involuntary muscular action" and undertook experiments to confirm his belief. He published his findings first in a letter to the *Times* of London (1853a) and then in greater detail in the *Athenaeum* (1853b).

For his experiments, Faraday gathered together a group of "sitters" whom he considered to be honest and who enthusiastically believed they could move the tables. By placing glued disks under the hands of these sitters, Faraday discovered that they were applying a force in the direction that the table eventually moved. From this he concluded that the tables rotated because the powerful wish and expectation of the sitters caused them unwittingly to push the table in the direction they expected it to move.

While Faraday concentrated on the physical aspect of table turning, James Braid, the inventor of hypnotism, took up the psychological. Like Faraday, he rejected the notion that a "fluid" force was involved. Braid (1853b) attributed the movement of the tables to "the extraordinary influence of dominant ideas in the minds of some individuals, in producing muscular action in accordance with those ideas, without any conscious effort of volition on the part of said subjects" (p. 38). He claimed that even the notion that there was *no* conscious volition may be called into question, since "to *will* the table to move from some impulse of mysterious nature streaming from our finger-points, held in contact with the body to be moved, and to will the voluntary muscles, at the same time, to be still and inoperative, seems to be a contradiction in terms" (p. 39).

William Carpenter, a noted British physiologist, came to the same conclusion: "The continued concentration of the attention upon a certain idea gives it a dominant power, not only over the mind, but over the body; and the muscles

become the involuntary instruments whereby it is carried into operation. . . .
The movement is favoured by the state of muscular tension, which ensues
when the hands have been kept for some time in a fixed position" (1853,
p. 547).

Michel Eugène Chevreul arrived at the same explanation for table rotation.
He described experiments in which he discovered that the sitters were applying lateral muscular force to the table without their being aware of that fact.
He also assumed that a similar explanation could be given for instances in
which the table tips or rocks. He did not attempt to explain other phenomena
included in the "spiritual manifestations," such as raps at a distance or divination, stating that "other disciplines" should be consulted in those matters
(1854, pp. 170–71, 214–24).

Jacques Babinet, a French physicist who specialized in developing experimental apparatuses, added his voice to those who held the phenomena to be
self-delusion. Writing in 1856, Babinet stated that all movement of tables may
be attributed to two factors: unconscious motion and "naissant" movements.
Unconscious motion referred to the pressure exerted on a table by a sitter
without awareness of that pressure. Naissant movements designated those
muscular movements that, though small, are very powerful: "It is easy to
observe in the movements of quadrupeds, reptiles, and fish many examples of
these primary, tiny movements, so strong and so rapid, yet hardly noticed.
One can call these movements *naissant,* and say that in the organization of
animals all naissant movements are, in origin, very strong and very rapid.
. . . Naissant movements are barely noticeable, but irresistible" (Babinet
1856, pp. 241, 242).

Babinet also responded to claims of table movement without contact. His
explanation was simple: since thought cannot act on matter, table moving
without contact is impossible; therefore it must not occur (pp. 37–41). Knocking sounds made in tables were accomplished by something he called "acous-
tic ventriloquism" (p. 51).

Fluid

While the explanation of Faraday et al. satisfied some investigators, others
felt that most of the phenomena were still unexplained. C. H. Harvey's
sardonic comments concerning Faraday's findings expressed the dissatisfac-
tion of many:

> Of course we must lay the crowning discovery of the nineteenth century,
> recently made by the great experimental philosopher of London, out of
> the account. That discovery, the reading public have already learned,

consists in this: A company of gentlemen and ladies seating themselves around a table and resting their hands upon it, can, if they all push together, and hard enough, actually move that table; and if it is varnished sufficiently smooth, they can move *pasteboard* or *paper* upon its surface, by pressing steadily all together in the same oblique direction. For the sake of those religious editors, and Rev. Professors, and the learned D. Ds., who need this item of philosophical information to complete their education, one can but rejoice that this great discovery has been made and published to the world. But it sheds no light upon the cause which moves a table under the circumstances we now have before us, for no hand nor any other visible substance, except the carpet upon which it stands, is in contact with it. Professor Faraday's late discovery therefore fails us here. (1853, pp. 177–78)

Less cutting, but equally critical, were the comments of S. R. Maitland (1855): "When the Professor came to supply an explanation of his own he was less happy. It was a pretty little key, but it did not fit the lock. However ingenious his illustrative apparatus might be, the solution itself did not meet the facts of the case" (p. 26).

The kinds of "facts" Harvey and Maitland were talking about included accounts of tables rising completely off the floor when the only contact was the sitters' hands resting on top of the table and tables moving without any human contact at all. Accounts of such phenomena were so numerous and appeared to be so well attested that other explanations had to be found. Those who were dissatisfied with fraud, delusion, and supernatural agency as an explanation looked to some natural force to account for what was going on. They posited a "fluid" (that is, an "influence" or "imponderable force") existing in nature but largely unexplored by natural science, as the table-moving agent. Some believed it was electrical fluid, others magnetic or even thermal. But in the end the fluid of animal magnetism and the fluid of Reichenbach's "odyle" were the most popular choices.

The Fluid of the Spiritualists

The existence of a strong mesmeric tradition in the United States made it inevitable that some of the concepts and nomenclature of animal magnetism would enter into the early attempts of spiritualists to explain their own phenomena. Many of the early spiritualists had themselves experimented with animal magnetism and easily drew connections between the concept of a fluid that acted at a distance and the invisible work of the spirits. William T.

Coggshall, for example, became involved with animal magnetism in 1845. His book *The Signs of the Times* (1851) reveals that terminology from mesmeric practitioners had entered into spiritualism's vocabulary. The book shows that the peculiar mixture of concepts drawn from animal magnetism and electricity found in mesmeric literature of the day (such as the "electrical psychology" of John Bovee Dods) was also present in spiritualistic thinking. Coggshall wrote that the spiritualistic "circle" where people sat in a ring was termed a "magnetic circle" by spiritualist mediums. The circle was also described in electrical terms: "The clairvoyant spoke of the rappings, calling them electrical vibrations, and said communications might be had from the Spirit-world if a battery was formed. Inquiry was made how a battery could be formed. The reply was, 'By sitting around a table'" (p. 33). Spirits were often called "electric beings" and the rapping out of messages was described in terms of the newly invented telegraph, with the "magnetic fluid" of Morse replaced by the electrical impulses emanating from spirits.

A "New Fluid"

Outside of the United States, there were many who looked for the cause of the table phenomena in some sort of fluid and left the spirits out of it altogether. Félix Roubaud called the moving agent a "new fluid" that was neither electricity nor heat nor terrestrial magnetism. He was satisfied to refer to it simply as an "unknown agent" whose properties could and should be investigated (1853, pp. 29–30). The anonymous author of *Practical Instructions in Table-Moving* (1853b) was also content to identify the phenomena merely as "the new fluid" (pp. 10, 14).

John Prichard, relying on secondhand information, began by favoring Faraday's explanation of table moving but did an about-face when he carried out his own experiments and several times saw a table completely lift off the floor. Convinced that what he had witnessed could not be accounted for by unconscious muscular action, Prichard (1853b) attributed the movements to "the agency of electric currents instantaneously permeating the spaces of matter" and thereby counteracting the forces of gravity (p. 15; see also Gentil 1854b, Goupy 1853, 1860).

Some researchers attempted to tie the table movements directly to the nervous system of the medium or sitters. Gentil (1854a), for instance, spoke of a kind of "animated electricity" present in the physical organism that disengages from the body and penetrates the table, causing its motion (pp. 282–83). They believed the table-moving force was an extension of the "nervous fluid" that passes through the body, mediating sensation and producing muscular movement. Probably the most outspoken of these authors was John Bovee Dods (1854), who wrote of "two brains" that operate within humans using an "electro-nervous force." The "back brain" mediates the involuntary or instinctual powers of the mind, while the "front brain" mediates the voluntary powers. An excessive buildup of nervous force in the back brain unduly charges the involuntary powers and "the result is those singular manifestations that are so confidently attributed to the agency of spirits" (p. 29).

For Dods, table rapping was "occasioned by too great a redundancy of electricity congregated upon the involuntary nerves, through passivity of mind, and thus imparting to them extraordinary nervous force." The rapping sound resulted when this "electro-magnetic discharge from the fingers or toes of the medium" came into contact with a table or other solid object (pp. 28–29). Table tipping and moving were also due to an excess of nervous fluid in the back brain. In this case the result was either an unconscious exertion of muscular force on the table by the medium or the buildup of an electro-magnetic charge in the table and movement by attraction or repulsion without physical contact with the table (pp. 86–87).

Animal Magnetism

The fluid of animal magnetism was considered by many experimenters in table moving to be a sufficient explanation for the phenomena. Ferencz Grof Szapary, who developed his own school of magnetic practice, wrote of a primitive human force called "gyro-magnetism" that caused the table phenomena. Szapary was already accustomed to sitting around a table with a group (called an *agape*) for magnetic purposes (Szapary 1854a). Since the late 1830s he had gathered people to generate group healing effects. When the table-moving fad hit Europe, Szapary easily incorporated the phenomena into his practices. His concerns, however, were not so much with the action of gyro-magnetism to move furniture as with the mystery of how phenomena of all sorts are produced in groups (Szapary 1854c, pp. 312ff).

One of animal magnetism's greatest proponents of the time, the Reverend Chauncy Hare Townshend, also looked to explain the table phenomena. Rejecting the notion of spirit communication as "contrary to the general Law of beauty and order" (1854, p. 208), Townshend stated that all communications come from man's own brain. The rappings themselves were mesmeric: "Whatever in it is contrary to sound reason, results either from fraud or error: while, whatever puzzling *residuum* may be left, is entirely accounted for by the Mesmeric Theory of an agent which sometimes produces material effects through cerebral motion, and, in rare cases, causes such increased mental perception as reaches to a knowledge of distant events or of the thoughts of other persons" (p. 209).

From 1853 to 1855 a number of articles and brief comments on table moving appeared in *The Zoist*. Experiments carried out by the various writers seemed to have convinced them that the table-moving phenomena were genuine. The explanation that spirits were involved was unanimously rejected; mesmeric fluid was asserted to be the only reasonable explanation for the movements and messages. In a letter to the editor of *The Zoist*, John Elliotson, published in the July 1853 issue, Townshend described his first experiments with table moving and his contention that it was due to mesmeric force. In turn, Elliotson wrote of his own experiences with the phenomena and, while admitting that they may be genuine, strongly condemned those who believed that they were due to the action of spirits. George Sandy, another well-known apologist for animal magnetism, affirmed in the same issue that "this action of the tables, induced by continued contact with a chain of human fingers, is nothing but simple mesmerism, developing itself in an unexpected phase" (p. 179). In a detailed examination of the phenomenon and its prehistory in American spiritualism, J. W. Jackson (1854) sought to explain in nonspiritist terms the "extraordinary mental capacity" for intellectual achievement that otherwise unexceptional men and women showed in trance. This "exaltation in the mental functions" he attributed directly to animal-magnetic fluid. He further asserted that the fluid of animal magnetism can cause physical objects to move and become the means of expressing this kind of unusual awareness, thus explaining the messages of table moving (pp. 1, 8–9).

Odyle

Edward Coit Rogers also accepted the fluid of animal magnetism as the cause of the table phenomena, but he believed that more questions about the action of the agent needed to be addressed. Examining recent experiments on the effects of electricity and magnetism on the muscles and nerves of the body, Rogers asked whether either of these agents could explain the physical phenomena of the spiritual manifestations. In a work published in parts and completed in 1853 Rogers concluded that, while some phenomena were attributable to electricity, for many a "new agent" had to be posited. This new agent had to emanate from the human body, be affected by the human brain, and be capable of moving physical objects at a distance. For this Rogers turned to Baron Karl von Reichenbach and his odyle force.

Von Reichenbach (1788–1869) was a highly respected scientist, metallurgist, and chemist (he was the discoverer of kerosene and famed for his study of meteorites). He spent twenty years experimenting with a force that he called "od" (variously translated into English as "od," "odyle," and "odic force"), a term derived from the name of the Norse deity Odin. He defined it as a power that permeates the whole of nature and emanates from every substance in the universe. It could be perceived only by people whose unique sensitivity caused odic emanations to register as vague feelings of heat or cold, or as a light of varying colors that could be seen radiating from odic sources. It could also be seen as a cloud collecting around the recently deceased—the visions of "ghosts" in graveyards. Reichenbach saw od as a general physical force, of which animal magnetism was simply a special case.

Believing that at least one-third of human beings could perceive the emanations to some degree, Reichenbach launched a series of experiments. He painstakingly wrote down his subjects' descriptions of what they perceived and noted the conditions in which each experiment took place. He concluded that od could be transferred from one body to another by touch, but that even proximity caused a buildup of odic charge. The principal sources of emission from the body were the mouth, the hands, the forehead, and the occiput.[4]

Rogers (1853a) believed that the odyle of Reichenbach was "identical with the animal magnetism of Mesmer" (p. 159). In the process of examining the facts that had to be explained, he spent some time on two contemporary cases that had puzzled many investigators: Angélique Cottin, the "electric girl," and the seeress of Prevorst.

A native of La Perrière, France, Angélique at age fourteen, in the year 1846, became the center of happenings that puzzled and frightened her family. Furniture in her presence began to shake and jerk in ways that could not be halted by muscular observers. It was assumed that the girl was bewitched or possessed, and she was brought to the parish priest. He, however, was inclined to think that there was some natural explanation and sent her to a physician. The phenomena increased—some objects would be propelled away from the girl with great force, others would be attracted, while at the same time some people around the girl experienced electric shocks. The physician had no explanation and decided to bring Angélique to Paris to be examined by the astronomer and physicist François Arago (1786–1853), who

4. Reichenbach produced a voluminous literature on his researches. His most popular work was the two-volume *Physikalish-physiologische Untersuchung über die Dynamide des Magnetismus, der Electrizität, der Wärme, Des Lichtes, der Krystallisation, des Chemismus in ihren Beziehungen zur Lebenskraft* (1849), which was immediately brought out in two English versions. Reichenbach's sixteen-hundred-page *Der sensitive Mench und sein Verhalten zum Ode* (1854) gave incredibly detailed results of a large number of experiments constructed to determine the nature and range of sensibility to od.

was known worldwide for his work with electromagnetism. Arago wrote a report about Angélique for the Paris Academy of Science stating that the left side of the girl's body sometimes exhibited attractive force, but most frequently repelled objects; any light body placed on a table would be forced away from her without contact, simply by the approach of her left hand; a table overturned when touched by her hand or by a thread held by her; the force of repulsion was strongest from her pelvic area; a chest held down by three men and a chair held by two men were thrown violently when Angélique attempted to sit on them.[5]

Rogers noted that all of the Cottin phenomena were random and reactive, with no sense of the presence of a mind behind them. He agreed with Arago that some hitherto unknown force was at work in the "electric girl" and speculated that this was related to the "nervous force" that operates normally in the body's nerve centers. But in the Cottin case the nervous system was in an abnormal state and mobilized the unknown but nerve-related force to produce the effects (pp. 52–62). Rogers believed that the force emanating from Angélique's nervous system operated without her conscious cooperation or awareness: "The Will and the Reason have no control of this force in its action from the nerve centres in their abnormal condition, and it acts from the person without cognizance of the consciousness" (p. 61).

In Friederike Hauffe, the seeress of Prevorst so closely observed by Justinus Kerner, Rogers saw what he believed were examples of the same blind nerve-based force operating to affect physical objects. The seeress could seemingly produce rappings and other sounds at will, even at some distance from her body. She was also able to move objects without contact, sometimes raising light objects to the ceiling (Rogers 1853a, pp. 78ff).

Rogers believed that the force identified and described by Reichenbach could fully account for the movements of objects without the use of physical force. But he held that to account for random sounds and motions was only half the job. More difficult was accounting for the phenomena indicating that some kind of intelligence was at work, particularly when messages were rapped or tapped out at the table. It was Rogers's conviction that here too Reichenbach's odyle provided the solution.[6]

5. For more information about Angélique Cottin, see Tanchou 1846.

6. Rogers (1853a) was not deterred by the fact that Reichenbach's "sensitives" did not manifest the phenomena Rogers hoped to explain: "comparatively few persons observe nature sufficiently to see that the same agent acting on the same substance, will vary its phenomena, just as the circumstances are varied, under which the one will act'upon the other. So it will be concluded by such persons, that the physical agent engaged in the so called 'spiritual manifestations' *cannot* be the Odyle of Reichenbach, because 'mediums' do not exhibit the same

Rogers believed that odyle force could build up in the body and then be expended in a sudden discharge. That discharge could be controlled by the sympathetic nerve system, as in the case of Angélique Cottin or Friederike Hauffe, or it could be controlled by higher psychological centers, as in the case of mediums and table sitters (pp. 137–38). For the latter, the resulting phenomena may produce physical effects that convey intelligent content, such as the messages produced at the tables (pp. 163ff).

Odyle force was also the explanation chosen by Asa Mahan, first president of Oberlin College. Mahan, a devoted Christian, wanted to refute the claim of spirit action in the phenomena and discredit spiritualism as a religion. Like Rogers, he looked for a natural explanation for what was going on. In the process, Mahan drew up a set of principles to govern theorizing about the phenomena that could well have been used as guidelines for the conduct of psychical research (1855).

De Gasparin

One of the most thorough experimenters with table turning was Count Agénor de Gasparin, a distinguished spokesman for French Protestantism and an elected member of the Chamber of Deputies. From his own experiences, Gasparin was not impressed by those who attributed table motion to delusion: "The tables turn in spite of M. Faraday; their death has been predicted, it has been demonstrated; but they continue to turn. They turn so well that their antagonists are beginning to revise their opinions, and the latest works written in opposition to them give evidences of a caution, a circumspection, I might almost say a respect, to which they are certainly unaccustomed" (1857, 1:xxi).

Gasparin believed that many instances of table moving involved errors in observation and testimony. In other instances he accepted the possibility of fraud. But he insisted that in many cases phenomena remained that had to be accounted for in another way, by positing the existence of a fluid or force or physical agent that is not supernatural and that "resides in the persons, not in the table" (1:93–94).

In giving his explanation, Gasparin attempted to define the role of consciousness in the production of the movements. He insisted that the "will" was involved, but without the awareness of the operator:

The table does not merely turn, it raises its feet, it strikes numbers, indicated by our thought, in one word, it obeys the will, and obeys so

_____

phenomena in all respects, that Reichenbach's patients did; neither did the latter receive the 'rappings,' or 'move tables'" (p. 21).

well, that the suppression of contact does not suppress its obedience. Lateral impulsion or attraction, which accounts for the rotations, cannot account for the elevations! Why not? Because the will directs the fluid, now upon this foot, now upon that. Because the table identifies itself in some sort with us, becomes one of our members, and executes the motions conceived in our minds, in the same way that our arm does. Because we have no consciousness of the direction communicated to the fluid, and govern the table even without representing to ourselves that a fluid or any force whatever is in play. (1:98)

In this passage Gasparin stated the problem that arose again and again for those who attempted to explain table moving as other than fraud or delusion: how does one account for the intelligent nature of some of the phenomena, with their answers to questions and messages of advice? If it is accepted that discarnate spirits are the source, there is no problem. But if that explanation is rejected, and fraud is not involved, how is it possible that intelligent communications could take place without the conscious intention or awareness of the operators? In attempting to answer this question, the investigators of table moving entered into some of the earliest speculations on the possibility of unconscious mental activity.

Unconscious Mental Activity

If, as many believed, "intelligent" table phenomena were being produced without participation of spirits or the conscious minds of the medium or sitters, from what did their intelligibility derive? The only possibility was that the medium or sitters were the *unwitting* source of the intelligent action. How this was possible became a subject of intense speculation.

Two explanations were offered. One was that the tables produced movements with intelligible content through the participation of the musculature of the medium or sitters, without their conscious awareness that their brains, functioning purely automatically, produced the muscular movement. On this hypothesis their brains were putting into action already existing mental states that corresponded to the spoken or unspoken expectations of the situation. No original thinking was involved. The other explanation was that the intelligent table movement resulted from the action of a fluid or force emanating from the nervous systems of the medium or sitters. It was directed by mental activity concurrent with but unknown to the conscious thought of those in whom it occurred. Here the content of the messages did involve original thinking.

Unconsciously Produced Muscular Activity

Although Faraday suggested that the expectations of the sitters produced unconscious muscular action that caused the table movements, he did not attempt to investigate the psychology involved in such an explanation. That was left principally to two British colleagues: James Braid and William Carpenter.

Writing in 1853, Braid developed a theory of suggestion and mental concentration that gave psychological flesh to Faraday's explanatory skeleton. He explained hypnotism as a state of mental concentration in which the mind of the subject is engrossed in a single idea or train of thought. In that state words or other impressions act as suggestions to the hypnotized person and these suggestions tend to be immediately translated into action (1853a, pp. 3–4).

In this way Braid formulated a notion of a "fixed dominant expectant idea" (p. 12), which he believed could explain many phenomena that had been attributed to animal magnetism or spiritualism. In this early elaboration of a psychology of fixed ideas,[7] he explained the rotation of tables as the result of unconscious effects of a suggested idea that becomes dominant in the persons involved and causes them to produce the physical movement without their being aware of it. The expectation is that the table will move. Participants and witnesses of the phenomenon are so engrossed in anticipation of the movement that they overlook its real cause (p. 38).

Braid had a similar explanation for the tapping out of messages by table legs:

> Considerable pressure on one side of a table, of suitable construction, would no doubt cause the table to incline in that direction, and thus to do its manners, or to elevate one leg from the floor and bump upon it alternately, according to the ideas in the minds of the experimentalists. . . . The influence of expectant ideas in the mind of persons pointing to the letters of the alphabet, in varying the action when they arrive at the letter which they allege may be the one to be indicated, and the change on the expression of the features of unwary subjects when doing this, and in the tones of their voice when calling letters, may all very readily enable the "medium" to make happy guesses. (p. 44)

7. It has been suggested that Janet derived his notion of "idée fixe" from Liébeault and Charcot (Ellenberger 1970, pp. 148–49). Braid's clear elaboration of his view of "fixed dominant ideas" suggests that Liébeault's formulation was not entirely original, for it is known that Liébeault was well acquainted with Braid's writings.

Braid acknowledged that he was not the first to recognize the influence of unconscious dominant ideas on muscular movements. He specifically cited the writings and lectures of William Carpenter and the term he coined to denote such influence: "ideo-motor." Nevertheless Braid emphasized, in an anonymously published letter in the *Manchester Examiner and Times* of April 30, 1853, that he was himself the first to relate this notion to table moving. Carpenter, however, was not so easily removed. Writing in the October 1853 *Quarterly Review,* he issued a critique of the whole of mesmeric phenomena and concluded with a rejection of the main phenomena of table moving. At the beginning of the article he laid down "well-known" principles as the foundation for such concepts as "dominant ideas": "The mind is liable to be seized by some strange notion which takes entire possession of it, and all the actions of the individual thus 'possessed' are the results of its operation" (p. 509). For Carpenter, the effects of such a dominant idea are far-reaching, for this "possessed" state of the mind is quite easily produced by suggestion. And since in his view "a large part of our ordinary course of thought, and consequently of action, is determined by direct suggestions" (p. 506), we should look to the theory of dominant ideas to explain many otherwise baffling experiences, such as the phenomena of mesmerism and the delusions of table moving. Mesmeric phenomena, according to Carpenter, may be reduced to the action of the power of suggestion on a mind that has, for the moment, lost volitional control as a result of fatigue produced by prolonged fixation of the eyes. In this view Carpenter relied heavily on the experiments of Braid, whose explanation he completely accepted (p. 503).

Carpenter held that the "spiritual manifestations" of American spiritualism in general, and the phenomena of table moving in particular, were due to the same kind of vulnerability to dominant ideas. While possessed by a dominant idea, the individual has no ability to doubt that idea or test its reality on the basis of previous experience. For that reason, when fascinated with the supposed marvels of table moving, the participant becomes unwittingly dominated by those ideas. In that state "the continued concentration of the attention upon a certain idea gives it a dominant power, not only over the mind, but over the body; and the muscles become the involuntary instruments whereby it is carried into operation" (Carpenter 1853, p. 547). Participants, with their hands resting on the table, mistakenly think the table moves on its own and even pulls them around after it: "Although the performers may most conscientiously believe that the attraction of the table carries them along with it, instead of an impulse which originates in themselves propelling the table, yet we never met with one who could not readily withdraw his hand if he really *willed* to do so. But it is the characteristic of the state of 'expectant attention,' to which the actors give themselves up in all such performances, that the power of volition is entirely subordinated to that of the 'dominant idea'" (pp. 548–49).

In the following years Carpenter elaborated his concept of involuntary action produced by dominant ideas into the notion of "unconscious cerebration." The word "cerebration" referred not to thinking but to the reflex action of the brain. Unconscious cerebration was intended as an explanation for automatic actions that seem to be intelligent but do not come into the agent's awareness. Despite the controversy surrounding the use of the terms (see, for example, Laycock 1876), Carpenter's formulation received broad acceptance among those who rejected the spiritualist interpretation of mediumistic phenomena.

Concurrent Unconscious Mental Activity

A number of writers described intelligible table messages as the manifestation of unconscious knowledge. Joseph Gentil (1854b), for instance, wrote that the "intelligent principle" of the human mind was operative in producing these communications, but in such a way that its operation did not come into conscious awareness. How is such a dichotomy possible? How must human psychological structures function to make sense of this explanation? Three writers attempted to answer these questions, and in the process they produced important early speculations about unconscious mental activity.

Samson

The first was George Whitefield Samson, president of Columbian College (now George Washington University) in Washington, D.C. Writing in 1852 under the pseudonym Traverse Oldfield, Samson attempted to account for the manifestations of spiritualism by positing a "nervous principle" intermediate between spirit and matter. This principle, not electricity but similar to it, was the means by which persons achieved thought transference and was the intermediary between mind and matter that made it possible to produce movements or sounds at a distance. What the ancients attributed to the action of a demon, wrote Samson, we can attribute to this spiritual medium, which can fittingly be termed "daimonion."

Given the reality of his nervous principle, the explanations for table rapping, automatic writing, and automatic speaking were identical. The raps were merely the thoughts of the sitters reflected back to them: "Mysterious *rappings* give response to our thoughts, uttered or merely conceived, as we sit around the table. . . . These responsive rappings are the working of our own nervous organism, echoing to our own thoughts" (pp. 38–39). The thoughts echoed in the rapped messages could be those of the medium, the sitters, or a combination of the two. Their minds utilized the nervous principle to produce the physical effect of rappings and spell out messages consonant with their thoughts. This process was unconscious, carried out without the knowledge of the principals.

That the messages merely reflected the thoughts of those involved was borne out, Samson believed, in the tone and content of the communications. One could see the temperament or moral philosophy of the medium or sitters clearly reflected in what was conveyed. Also, Samson insisted, no information was ever received in these séances that did not exist in the minds of someone present.

Samson held that the same explanation applied to automatic writing and automatic speaking produced by mediums. All information could be traced to the minds of either medium or inquirer. If the information existed in the mind of the inquirer, it could easily be communicated to the medium by a kind of thought transference accomplished by means of the nervous principle. These thoughts would then be converted into an intelligible written or spoken message without the medium's awareness. Addressing himself to a friend, Samson asked him to experiment and find out for himself:

Bring a man to your table, a part of whose name you know; and when that part is written, ask the spirit whom you may imagine guides your pen to write the other part. Most assuredly you will find that only your *own knowledge* will be responded to. Prepare, then, to watch more closely your own mind's working, go on and observe the other responses you receive. You may not at first be able to trace all you write to your own positive knowledge, your once known and forgotten, but in that moment of intense mental action remembered thought and realized imaginings. . . . No man, under such circumstances, utters or writes anything but his own thought. . . . Once again a reporting medium mentions facts and thoughts, or imaginings, which are not in her own mind, but in that of the inquirer. She receives by the rappings, or she writes with the pen, or she utters with the voice, not her knowledge, or surmise, or impression, but that which belongs to the mind of him put in communication with her. . . . Cases, indeed, are reported, in which inquirers have been informed, by the medium, of circumstances in the lives of relatives of theirs, and of other facts of which they suppose themselves never before to have had knowledge. . . . [But] who can say certainly that any fact which is thus reported from the medium never was known to [the inquirer]? . . . Who knows what facts, casually mentioned in his hearing in childhood, entirely uncomprehended and not noted in memory, are yet fast adhering in his mental organism; and who can say, positively, that the mysterious communications of the spiritual "medium" are not those deep-hidden impressions brought out under a strong nervous excitement? (pp. 41–43)

Samson here theorizes about a phenomenon later to be termed "cryptomnesia" by psychical researchers (Thalbourne 1982, p. 18). This involves the memory of some event or experience that has been forgotten by the conscious mind but later appears in awareness without the person's recognizing it as memory. This notion that knowledge can be active in the mind but not conscious is key to Samson's explanation for the origin of intelligible table phenomena. It is also significant in the history of psychological thought, for it supplies a basis for speculation about unconscious mental activity taking place concurrently with conscious thinking.

Rogers

In his book "To Daimonion" Samson referred with enthusiasm to a book that had "just come from the press:" *Philosophy of Mysterious Agents, Human and Mundane* (1852), by Edward Coit Rogers.[8] Rogers's book, attempting to expose all the psychological dimensions involved in table moving, proved to be one of the most thorough contemporary studies of the phenomena. He proposed a fluidic explanation for the "spiritual manifestations," among which he included the movement of objects without touch, the production of sounds without physical means, and the communication of intelligible messages through table moving.

Rogers made an emphatic distinction between phenomena produced with characteristics of intelligence (such as message rapping) and those produced without such characteristics (such as random movement of furniture). The latter were, according to him, the result of a force emanating from the human organism without any participation of the psychological or spiritual aspect of the agent. As an example he cited the phenomena of Angélique Cottin: "We say that the difference between this case and the 'mediums' of the present day, in whose presence tables are made to move, sometimes without contact, is,

---

8. This was to be the first of five parts, but in the following year Rogers completed the work with a single publication under the same title (1853a). Later that same year Rogers defended his position against critics in a pamphlet titled *A Discussion of the Automatic Powers of the Brain* (1853b).

that the force in the case of Angélique discharged itself by causes acting below the psychological centres; whereas the discharge of the force from the organism of the 'mediums' is more at the command of the brain-centres" (Rogers 1853a, p. 138).

To Rogers's way of thinking, there could be a participation of brain centers in an activity without conscious awareness on the part of the subject. That was possible because, for him, mind and brain were two different things, and only mind was essentially linked to consciousness. For that reason, it was possible to have mediumistic action in which intelligent communications took place without the medium's being aware of their content or meaning. In this case brain centers were active, but the mind was not involved:

> To *know* a word is to be *conscious of it.* Should the *mind,* therefore, express the word, it would do it *knowingly, consciously.* It follows, therefore, that if a word is expressed without knowledge, without consciousness, it did *not* come *from the mind.* That is, it did not come from the intelligent, thinking, self-conscious, self-determining, responsible agent. Either, therefore, there must be two totally distinct minds to one person, one self-conscious, &c., and the other unconscious,—which is a solicism,—or a mind which is always necessarily conscious of its *own acts,* and therefore, *responsible for them,* and *another* part which is *not* the *mind,* and is *not necessarily conscious of its own acts,* and is *not, therefore, a responsible agent.* (p. 168)

By the "other part which is not the mind" Rogers meant the brain. In his framework, the brain is not conscious of its own acts and may operate in an automatic fashion. Because the brain is the instrument of the mind, it stores mental information and may on its own produce intelligible communications. It would be a mistake, however, to assume that consciousness is involved in these communications and attribute them to the mind.

Rogers believed that information could be taken in from the environment and stored in the brain without the participation of the mind. This information could then be given external expression by the brain, again without participation of mind. For example, "unconscious impressions received in *childhood may, under peculiar conditions of the brain, be thus automatically displayed*" (p. 169).

This split between mind and brain could manifest itself in a yet more striking way: "The brain, under peculiar conditions, will act not only independent of the mind's control, but directly opposite to the decisions of the latter" (p. 223). This state of affairs makes it possible for a person to seem to be divided against himself: "'A bias, a prejudice, a predilection' of the brain may or may not become known to the mind of the person. Most persons will deny their existence in themselves. Even the honest do this; and it is because *their mind does not take cognizance of unconscious impressions*" (p. 173).

For Rogers, the split between mind and brain was a problem. Ideally, the mind imposes its discipline on the brain, and the individual lives a responsible, rational life. What is communicated to others has the full participation of consciousness. But maintaining this discipline is not easy, and some people, such as mediums, allow free play to the brain:

> It is only by the severest mental or spiritual discipline that a person becomes a *master* of his brain. And even with this, he never can be able to *prevent* the impressions which some objects will make upon the delicate brain-centres. It is, indeed, the property of the brain to receive impressions; but it is the prerogative of the *self-conscious, self-determining, disciplined mind, to reject or to receive their influence*. And this is the reason why a highly disciplined *mind* prevents a person from *becoming a medium*. Because an undisciplined mind has not a control of the brain, it cannot prevent the influence of others in making impressions upon it; and, when made, it cannot prevent their reflex action or reflection back upon the outward world (p. 173).

This "reflection back upon the outward world" included not only manifestations through the bodily action of the medium, such as automatic writing and automatic speaking, but also intelligent, physical, fluidically produced manifestations at a distance, such as table moving.

According to Rogers, then, a medium was a person who could suspend the discipline of the mind to allow the action of the brain to come through. Mental passivity and physical responsiveness were the qualities most desirable for successful mediumship (pp. 173–176). The medium was a kind of automaton easily played upon by others. In the mediumistic state, will and reason are more or less dormant, and the table-moving medium acts from the "involuntary imagination" and unconsciously directs the odyle fluid in a manner that gives expression to the ideas from the brains of others (p. 125).

Unlike mesmeric practitioners, who believed that abnormally *superior* mental qualities sometimes manifest in the trance state, Rogers equated brain action with lower, animalistic functions. He divided psychological functioning into two classes of phenomena: those that are determined by the person as a self-conscious, self-reasoning, self-governing agent and those that the person does not determine and are beyond his control or even contrary to his conscious intentions. Rogers counseled the reader to "ask your *self* wherein *you* are different from a mere animal. To see the grand difference, notice that the psychological nature of the animal is *controlled by outward objects acting upon internal senses* and propensities; that it has no self-judging, self-deciding, self-governing, self-conscious personal identity" (p. 29). It is the very fact that a human being experiences both classes of phenomena that makes him human; "the former makes him a governor of himself, the latter makes him an automaton—the tool of any sensuous influence that may preponderate at the time" (p. 29). Because humans are also animals, they are subject to automatic action. When their self-conscious personal identity is suspended, they can be made to assume *any* sense of identity, "from the supreme Divinity to that of a toad" (p. 29). Suspension of personality identity occurs, says Rogers, in insanity, sleep, and the mesmeric trance.

Since the brain, according to Rogers, is associated with the animal functions, it plays its part automatically, without benefit of rational processes. To support this view, he cited instances of mediumistic and automatic action that, while having the appearance of intelligent action, merely reflect previous mental thinking incorporated into the brain's memory and produced reflexively. He assigned quite remarkable abilities to the action of the brain centers, such as the power, exercised through the agency of the odyle fluid, to move physical objects and produce sounds. Another is the brain's ability to "represent the *identity of another person*"—that is, pretend to be a discarnate spirit (p. 182).

Rogers had to attribute this power to the brain in order to account for the predominantly spiritistic nature of the communications he was studying. Since he rejected the hypothesis that spirits produced such communications, he had to find in the brain a capacity to impersonate intelligent entities: "There is another part of the medium, that, under peculiar conditions may be made to act *like a secondary personality,* but has none invariably of its own. . . . The personality . . . must act independent of the *medium's real personality*; namely, independent of his consciousness, reason, and will" (p. 184). Rogers identified this phenomenon of impersonation by the brain with that demonstrated in mesmeric trance. He cited William Gregory's *Letters on Animal Magnetism* (1851) as a source for a description of the same experience: "He often loses, in the magnetic sleep, his sense of identity, so that he cannot tell his own name, or gives himself another, frequently that of the operator" (p. 85).[9]

9. Rogers did not deal with the fact that according to Gregory's description of double consciousness in magnetic sleep (as well as those of Puységur, Deleuze, and others), the trance consciousness is often shown to be more elevated and intelligent than that of the waking consciousness. This would not fit in well with Rogers's notion of the brain as source of nonreflective, animalistic activity.

When the mind loses its hold over the brain, various influences from the internal or external environment can intrude and give personal form to the unreflective productions of the brain centers. Among the external influences often at work in table-moving sessions, "the influence of surrounding persons upon the character of some of the responses has been too palpable and evident to be denied" (Rogers 1853a, p. 189).[10] The mind, wrote Rogers, is the personal self, the *I*. It is the consciously thinking, consciously willing, consciously loving, consciously responsible person. The brain, on the other hand, is the material substance that has the capacity to represent faithfully the characteristics of the mind. According to Rogers, it may not be easy to distinguish between the person who is functioning automatically or from the brain alone (such as in automatic writing or automatic speaking) and the person who is functioning from the brain under the controlling action of the mind. There may be no obvious difference in the content of what is produced: the only distinguishing feature may be "*consciousness of personal action,—a perfect sense of selfhood in it*. Thus, 'I think—I am *conscious* that I think; I write,—I *know* that I write;'—that is, the mind has the guidance and control, and is always conscious of her own acts" (p. 209). Rogers is able in this way to reduce all non-normal psychological functioning to this separation of brain from mind: "The psychological phenomena of mesmerism, pathetism, spontaneous somnambulism, clairvoyance, insanity, spiritual manifestations, &c. &c., *are not the phenomena of mind,* but of the *brain without* the *mind*" (p. 211).

The suspension of the controlling action of the mind over the brain, Rogers asserted, may sometimes occur without the loss of the mind's consciousness. This is most clearly illustrated in automatic writing. Here the brain is thrown into unconscious action and controls the muscles of the writing arm. At the

---

10. It is interesting that there were those who theorized that the group participating in the experiment formed a kind of unified mental organism that united and sifted the thoughts of the group members. According to this view, the medium simply channeled the "mind" of the group entity. In attempting to explain how one person in the table-moving group becomes the interpreter of the thoughts of the others, Alcide Morin (1854) posited the existence of a "new being created from their individualities" (p. 57). This "being of reason" (p. 60) makes it possible for the many to have one common intention and apply the force that moves the table. The group thus can respond to questions from the "instincts of the many" (p. 74) and, since instinct is more reliable than reason, produce answers that are quite accurate. According to Morin, however, it can also happen that the prejudices and fantasies of a group can produce group illusions and hallucinations that have no basis in reality (pp. 93–94). Another writer, Jobard, also believed that a communal entity was at work in table moving. In his view, the intelligent messages emanated from a collective being formed by the participants that had an individuality distinct from theirs and existed only for the duration of the group experience (see Morin, p. 382).

same time the writer may be perfectly conscious and engaged in intellectual activity with no knowledge of what the arm is writing (p. 226).

Rogers spoke of the action of the brain without the mind as "automatic" and as a "suspension of spiritual action," so that without the controlling influence of the mind it operates according to whatever impressions most strongly affect it at the moment—"a glass of wine, or a cup of tea, or the irritation of a disease . . ." In a complete suspension of the mind, the person is not responsible: "When the *mind* produces the action of the cerebral machinery, and so of other parts, it is *conscious* of it; but if the other causes have set the machinery in motion, the mind is not conscious of the act and is not responsible. Thus, for instance, the machinery may so operate as to take the life of another. If this is done without the design or government of the mind, as in a state of somnambulism, dreaming, insanity, religious ecstasy, who is responsible? who is the criminal?" (pp. 215–16).

Rogers's attempt to explain intelligent communications occurring without conscious awareness led him to an intricate view of the mechanics of human mental processes. Human psychological functioning, in his account, arises from two separate sources: the mind and the brain, the mind involving conscious mental action and the brain involving reflexive actions without self-awareness. The proper state of human beings is for the mind to exert a controlling action over the brain. In abnormal conditions, however, the brain can function without the controlling action of the mind—operating automatically and without self-awareness (such as in mesmeric trance or mediumship). The brain, being in sympathy with all external nature, also functions as a receptor for mental impressions from persons surrounding it. In certain circumstances, it can direct the action of odyle fluid and produce physical movements at a distance, and convert internal states or impressions received from the outside into reflexive action manifested in fluidically caused physical movements with intelligible content, such as table movements and raps (pp. 202–8, 286–87).

Rogers's insistence on the automatic nature of the productions of the brain seems straightforward and unambiguous. But when he insists that the brain is analogous to the mind and that one can be distinguished from the other only through the added quality of self-awareness, his position becomes much more complex. Then the brain seems every bit as capable as the mind of functioning intelligently and purposefully. In this framework, the brain is like a second intelligent agent, capable of operating alongside the mind and of having ideas and values that may be at variance with those of the mind. This means that human beings are divided, with a self-conscious mind and a brain with all the capabilities of the mind except self-awareness. The brain operates largely outside the awareness of the conscious mind, every once in a while putting forth thoughts or impulses that intrude into the mind's consciousness and which the conscious mind may or may not be able to control.

From this point of view, Rogers's concept of the brain hints at a subconsciously functioning psychological stratum without quite arriving there. Two years later, another writer took that final step.

Lettre de Gros Jean

Anonymously published in 1855, *Seconde lettre de gros Jean à son évêque au sujet des tables parlantes, des possessions et autres diableries* broke new psychological ground with its speculations about the inner structure of the psyche.[11] It provided a framework for understanding table moving, possession, automatic writing, and other automatic phenomena and presented a novel picture of unconscious mental activity.

According to the author, in human beings ideas and knowledge are given a unity by personality. There is an "I" that thinks and knows, and from this "I" the individual makes judgments and takes action. As a person thinks new thoughts, they are attached to the personality and said to belong to the "I." This sense of personal oneness is provided by the will. Thoughts, however, can separate into distinct streams, each with a life of its own. We see this, for instance, in the distinct lives of sleep and wakefulness. In certain conditions—when the will weakens—a person can find himself separated into distinct parts, each with its independent stream of thought and its separate identity.

This weakening of the will makes it possible for mediums, who regularly experience it, to have more than one personality, and that is precisely what happens when the table-moving medium is not conscious of the response produced through her:

> The young woman has no internal knowledge of the response which is formed in her intelligence outside of her "I." She only knows the response through the movement of the table. Intellectual division is complete. At the same time the dissident thought expands its domain. It no longer simply deals with questions addressed to the table; on the contrary it now asks questions of the persons present, one after another, about such and such a subject, placing itself in such and such an order of

11. The author may be Paul Tascher. This work is so scarce that there does not seem to be any public facility that possesses it. The quotations and references used here are taken from Janet 1889, pp. 397–401. There is an equally scarce first letter, *Grosjean à son évêque au sujet des tables parlantes* (1854), not mentioned by Janet.

ideas: old reawakened memories of which the young girl has no awareness, romantic fabrications, sentimental fantasies, digressions—all
things that can be produced when intelligence and imagination are
abandoned to themselves. (Tascher 1855, p. 11)

For the author of the *Lettre*, these fabrications are produced by a second
personality separate from the normal personality. The same kind of division of
consciousness is present in other mediumistic phenomena, such as automatic
writing, and in somnambulism. Writing mediums may exhibit a second personality that is "inflamed, passionate, and unrestrained" and momentarily
blots out the normal personality. Or the medium may exhibit "two simultaneous currents of thought, the one which constitutes the ordinary person, the
other which develops outside of it." On the other hand, with somnambulism
"we are now in the presence of the second personality only, the other being
annihilated in sleep" (pp. 23, 44).

The author's notion that a second personality exists concurrently with the
ordinary personality was novel. This second personality is not merely the
manifestation of mechanical automatism, as Braid and Carpenter would have
it. Neither is it the manifestation of animal thinking operating as a reflex
action of the brain, as Rogers proposed. It is an intelligence separate from the
ordinary personality, but capable of "romantic fabrications" and "digressions"
that are signs of an independent but simultaneous "current of thought." For the
anonymous author, a psychological examination of mediumship demonstrates
that human beings can carry on complex mental activity outside the awareness
of their normal selves. This formulation of a theory of unconscious mental
activity was far ahead of its time, and is not seen again until the work of Janet
and others in the 1880s.

Chapter 13

Animal Magnetism and Psychical Research

Psychical research, the scientific study of the paranormal and the precursor of parapsychology, had its official beginning on February 20, 1882, with the establishment of the Society for Psychical Research in England. There were, however, three decades of preparation, during which scientists and researchers evolved increasingly accurate techniques for observing paranormal phenomena. This was made possible and even necessary because from around 1850 paranormal phenomena began to become available in abundance.

Animal magnetism played a central role in the birth of psychical research because it was crucial in creating the conditions that produced the surge of apparently paranormal occurrences. Paranormal happenings were associated with animal-magnetic processes from the beginning. Most important, animal magnetism helped prepare the ground for American spiritualism, the movement that produced a plethora of paranormal phenomena in the séance room, in the home, on the stage, and in the church. When spiritualism spread to England, Europe, and other parts of the world, its fame was associated specifically with table moving, a phenomenon that brought the experience of the paranormal within reach of almost anyone who wished to experiment.

It is not surprising, then, that when the first group of serious-minded men and women was formed to investigate the paranormal systematically, the study of animal magnetism and its effects was among their principal initial projects. The events that led to the formation of this association are well worth noting, however briefly.[1]

Early Researchers

Michael Faraday, with his examination of table turning, was the first scientist of note to attempt to deal with the paranormal phenomena arising from

1. There are a number of excellent histories of psychical research, especially Moser 1935, Gauld 1968, Inglis 1977, Moore 1977, Cerullo 1982, Grattan-Guinness 1982, and Haynes 1982. Mauskopf and McVaugh 1980 and Inglis 1984 cover later developments and the rise of modern parapsychology.

spiritualism. As we have seen (chapter 12), his work was cursory and limited in scope, and his findings were negative.

The next attempt to observe and explain the phenomena scientifically was undertaken by Robert Hare, a respected medical doctor and professor of chemistry at the University of Pennsylvania who was considered an expert on epilepsy and had written more than 150 articles on scientific subjects. Hare had read accounts of a "spiritual" force that could move tables and objects without contact, but was skeptical. In 1853 he conducted tests to determine the matter and to his surprise obtained positive results. His disbelief changed to a complete acceptance of spiritualism. Hare published his findings, along with detailed diagrams and descriptions of his experiments, in 1855. Despite Hare's scientific reputation, there seems to have been little response to his study. It was nearly twenty years before experiments of this kind were again attempted.

The evolution of psychical research into paranormal phenomena centered in England.[2] Spiritualism came to England in the early 1850s and within a short period of time had made a powerful impression on the public. Two American mediums, Daniel Dunglas Home and Mrs. W. Hayden, led the way. Their success encouraged the development of home-grown talent, and by the early 1860s spiritualism was thriving.

The impressive occurrences connected with mediums along with the active support of spiritualism by the renowned naturalist Alfred Russell Wallace spurred a new and more thorough investigation of the phenomena than had occurred so far (Wallace 1866, 1874, 1875). The London Dialectical Society, a sophisticated debating club made up largely of lawyers, physicians, and other professionals, also contributed to the progress of psychical research. The purpose of the society was to afford a hearing for subjects that were shunned elsewhere. In 1869 it formed a committee to study spiritualistic

---

2. There was significant psychical research in other countries as well. In Germany the writings of Maximilian Perty (1861, 1863, 1869, 1877) were a serious attempt to provide an orderly study of mediumistic phenomena in the framework of a neoplatonic philosophy. In the context of their scientific works, Johann Zöllner (1878–1881, 1880) and Gustav Fechner (1851, 1861) also tried to develop an explanation of paranormal phenomena. The renowned French astronomer Camille Flammarion (1862–1863, 1866) attempted to classify mediumistic phenomena. The Russian Alexander Aksakov, imperial counsellor to the czar, carried out numerous investigations of mediums and in 1874 founded and edited *Psychische Studien,* the first great psychical research periodical, which continued into the twentieth century as *Zeitschrift für Parapsychologie.* In the United States, William and Elizabeth Denton performed experiments with "psychometry," the application of clairvoyant sensitivity to ancient artifacts, to determine whether there was a correlation of psychic intuition and archeological fact.

phenomena and proceeded to gather evidence from testimony and direct experimentation both for and against the genuineness of the phenomena. Contributions were received from some of the great intellectual lights of the day and some of the best-known spiritualist mediums. A subcommittee, formed to carry out direct experimentation, reported that sounds were produced and heavy objects moved without any physical action or mechanical contrivance. The report was not accepted by the Dialectical Society as a whole and so was published by the committee on its own authority (London Dialectical Society 1871).

In 1871 the cause of the paranormal was boosted by publication of the initial experiments of William Crookes, one of the most respected scientists of the nineteenth century. He first gained renown when, at the age of twenty-nine, he discovered the element thallium. He went on to become the inventor of the radiometer, the spinthariscope, and the Crookes vacuum tube, which was instrumental in the discovery of X-rays. He was founder of the *Chemical News* and editor of *The Quarterly Review of Science*. In 1871 Crookes began experiments, with the aid of the American medium Daniel Dunglas Home, to discover if it was possible to cause objects to move without physical force. He eventually proved to his complete satisfaction that there existed a "psychic force" that could move objects and apply pressure at a distance. In his published report (Crookes 1874) he gave a detailed account of the construction of the apparatus used and the physical circumstances of the experiments. Crookes's report made a powerful impression on intellectuals of his day and contributed greatly to the climate that led to the formation of the Society for Psychical Research.

Research in the same area was carried on by the barrister Edward William Cox and published in two works (Cox 1871, 1873). Cox founded the Psychological Society of Great Britain to carry out intensive research into the nature of psychic phenomena, but it dissolved at his death in 1879, having accomplished very little of that task.

Among the most important British mediums of this early period was William Stainton Moses, an Oxford graduate and an ordained minister. Moses became interested in spiritualism in 1870 and began to experiment on his own. In a private circle that met at his house, Moses was able to bring about striking physical phenomena including the movement of objects without touch and the appearance of mysterious lights. Moses also practiced automatic writing and produced a remarkable set of purported communications from spirits. Because of his impressive intellectual and personal qualities, his phenomena were taken seriously by many academics.

The Foundation of the Society for Psychical Research

In 1874 Moses reported his experiences to two Cambridge scholars, Frederic W. H. Myers and Edmund Gurney. For some time Myers had been discussing his interest in psychical investigations with Henry Sidgwick, a respected Cambridge professor with broad philosophical and spiritual interests. Many years later Myers recalled what was to be a most important conversation for the development of psychical research:

> In a star-light walk which I shall not forget . . . I asked him, almost with trembling, whether he thought that when Tradition, Intuition, Metaphysic, had failed to solve the riddle of the Universe, there was still a chance that from any actual observable phenomena—ghosts, spirits, whatsoever there might be,—some valid knowledge might be drawn as to a World Unseen. Already, it seemed, he had thought that this was possible; steadily, though in no sanguine fashion, he indicated some last grounds of hope; and from that night onwards I resolved to pursue this quest, if it might be, at his side. (*Proceedings of the Society for Psychical Research* 15 [1901]: 454)

In 1873 Myers suggested to Sidgwick that they begin investigating spiritualistic phenomena. Edmund Gurney and others were drawn into the project, which continued with increasing intensity up to 1880, when it lost momentum.

Meanwhile, William Barrett, professor of physics at the Royal College of Science in Dublin, having developed an interest in mesmerism, began a series of experiments in the physical rapport of the mesmeric subject and the mesmerizer. He also focused his attention on table turning and the "willing game," where the silent will of one person is supposed to cause another to act. In 1881 the British National Association of Spiritualists proposed to Barrett that a society be formed to investigate the phenomena of spiritualism, and in January 1882 Barrett assembled a mixed group of spiritualists and academics, including those who had been working under Sidgwick's leadership at Cambridge. Myers and Gurney agreed to participate in such a society only if Sidgwick were president. The group agreed to this stipulation, and on February 20, 1882, the Society for Psychical Research had its founding meeting.

The society set itself six aims:

> 1. An examination of the nature and extent of any influence which may be exerted by one mind upon another, apart from any generally recognised mode of perception.

2. The study of hypnotism, and the forms of so-called mesmeric trance, with its alleged insensibility to pain; clairvoyance and other allied phenomena.

3. A critical revision of Reichenbach's researches with certain organisations called "sensitive," and an inquiry whether such organisations possess any power of perception beyond a highly exalted sensibility of the recognised sensory organs.

4. A careful investigation of any reports, resting on strong testimony, regarding apparitions—at the moment of death, or otherwise, or regarding disturbances in houses reputed to be haunted.

5. An inquiry into the various physical phenomena commonly called Spiritualistic; with an attempt to discover their causes and general laws.

6. The collection and collation of existing materials bearing on the history of these subjects. (*Proceedings* 1 [1882–1883]: 3–4)

It soon became clear that the alliance between the spiritualists and the Cambridge group was shaky. Many of the spiritualists viewed the society mainly as a tool for the vindication of the claims of spiritualism and were reluctant to take part in research that might ignore the role of spirits or even disprove certain phenomena. When in 1885 Myers wrote an article for the *Contemporary Review* concluding that the phenomena of automatic writing could be adequately explained through the action of the writer's unconscious mind combined with a telepathic ability to pick up the thoughts of others, the issue came to a head. Resignations of spiritualists from the society grew, and when in 1886 the leaders of the spiritualist segment of the society—C. C. Massey, William Stainton Moses, and E. Dawson Rogers—resigned, spiritualist involvement effectively ended (Cerullo 1982, pp. 70–84).

The Investigation of Animal Magnetism and Hypnotism

Particularly from 1880 on, the question of the reality of animal magnetism as distinct from hypnotism was seriously debated. Many denied the existence of animal-magnetic phenomena outright. Others found some of the phenomena credible and identified them with hypnotic phenomena. Still others believed that animal magnetism dealt with a life force that could be controlled and used in healing. The debate centered around the issue of action at a distance. Can a magnetizer induce somnambulism or cause healing at a distance? Can a somnambulist perceive thoughts, feelings, or objects at a distance without the use of the senses? In other words, does animal magnetism produce distant effects by means of some quasi-physical influence, or are all magnetic phenomena merely the result of suggestion and imagination? Because the problem was phrased in these terms, it was viewed by many as an issue of psychical research.

To carry out the study of hypnotism and mesmeric trance, the Society for Psychical Research formed an investigative group called the Committee on Mesmerism. It included William Barrett, Edmund Gurney, Frederic Myers, Henry Rudley, W. H. Stone, George Wyld, and Frank Podmore, who served as secretary. In its first report (Barrett et al. 1883a) the committee noted that although it was quite common for scientists to accept the reality of hypnotic phenomena and deny the reality of anything that could not be explained in terms of this psychological state, there were persons of sound judgment who accepted the action of a "special influence or effluence passing from the operator to the subject." In its investigations the committee would attempt to discover if there was evidence of effects that could not be adequately explained through hypnotism and that would therefore have to be attributed to mesmeric action. It initiated a series of experiments using hypnotic and mesmeric techniques.

In its second report (Barrett et al., 1883b) the Committee on Mesmerism called attention to a number of strong indications that mesmeric influence was not explainable in terms of hypnotic states. Among the evidence derived from the committee's own experiments was a commonality of sensation between mesmerizer and subject, the apparent ability of the subject to obey the unspoken will of the mesmerizer, and the lingering presence of some mesmeric influence in inorganic objects detectable by mesmerized subjects. The experiments had been designed to remove the possibility of suggestion or subliminal influence, and the preliminary conclusion of the committee was that some sort of physical or concrete influence seemed to pass between mesmerizer and mesmerized.

In an appendix to the second report (Barrett et al., 1883c) the committee noted five external manifestations of mesmerism: changes in sensibility to pain, in sensory and supersensory perception, in the current of consciousness, in memory, and in emotional disposition or character. They noted that all these phenomena occur spontaneously in natural sleepwalking or somnambulism and rejected objections to the reality of mesmerism based on the "incredible character" of the phenomena. Although the committee concentrated its attention in these reports on the issue of mesmerism versus hypnotism, it had a broader study in mind for the long run: an investigation of "clairvoyance, phreno-mesmerism, mesmeric healing, and mesmeric effects produced without either fixation, manipulation or expectancy" (*Proceedings* 1 [1882–1883]: 284).

The committee's third report noted its experiments with the transfer of sensation from mesmerizer to subject and in an appendix cited similar experiments performed by earlier mesmerizers, such as Esdaile, Townshend, and Gregory. This report was published in volume 2 of the *Proceedings* (1883–1884). This volume also contained the first two of Edmund Gurney's important studies on hypnotism, "The Stages of Hypnotism" and "The Problems of Hypnotism."

In 1885 Gurney and Myers collaborated on "Some Higher Aspects of Mesmerism" (*Proceedings* 3 [1885]: 401–23), in which they stated, among other things, the criteria that should be applied to determine the efficacy of mesmerism in healing:

1. The case should be reported throughout by a medical man; or, at the very least, there should be a medical man's *diagnosis* and *prognosis* of the patient's malady before mesmerism is resorted to, and satisfactory evidence of the restoration to health.

2. The case should be reported, as nearly as may be, at the time, and publicly, so that objections may be taken to it before the circumstances are forgotten.

3. The case must be one in which no other form of medical treatment has been concurrently employed.

4. The recovery should be such as cannot reasonably be attributed to the *vix medicatrix naturae*.

5. The influence of imagination should be, as far as possible, excluded. (*Proceedings* 3 [1885]: 405)

By this time it had become clear that the usefulness of the Committee on Mesmerism had passed, and the investigative mantle had been taken up by Gurney and Myers. From his interest in hypnotic phenomena, Myers moved naturally into the study of automatic writing, hysteria, multiple personality, strata of personality, and finally an overall theory of personality highlighted by his notion of the "subliminal self."[3]

Gurney's contributions to the *Proceedings* relating to hypnotic and mesmeric investigations were also notable, but were brought to a sudden end when he died in 1888 at the age of forty-one. His studies of mesmerism and

---

3. Myers's early articles in *Proceedings* include "Human Personality in the Light of Hypnotic Suggestion" (4 [1886–1887]: 1–24), "On Telepathic Hypnotism, and Its Relation to Other Forms of Hypnotism" (4 [1886–1887]: 127–88), and "Note on Certain Reported Cases of Hypnotic Hyperaesthesia" (4 [1886–1887]: 532–39). See also Myers's articles in *Proceedings* from 1886 to 1900, as well as Myers 1903. The development of Myers's theory of human personality will be dealt with in greater detail in chapter 16.

hypnotism, along with those of Myers, form a corpus of research that is well worth consulting even today.

Animal Magnetism, Hypnotism, and Early Psychical Researchers

The literature of psychical research in the first decade after the foundation of the Society for Psychical Research reveals how intimately that endeavor was intertwined with issues relating to animal magnetism and hypnotism. Hundreds of articles and books were written on the paranormal phenomena arising from mesmeric or hypnotic states (see Crabtree 1988). Many of them repeated the question: does action at a distance occur in magnetic phenomena and is there an essential difference between animal magnetism and hypnotism?

In 1886 a number of articles appeared in England and France describing experiments to determine whether the induction of somnambulism at a distance was genuine.[4] This flurry of activity seems to have begun with experiments carried out by Pierre Janet at Le Havre in 1886 (Kopel 1968) with Mme B, who was particularly susceptible to somnambulism. Janet had her magnetizer, a Doctor Gilbert, attempt to induce the somnambulistic state simply by thinking about it. When this proved successful, he had the doctor attempt the same thing from another location, with the same positive result. In a second series of experiments a new element was added: the subject was placed in a somnambulistic state from a distance and given commands to carry out. These experiments were also successful (Janet 1886a, 1886b).

Frederic Myers, who was present at the first set of experiments, spoke with Dr. Gilbert and his family about Mme B and wrote an article (Myers 1886b) agreeing that the induction of somnambulism at a distance had indeed taken place. He preferred to call it "telepathic hypnotism," believing that the basis of the induction was communication by telepathy (a word coined by Myers) between operator and subject.

One of those who came to Le Havre to observe the second set of experiments was Julian Ochorowicz, the leading Polish psychical investigator of animal magnetic phenomena. In 1887 he published a major work, *De la suggestion mentale,* describing the occurrences at Le Havre and his own experiments with effects produced at a distance (pp. 118–44).[5] Among the influences that could be transmitted at a distance Ochorowicz included physical ailments or debilities, emotional states, sensations, ideas, and will (pp.

---

4. See Gley 1886, Héricourt 1886, Janet 1886a and 1886b, Myers 1886, and Richet 1886; also Dufay 1888 and Richet 1889.

5. An English translation, *Mental Suggestion,* appeared in 1891; apparently no Polish version was ever published.

157–325). He believed that these distant effects occurred in animal magnetism and were different from those of hypnotism, for they involved the presence of what the magnetizers termed an "emanation" of some sort and what Ochorowicz called a "certain physical action" (p. 179).

Having established to his satisfaction the reality of these various kinds of transmission, Ochorowicz concluded that they may explain many phenomena that have puzzled observers. Among these he lists:

> 1. Certain cases of instinctive appreciation of illnesses, 2. certain cases of direct nervous contagion, 3. certain illusions by observers who did not guard against mental influence, 4. certain cases of alleged vision at a distance, 5. certain phenomena of veridical hallucination which are incredible but well attested, 6. the communication of certain sensations during dreams in normal sleep, 7. alleged divinations by the "spirit rappers," 8. the mystical influence of certain persons, 9. the personal difference between "hypnotizers," and the different characteristics of the effects they obtain, 10. many facts registered in the history of civilization and attributed to demons, oracles, sorcerers, the possessed, etc. (p. 538)

Another important psychical researcher into the phenomena of animal magnetism and hypnotism was Eugène de Rochas d'Aiglun, onetime administrator of the Ecole Polytechnique of Paris. Initially Rochas began his researches from an interest in the findings of Baron von Reichenbach (Rochas 1866) in regard to the od (see chapter 12). Rochas related this effluence to the fluid of animal magnetism and conducted experiments aimed at detecting its presence and defining its properties (Rochas 1887, 1891).

Rochas delineated what he believed to be the principal "profound" (1892) and "superficial" (1893) stages of hypnotism. Like Ochorowicz, he believed that a physical influence was involved in the induction of hypnotism and the rapport maintained between the hypnotist and subject. He took this concept one step further in describing an "exteriorization of sensibility" (Rochas 1895)—a phrase originally used by Joire (1892)—and an "exteriorization of motricity" (Rochas 1896), in which the normal abilities to receive sensation and affect objects physically are extended beyond the limits of the body. These two notions were employed to explain certain types of clairvoyance and certain physical phenomena of mediumship. The exteriorization of sensibility could explain how some psychics are able to feel the sensations of others and perceive the interior of their bodies to diagnose illnesses; the exteriorization of motricity explained how mediums can move physical objects without contact.

In the 1880s a number of works appeared supporting the view that animal magnetism and hypnotism were distinct. In 1882 A. Baréty, a member of the Faculty of Medicine at Paris, published a small treatise to this effect and followed it up five years later with a massive tome, *Le magnétisme animal étudié sous le nom de force neurique*. He posited an identifiable force, analogous to electricity, that circulates through the body and categorized its physical and physiological characteristics. He believed that this "neuric force" was produced in the nervous system and radiated from the eyes, the ends of the fingers, and the lungs. Baréty held that sensitive individuals may perceive the force at some distance from the body, varying from a few centimeters to several meters, and that some substances block this force as it radiates from the body while others transmit it. Baréty equated the force with Mesmer's animal magnetism.

In *Le nouvel hypnotisme* (1887) Lucien Moutin compared hypnotism to animal magnetism and found them to be different phenomena. He noted that the induction of "braidism" demanded complete passivity on the part of the subject, whereas mesmerists were able to bring about their phenomena while the subject was in certain ways active. Moutin insisted that without inducing sleep or drowsiness, magnetizers were able to produce effects that Braid never could. He noted that Du Potet brought about magnetic somnambulism at a distance and cited his own successful experiments of this type. He detailed cases in which subjects could not resist falling into somnambulism when touched in a certain spot, no matter how hard they tried. Moutin wrote that there are phenomena, such as second sight, produced through animal magnetism that Braid never attained through hypnotism.

Turning to the subject of healing, Moutin saw great differences between hypnotism and animal magnetism. For individuals to be successfully treated by hypnotism, they had to be susceptible to hypnotic trance. With animal magnetism, however, healing could take place without trance, and to those who said that it was through suggestions that such cures took place, Moutin answered that he had himself cured infants and animals, who could not be affected in that way (1887, pp. 58–73).

In 1889 Albert Moll wrote a study of hypnotism that would become a classic and go through numerous editions and printings. Moll worked with August Forel, Max Dessoir, and other researchers to define the nature of limits of the phenomena associated with hypnotism and animal magnetism. He published his first findings as *Der Hypnotismus*. Beginning with a history of hypnotism in France and Germany, he went on to describe the characteristic symptoms of the hypnotic state, the role of suggestion, possibilities of simulating the hypnotic state, and its uses and legal implications. Moll also inquired about the status of animal magnetism and whether it possessed a reality above and beyond hypnotism. He criticized experiments done to prove the existence of the magnetic fluid and generally expressed skepticism about its reality.

Three years later Moll took up the issue of animal magnetism more directly. In *Der Rapport in der Hypnose,* based largely on Moll's original experiments, he pointed out that although the term "rapport" in its most general sense referred to a special connection between hypnotist and subject, it had assumed a bewildering variety of nuances over the previous hundred years. In the process of discussing the various kinds of rapport, Moll examined an issue that he considered essentially related: whether there was such a thing as animal magnetism and whether an animal-magnetic fluid existed. Moll said that he had not been able to observe the phenomena so often cited by the mesmerists as conclusive proof, but he could not bring himself to deny conclusively the reality of animal magnetism.

In 1890, H. R. Paul Schroeder wrote a work on the healing technique of magnetism in which he concluded, quite unlike Moll, that animal magnetism and hypnotism were two different phenomena with distinct characteristics. Hypnotism, he said, was a state brought about either by fixation of the gaze on some object or by suggestion through words. Hypnotism from a fixed gaze results in catalepsy or stiffening of the limbs, along with a state of passivity or lack of will. Hypnotism through suggestion produces a state in which the subject can be brought to carry out some action or enter a special mental state and, for instance, become unaware of pain. Schroeder claimed that the latter process was identical with faith healing except that the subject invests faith in the hypnotizer rather than in God. For Schroeder, the highest state of hypnotism united both fixation and suggestion, producing powerful phenomena in which the cataleptic subject carries out the commands of the hypnotizer.

Animal magnetism, on the other hand, was for Schroeder a process in which the operator manipulates the "nerve-ether" or "nerve-force," the life principle of the body through the imposition of a healthy nerve-force on the unhealthy nerve-force. He likened this operation to a transfusion. Schroeder pointed out that sometimes the patient at first seems to worsen because of the powerful reaction caused in the unhealthy nervous system, but this "crisis" passes and then the beneficial effects of the infusion of nerve-force are perceived.

Schroeder contended that the mental state produced by the two processes was also distinct. He held that in magnetic somnambulism the subject maintained his own mind and spirit, whereas in hypnotic trance the subject became a virtual machine to be operated by the suggestions of the hypnotizer.

Julian Ochorowicz made his own contribution to the issue. In experiments on hundreds of subjects he believed he had discovered a relationship between sensitivity to magnets and hypnotizability. This led to his invention of the "hypnoscope," a magnet in the shape of a slit tube that, when placed on the finger of the subject, would indicate hypnotic susceptibility (see Dingwall 1968, 3:108–35). Ochorowicz found that some hypnotic or magnetic subjects were able to view objects, persons, and events at a distance, and he believed that thought transference often occurred in this state. His experiments with "eyeless vision" were, in his opinion, conclusive, and, seeing this ability as completely natural, he urged the serious scientific study of all phenomena relating to distant influence.

In 1897 Ochorowicz published *Magnetismus und Hypnotismus,* which contained a history of animal magnetism, a history of hypnotism, and a comparison of the two phenomena. From his studies and experiments, he concluded that magnetic sleep and hypnotic sleep were distinct, and that although hypnotic sleep had an important place in scientific studies, magnetic sleep had a much greater practical value (p. 135). Ochorowicz pointed out that the two states were produced in different ways—hypnotic sleep by a kind of fascination produced by concentrating on a piece of "dead matter," such as a bright object, and magnetic sleep by the action of one living system on another. Calling attention to current hypnotic experiments centered around automatism, Ochorowicz believed that the contrast could not be clearer:

> The modern hypnotizer develops automatons who can be maneuvered this way and that as a blind tool of various influences, while the magnetizer depicts with religious respect the highest states of the magnetized. Magnetizers pay heed to their mediums, in that they take counsel with them as with seers, while hypnotists only order and command, disregarding the spontaneity of the medium. Magnetizers subjugate themselves to nature, while hypnotizers want to conquer it. The former recognize a possible influence [on the magnetized person] and are careful not to get mixed in with it. The latter recognize no personal influence and permit anybody to try their hypnotic jokes [on the subject]. Which side is right? As usual truth lies in the middle. Yet the moral and therapeutic aspects of these two are not quite the same. And although I recognize the scientific value of hypnotic research, I do not hesitate to ascribe to magnetism the greater practical worth. (pp. 134–35)

For Ochorowicz the physical action of one organism on the other, in magnetic sleep, produces phenomena that are not found in hypnotic sleep, including communication of thought without use of the senses and the reception of distant impressions. He noted that hypnotizers did not come across these phenomena in their experiments (pp. 136–37).

Hudson

In the same period Thomson Jay Hudson (1834–1903) provided a most original and thought-provoking analysis of the distinction between animal magnetism and hypnotism. Hudson, an American lawyer turned journalist, in 1893 published *The Law of Psychic Phenomena*. Although this book was destined to enjoy tremendous popularity in the United States, it seems to have been largely overlooked by the more serious investigators of mesmerism and hypnotism.

Hudson introduced the notion that the human mind is dual, consisting of an objective mind and a subjective mind. The objective mind deals with the objective world and observes reality by means of the five senses. It is the "outgrowth of man's physical necessities. It is his guide in his struggle with his material environment" (Hudson 1895, p. 29). The subjective mind comes to know its environment by means independent of the physical senses—by "intuition." It is the seat of the emotions and the storehouse of memory, and operates most effectively when the five senses are in abeyance. It is the subjective mind that manifests in somnambulism.

The subjective mind sees without the use of the eyes, can leave the body and travel to distant places, and can read the thoughts of others. Whereas the objective mind is merely the function of the physical brain, the subjective mind is a "distinct entity, possessing independent powers and functions . . . and capable of sustaining an existence independently of the body" (p. 30). Whereas the objective mind is immune to the power of suggestion, the subjective mind is highly susceptible. According to Hudson, the subjective mind as an individual entity is subject to suggestions both from the objective mind of another person and from its own objective mind (pp. 30–31).

Hudson posed the question about mesmerism versus hypnotism in provocative terms. He asked why the peculiar effects so commonly reported by mesmerists—the "higher phenomena," such as telepathy, clairvoyance, and distant healing—were more or less absent in the experiments of hypnotists. In answering he noted that the higher phenomena first showed signs of "decadence" when James Braid came on the scene, since Braid was determined to explain away the higher phenomena at all costs. Hudson accepted the basic facts adduced by Braid. He agreed that a subject could be hypnotized without any involvement on the part of the operator and that intense gazing at an object could displace the threshold of consciousness. To Hudson's way of thinking, with this displacement the subjective mind was elevated above the threshold, bringing the subjective powers of the mind into play. The subjective mind was then able to operate independently or synchronously with the objective mind, depending on the depth of hypnosis: the deeper the hypnotic trance, the more forcefully and independently the subjective mind could exercise its powers (pp. 107–8).

At this point Hudson introduced a novel and ingenious analysis of the method of Braid. He pointed out that Braid had shifted the action from the magnetizer to the subject and that it was because of this shift that the phenomena had changed. With the induction of hypnotism, although the subject may gain access to his subjective mind, the hypnotizer is left untouched by the process. On the other hand, "when a mesmerist employs the old methods of inducing the subjective state,—passes, fixed gazing, and mental concentration,—*he hypnotizes himself by the same act by which he mesmerizes the subject*" (p. 108). Both are placed in a state in which their subjective minds are accessible. Now since telepathy takes place between two subjective minds, thought reading can only occur when the operator is also in a trance state. Clairvoyance that depends on the operator knowing the hidden information is likewise dependent on the dual-trance state.

Healing too was dependent on this state of double subjectivity. Hudson believed that healing took place through the "force" of one subjective mind acting on another. Distant healing required the telepathic transmission of this "force." For these reasons both subjective minds must be engaged (p. 110).

It is clear that for Hudson the higher phenomena and magnetic healing could only be brought about by using the old mesmeric approaches. If some higher phenomena did develop in hypnotic experiments, it was only because unwittingly the hypnotizer also entered the subject state. So for Hudson, hypnotism could bring about only those effects produced by the sole action of the mind of the subject. Mesmerizing, however, supplements those effects by a constant force emanating from the subjective mind of the operator and by mental impressions conveyed telepathically from operator to subject.

But is a magnetic fluid involved in this process? Hudson insisted that the force produced in mesmeric operations was mental in origin, and "whether this mental action creates or develops a fluid akin to magnetism is a question which may never be solved. . . . When mesmeric passes are made over a patient, a fluid appears to emanate from the hands of the operator. . . . Is it not a fact, nevertheless, that the passes are principally useful as a means of controlling the mind of both the subject and the operator? There are many facts which seem to point unmistakably in that direction" (p. 112).

Hudson illustrated the importance of the operator's mesmerizing himself by describing hypnotic control of animals. He stated that trainers who utilize mesmeric or hypnotic power over animals employ methods that, wittingly or unwittingly, involve self-mesmerization. The result is that the trainer's subjective mind comes to the fore and dominates the subjective mind of the animal which, having very little objective mind, is then easily controlled (pp. 114–15).

Although Hudson did not decide the issue of the existence of magnetic fluid, he certainly did insist on the reality of action at a distance. His inclusion of the subjective mind of the operator in the equation was thought-provoking. Unfortunately, little was done to investigate the implications of the idea. His dual-trance theory of mesmerism was striking, and his two-mind theory of mental interaction found an echo in the psychological ideas of Frederic Myers (Chapter 16). Overall, Hudson introduced a unique and sophisticated complexity to the problem of mesmerism versus hypnotism.

By the mid-1890s neither the Society for Psychical Research nor other investigators had decided on the reality of animal magnetism as distinct from hypnotism or on the question of action at a distance by means of unknown forces. Nevertheless, the reality of psychological phenomena connected with both animal magnetism and hypnotism had gained increasing acceptance. In mesmeric and hypnotic trance states the mysterious world of the mind began to open up to investigators in a fresh way, making it possible to bring scientific thinking to bear on phenomena that had largely been reserved for philosophical and theological speculation. And this in turn allowed the development, for the first time, of systematic and effective psychological healing.

Part IV

Psychological Healing

Chapter 14

Dual Consciousness and Dual Personality

Mesmerism holds a place of preeminence in the history of psychological healing because it made possible the development of the alternate-consciousness paradigm. This paradigm was based on the discovery of a second consciousness within the human psyche revealing itself in magnetic sleep. Divided consciousness or double consciousness quickly became the subject of intense curiosity and experimentation. Although for decades the psychotherapeutic dimensions of this discovery were largely overlooked, magnetizers and those interested in "mental physiology" soon noted that double consciousness occurred not only during magnetic sleep but also spontaneously in sleepwalking states. Investigations of both natural and artificial double consciousness contributed significantly to the evolution of the alternate-consciousness paradigm through the early and middle decades of the nineteenth century.

Double Consciousness and Memory

Deleuze (1813) pointed out, following Puységur, that the somnambulist, upon waking, loses all memory of "the sensations and ideas that he has had in the state of somnambulism. These two states are so alien to each other that the somnambulist and the waking man seem like two different beings." Some somnambulists even share this view of themselves as two separate selves. Deleuze cites an example. "Madame N. . . . , who had a distinguished education, having lost her fortune following a legal battle, decided, on the advice of her husband, to enter the theater, where her talents assured her of success and a good salary. While engaged in this project she fell ill and became somnambulistic. Since in the somnambulistic state she was in favor of the opposite decision, her magnetizer sought an explanation from her. . . . 'Then why do you want to enter the theater?' 'It is not me, it is her. . . .she is mad'" (1:176).

Bertrand (1826) spoke about this separation of states in terms of two memory chains. He pointed out that not only does the subject on awakening lose all memory of what had occurred during somnambulism, but when again becoming somnambulistic he regains the memory of what had occurred in all previous episodes of somnambulism. This produces a kind of separate life, renewed at regular intervals and bound together by a separate memory chain. Although the waking person does not remember what has occurred in somnambulism, the somnambulistic person does recall the events of the waking state (pp. 409–10). Pigeaire (1839) added his voice to those who recognized this divided state, noting that magnetizers again and again observed "two very distinct lives, or at least two ways of being in the life of somnambulists. One sees them speak of one another as if their individuality were composed of two persons" (p. 44).

This separation of consciousness came to be called "double consciousness" (*Doppelschlaf;* see Brandis 1818, p. 127). Braid (1844), who called the phenomenon "second memory" as well, wrote: "By double consciousness I mean that there is a state of the sleep, when a patient may be taught anything, and be able to repeat it with verbal accuracy as often as in that stage again, whilst he may have no ideas either of the subject or the words when in the waking condition" (p. 32).

In 1840 Chauncy Hare Townshend, a British clergyman and mesmerist, attempted to explain the puzzling phenomenon of double consciousness. He attributed the inability to recall somnambulistic occurrences in the waking state to the lack of "introspective consciousness" in the state of magnetic sleep. He explained the memory chain within a succession of somnambulistic episodes as proceeding "naturally" from the principle that "a former series of thoughts should recur to us then, and then only, when we are under conditions of consciousness similar to those in which they were conceived" (1851, pp. 306–7). This is a remarkably clear formulation of the modern notion of state-related learning and state-related memory. Today this concept is a widely accepted explanation of somnambulistic amnesia.[1]

1. Chardel (1826) seemed to have hinted at the same explanation by relating lack of memory in the waking state to the absence of the affect present when the experience occurred in somnambulism (pp. 188–89). Puységur had also noted this total inability to recall events that occurred in magnetic sleep (Puységur 1784, p. 90). The question often arises among modern hypnotherapists as to why today's hypnotic subjects seem to display this characteristic so infrequently. The theory of state-dependent learning as first formulated in the 1960s (see Crabtree 1992) may provide an answer. In this framework, memories are attached to specific states of consciousness, and gaining access to memories depends on returning to the state of consciousness in which the information was first obtained. If the state in which the learning originally occurred is too disparate from the state in which the person is trying to remember, retrieval will fail. It seems that during trance states induced in hypnotherapy the patient deals with issues and concerns that have become broadly accepted in popular culture. The awareness that people have a subconscious or unconscious mind and that repressed or dissociated memories may surreptitiously affect one's actions, awareness of meditative states and of exercises involving guided imagery, and awareness of other psychologically sophisticated concepts re-

William Gregory, writing in 1851, emphasized not only the double memory chain but also the contrast of personality traits in the two consciousness states. Under somnambulism the subject's voice, facial expressions, and general bearing may change radically to reveal "a person of a much more elevated character than the same sleeper seems to be when awake. . . . He often loses, in the magnetic sleep, his sense of identity, so that he cannot tell his own name, or gives himself another, frequently that of the operator; while yet he will speak sensibly and accurately on all other points" (pp. 83–85). In emphasizing the contrast of personality traits in the two states, Gregory touched upon the basis in magnetic sleep for an explanation of a phenomenon that had been noted for some time: the doubling of personality.

Double Personality

Magnetizers sometimes noted a doubling phenomenon that indicated a separation of the person into two distinct personalities. Roullier (1817), for instance, described a dissociated voice that spoke to the somnambulistic person from within. He believed that this voice was an alienated part of the somnambulist and refused to see it, as some of his contemporaries might, as a spirit or entity invading from without (pp. 120ff). Friedrich Fischer (1859) noted the same phenomenon, calling it the "inner voice," and said it was a common occurrence in somnambulism (3:314ff).

Strombeck's somnambulist, Julie, while being cared for by Strombeck and his wife, spoke of seeing a dissociated body accompanying the dissociated voice:

"When seated on the sofa, I heard a rapping. I looked around and saw a body, which is difficult to describe. It told me that I must change certain things regarding the conduct of my routine planned for tomorrow.". . . My wife and I did not object to these changes, and we consented. . . . A quarter hour later my wife returned and said, "The speaking body has come back again near the bed of our patient and has prescribed that tomorrow at dinner she have a small plate of oatmeal." . . . I asked [Julie], "What is the phantom you saw last evening?" She answered, "It is not a phantom. I erred in believing it was. It was a voice that speaks within me. I thought it was outside of me, but it is inside." . . . I

---

duce the gap between the introspective states of hypnotherapy and everyday ideas. Since the gap between these states is not large, memory retrieval is generally easier. In an era in which pop psychological concepts that we take for granted were unheard of, the gap between magnetic or hypnotic trance states and ordinary consciousness would tend to be much greater. This would make retrieval of information from the state of magnetic or hypnotic trance on the whole more difficult.

inquired, "Under what form does this 'speaking body,' as you have called it, appear to you?" She replied, "It is as though a white cloud rises from the earth, from where, when it has risen, a voice emanates, the echo of which resonates in me. It is necessary to obey this voice; it is the equivalent of the inner sense [intuition] that occurs in sleep. (Strombeck 1814, pp. 65–70)

Julie thus confirmed Roullier's contention that the inner voice is the dissociated manifestation of a part of the magnetic subject.

It was not a big step to move from double consciousness and the division of the magnetic sleeper into two distinct parts to the notion of double personality. Eschenmayer (1816), for instance, stated that double personality (*doppelte Persönlichkeit*) referred to the fact that the sense of self (*Selbstgefühl*) and awareness of self (*Selbst-Bewusstsein*) in the normal condition and in the magnetic state were distinct: "Each [somnambulist] has a double being, one in the waking state and another in the crisis" (pp. 56–57).

Joseph Ennemoser claimed that this double personality manifested itself only in the profoundest and highest stage of magnetic sleep, which he termed "double sleep." In that state the subject

speaks of himself in the third person and considers himself both as his usual being of daily life and at the same time the extraneous expression of the enhanced spiritual "I," so that body and soul, thus separated, stand in relation to each other as two personalities. (Ennemoser 1852, p. 486)

Although the term "double personality" was being used by some authors, the exact meaning of "personality" in this context was not clearly enunciated. The need for clarification was addressed in a series of articles on artificial somnambulism by the physiologist Charles Richet (1850–1935). Richet, a professor of physiology at the University of Paris, a member of the Académie de Médecine and the Académie des Sciences, and a winner of the Nobel Prize in physiology and medicine, early in his career became interested in artificial somnambulism and began a series of experiments with it. The results were published in the *Journal de l'anatomie et de la physiologie* and *Revue philosophique* (Richet 1875, 1880, 1883).

Richet explained the doubling of personality in the two consciousness states as proceeding from the "destruction of memory" when the subject passes from somnambulism to wakefulness. "What makes the *I* is the collection of memories we have. And since they are reserved for a special physical state, one can rightly say that, theoretically speaking, the person is different, since in sleep it recalls the whole series of acts it absolutely does not know about in the waking state" (1875, p. 362). In Richet's view, it is memory that makes personality possible. The remembrance of past experiences, a remembrance that is always available to the mind, gives one a sense of being a person, a continuing unity. One remembers one's age, sex, nationality, home, social position, way of feeling and thinking, and the general context in which one exists. If memory is lost, so is the sense of personality (1883, pp. 227, 232).

Richet noted that if you ask somnambulists what they are thinking, they will reply, "Nothing." This "psychic inertia" carries with it a corresponding physical inertia. Into this psychic void the magnetizer can introduce suggestions that the subject will take up. Now the mind and the body become completely occupied with these ideas; the individual is subject to a "psychic automatism" and a "somatic automatism" (1880, pp. 360, 479–80). When the subject is in the passive state, a state without memory and without thought, the magnetizer can in this way create an artificial personality. Richet used suggestion to cause somnambulistic subjects to believe that they were, for example, a peasant, an actor, a general, a priest, a sailor, or an old woman, and he called this phenomenon the "objectivization of types." The experimental subjects took on the suggested personality readily and in every way seemed to believe they *were* that person. Their actions and words were those appropriate (from their own knowledge) to the type of person they imagined themselves to be. Their involvement in their new personality was complete. The sense of being an "I" remained, but the personality of that "I" was new (1883, pp. 228–33).

While magnetizers were trying to describe and understand double consciousness and double personality as it manifested in magnetic sleep, a similar phenomenon was being noted in the annals of psychological pathology. Beginning in the 1790s observers began to report about individuals who alternately manifested two or more personalities in their lives. These spontaneously occurring instances of dual and multiple personality were seen as analogous to artificially induced double states and were often interpreted in the light of the investigations of magnetic sleep.

Multiple Personality Disorder: The Beginnings

In the late eighteenth century investigators began to notice a condition in which someone would seem suddenly to become an entirely different person. The individual would remain in this new personality for some time, then return to "normal." This phenomenon would eventually be followed by another switch back to the new personality, subsequent return to normal, and so forth. It was at first labeled alternating personality or dual personality and is today known as multiple personality disorder (see Braude 1991).

In multiple personality disorder the individual spontaneously manifests more than one cohesive personality, and the extra personality (or personalities) makes no claim to an independent existence outside the body of the individual. The second element clearly distinguishes multiple personality disorder from the possession syndrome, in which the alternate personality presents itself as an intruding intelligence, a being that exists in its own right elsewhere (Yap 1960, Roullier 1817).

Although multiple personality disorder as defined here has gained recognition only in recent years, cases that meet this definition have been reported for at least two centuries. Eric Carlson (1981) has brought to light a case that was originally described by the Reverend Joseph Lathrop of Springfield, Mass., in a letter of 1791 to Ezra Stiles, president of Yale University. This case was cited in the same year by Benjamin Rush in his lectures to medical students. It concerned the son of Captain Joseph Miller of West Springfield. For some years the young man had been "visited with a peculiar kind of disorder, which operated by paroxysms." Each of these fits would last from two to three hours and then pass, apparently without ill effects. What intrigued Lathrop was that while in a fit, the captain's son "perfectly remembered things which occurred in the preceding fits; but nothing which had happened in the intervals, all his fits and everything which had passed in them were totally obliterated; but he could distinctly recollect the occurrences of the former intervals. The time of his fits appeared to him in continuity—as did also his healthful periods—when one was present the other was lost." It seemed to Lathrop that the young man displayed "two distinct minds, which acted by turns independently of each other" (Carlson 1981, p. 669).

A second instance, described by Eberhard Gmelin in his *Materialen für die Anthropologie* (1:2–89), involved a twenty-one-year-old Stuttgart woman who suddenly exhibited a personality who spoke perfect French and otherwise behaved in a manner typical of a Frenchwoman of the time. She would periodically enter these "French" states and then return to her normal "German" state. In the French states she could remember everything she had said and done in previous French states whereas in her normal state she had no knowledge of the French personality. In her French personality she believed herself to be a native of Paris who had emigrated to Stuttgart because of the French Revolution. In that altered state she believed people around her to be personages other than they were, incorporating them into her fantasy. She spoke in elegant, idiomatic French, and when she attempted to speak German (her native tongue) it was labored and hampered by a French accent.

Gmelin decided to see if he could discover something about her peculiar condition by magnetizing her. She had never been magnetized before, nor had she seen anyone else put into the magnetic state. Gmelin found that he could put her into her "altered personality" by magnetizing her and restore her to her "original personality" by using the usual methods of bringing a person out of the state of magnetic sleep. In this way he theorized that her pathological condition was in some way related to magnetic sleep. Gmelin also attempted to explain philosophically how a person could have more than one personality and spent considerable space exploring the psychological factors involved in switching back and forth from personality to personality.

Multiple Personality and the Alternate-Consciousness Paradigm

These early cases of multiple personality disorder appeared less than ten years after the discovery of magnetic sleep. Just after Puységur had come across the phenomenon of double consciousness and a second stream of thought and memory, pathological instances of second consciousness began to be noted. Multiple personality seemed to be the perfect spontaneously occurring illustration of the reality of the artificially induced divided consciousness that magnetizers were so diligently probing. It revealed a second consciousness secretly existing below the surface in an individual, which can suddenly break through into waking awareness. It is as though the victim of multiple personality alternates between normal consciousness and the consciousness of somnambulism, in the latter state manifesting a new personality that is the personification of that somnambulistic consciousness.

There appears, then, to be a connection between the rise of multiple personality disorder and the discovery of a second or alternate consciousness in magnetic sleep. The orientation of this new, alternate-consciousness paradigm was quite different from those of the intrusion and organic paradigms (see chapter 5). In the intrusion paradigm the intervention of some intelligent outside agent, such as a possessing spirit, is used to explain psychological disturbance. In the organic paradigm, the explanation is formulated in terms of chemical agents or physical lesions. In the alternate-consciousness paradigm, however, humans are viewed as divided beings. We have our ordinary consciousness, which we normally identify as ourselves, and we have a second consciousness, which reveals itself in the magnetic trance and can seem quite alien to our ordinary consciousness. The feeling of alienation from this second consciousness is due in part to the memory barrier that accompanies manifestations of the second consciousness, and in part to the experience of a distinct sense of identity. This alienation is the basis on which the alternate-consciousness paradigm explains disturbances of consciousness. For if there is a second, alternate consciousness within human beings, that consciousness may develop thoughts or emotions very different from those of the ordinary self. Such thoughts and feelings may impinge on the ordinary consciousness to varying degrees, causing the person to think, feel, and act in ways not characteristic of the ordinary self.

Before Puységur, the notion of an alien element producing thoughts, feelings, and acts contrary to a person's usual behavior was explained as possession (the intrusion paradigm).[2] After him, a new understanding of this kind of disturbance became possible.

I believe that the discovery of magnetic sleep and the appearance of multiple personality disorder are directly related. It seems to me that in non-organic mental disturbance there are two elements: the disturbance itself and the phenomenological expression of that disturbance, the symptom language of the illness. The symptoms are a message to others telling them what is going on inside the individual. That is, the symptoms are the language of the inner disturbance. How clearly that inner disturbance will be expressed depends on the adequacy of the language available, and that in turn depends on what categories for understanding humans and the world are current in society.

Until the emergence of the alternate-consciousness paradigm the only category available to express the inner experience of an alien consciousness was possession, intrusion from the outside. With the rise of awareness of a second consciousness intrinsic to the human mind, a new symptom language became possible. Now the disordered person could express (and society could understand) the experience in a new way: it was the second consciousness acting at odds with the normal self.

This means that when Puységur discovered magnetic sleep, he contributed significantly to the form in which mental disturbance could manifest itself from then on. For he helped make possible a symptom language through which the experience of an interior alienation of consciousness could be expressed without resorting to the notion of intrusion from the outside—in other words, without experiencing that condition as possession.

But how direct a link can one make between the work of the magnetizers and the rise of multiple personality? Several factors have to be considered. First of all, some instances of spontaneous double consciousness were reported before the discovery of magnetic sleep (Gauld 1992a, p. 629; 1992b).

2. For a discussion of the history of the spirit possession, see Lhermitte 1963, Montgomery 1976, Nevius 1893, Oesterreich 1974, and Summers 1966. A comparison of multiple personality disorder and the possession syndrome can be found in Crabtree 1985a. For a view of both possession and multiple personality as culturally originated, see Kenny 1981.

Although that indicates an emergent alternate-consciousness symptom language, there is no reason to think that those symptoms were explained at the time in terms of a second consciousness inherent in the human psyche. These instances could be considered scattered indications of an awareness not yet come to fruition in the consciousness of society.

Another factor concerns how directly the influence of knowledge of magnetic sleep relates to specific instances of dual or multiple personality. It would be difficult to show that animal magnetism was known in the local contexts of the earliest cases, and that kind of linkage might be too simplistic in any case. Just how a symptom language becomes current in a culture is not particularly clear. It is debatable that direct conscious knowledge of an idea is needed for that idea to affect people at large. For that reason, the notion that a new symptom language came into its own around the time of the discovery of magnetic sleep can still stand, even if that new symptom language were more broadly based. Perhaps the context for viewing the human psyche as divided was already prepared in society at large when Puységur made his discovery. Perhaps an incipient awareness of this new way of looking at human consciousness was already "in the air" at the time. Perhaps Puységur's work simply crystallized and gave an experimental basis for a notion that was, so to speak, working its way to the surface of cultural awareness. Perhaps inquiries into the phenomena of dreams and natural sleepwalking in combination with a growing desire to find natural rather than supernatural causes for human experiences had sufficiently prepared the ground for the multiple personality symptom language to develop. In any case, it is my belief that no matter how it developed, the emergence of the alternate-consciousness paradigm was necessary for the development of multiple personality disorder as a visible pathology.

Multiple Personality: The Evolution

Twenty years passed after the two multiple personality cases of 1791, before the next instance came to light. In 1811 a thirty-six-year-old woman living in the Allegheny Mountains of Pennsylvania began to show symptoms of the disorder. Mary Reynolds experienced a severe convulsion in the spring of that year and was found lying unconscious. When she came to, she was blind and deaf, and it took her some weeks to regain her sight and hearing. Later the same year Mary fell into a deep sleep from which her family could not awaken her. After about twenty hours she awoke in a new state, with a bright, witty, and engaging personality very different from her usual quiet, sad self. She seemed to have lost the memory of all persons and skills, even the ability to speak. She was like a young child and had to be taught everything anew. After five weeks she woke up one morning back in her normal state of consciousness and her original personality. She could recall nothing of the previous five weeks and became depressed when told what had occurred. A few weeks later she again switched to her second personality after a prolonged sleep and continued where she had left off in that state, retaining the skills she had developed before and recalling all that had occurred during that five-week period. The switching between states now occurred frequently and continued until 1829, after which Mary remained permanently in her second state. She never married but lived an active life, even working for some time as a schoolteacher. She died in 1854 at age sixty-nine.[3]

A number of other significant but less well known cases of multiple personality disorder were reported in the first half of the nineteenth century.[4] H. Dewar described an instance of double personality in the *Transactions of the Royal Society of Edinburgh* (1823). The case, related to him by a Dr. Dyce, was that of a girl of sixteen who suffered from a spontaneous but temporary period of "divided consciousness" or "double personality." The onset of the condition was marked by disturbed sleep with sleep-talking and the singing of songs. One evening she took on the role of an Episcopal clergyman and went through the ceremony of baptizing three children and uttering an extemporaneous prayer. When awakened from this sleep, she recalled nothing of what had happened. This event was followed by a number of incidents in which she performed complex household chores with her eyes closed. Eventually the occurrences began to happen on other occasions than when in bed, and she would switch into her altered state when engaged in various activities. While in this state she was able to communicate with those around her to some degree, but her perception of her environment was defective or distorted and she sometimes had hallucinations. In her altered state she played with children, did housework, and was attentive to sermons at church. She retained no

3. The Mary Reynolds case came to be well known in the nineteenth century. It was described in detail by a number of authors and referred to by many. The first account of Reynolds's dual personality was written by Samuel L. Mitchill (1816). For a recent account of the case and how it was reported, see Carlson 1984.

4. See Goettman, Greaves, and Coons 1992. Earlier lists of cases may be found in Taylor and Martin (1944) and Greaves (1980). Michael Kenny (1986) has published a work on multiple personality in the United States in the nineteenth and early twentieth centuries. For a discussion of the more recent literature on the history of multiple personality disorder, see Alvarado 1990. One might include among these early instances of multiple personality disorder the case of John George Sörgel, "The Idiot Murderer," described by Feuerbach (1846). Although the information given in this narrative strongly suggests multiple personality, it is insufficient to make a certain determination.

memory of any event that occurred during her doubled state. This condition cleared up after three months, and at the time of the report, seven years after it had begun, there had apparently been no return of the problem.

A similar case occurred in the United States in the 1830s. A seventeen-year-old girl from Springfield, Massachusetts, Jane C. Rider, was closely observed by a number of doctors, and her story was published by one of them, L. W. Belden (1834). Jane suddenly began to have sleep-walking episodes which became more and more elaborate. She would do complex household chores while in a somnambulistic state, seemingly with her eyes closed. Sometimes the "paroxysms" occurred while she was in bed, during which she would talk, sing, and recite poetry. If they happened when she was up and around, she was invariably engaged in industrious tasks, such as sewing or preparing meals. Her skill and efficiency in these states often exceeded her normal talents. Her entry into the somnambulistic state was preceded by a headache, and while in the state she often complained of a severe pain in the left side of her head. The young woman seemed to have the ability to sense rather than see objects around her, and apparently perceived objects and events at a distance, so that she came to be called the "Springfield Clairvoyant." At first her interaction with the people around her was limited in these states, but over a period of time she became more and more responsive. When she returned to her normal state, she experienced complete amnesia for events that had occurred while in her altered state.

Whereas the cases narrated by Dewar and Belden seem to show elaborated somnambulism bordering on dual personality, other authors described individuals who seemed to have more fully evolved alternate personalities. David Skae, in 1845, described a middle-aged British man who alternated from day to day between a state of depression accompanied by suicidal tendencies and a state of normal functioning, with amnesia in each state for the other (Carlson 1982, p. 10). In the same year Thomas Mayo reported a case in the *London Medical Gazette* involving an eighteen-year-old girl named Elizabeth Moffat. Like Mary Reynolds, Elizabeth was "dull and quiet" in her normal state but lively and spirited in her second state. She passed from one state to the other suddenly and sometimes remained in her second state for weeks at a time. She experienced a reciprocal amnesia in the two states and developed two completely separate lives.

Of the cases of multiple personality reported in the first half of the nineteenth century, few were more fully described than that of Estelle L., a young girl treated by Charles Antoine Despine at Bains d'Aix-en-Savoie in France in 1836 (see Fine 1988). Writing in 1838, Despine reported that when Estelle was age seven, her father died in an epidemic that nearly took her mother and sister as well. Signs of hypersensitivity began to appear at that time, and after a fall at age nine, she began to show further symptoms, such as headache, the sensation of weight on the chest, and stomach pains. She gradually lost her mobility and could never get warm. Estelle also began to experience hallucinations, anxiety attacks, and disturbing dreams. In 1836 her mother took her to the baths at Aix-en-Savoie in the hope of some physical remedy. There Despine, a physician at the baths, was called in to treat the young girl (Despine 1838, pp. 2–9).

Despine learned from Estelle's mother that when the girl was alone she had conversations with angelic beings and that she experienced periods of amnesia. He decided to treat Estelle with magnetic sleep. He convinced the mother of this approach, but it took some time to get the daughter to agree. Estelle was very bright intellectually, and Despine was struck by the contrast between her lively mind and her feeble physical state. As treatment progressed, she regained her strength. Though Despine had been pessimistic at the start, he was delighted to see that magnetic somnambulism could be quite effective for this kind of condition.

Despine discovered that Estelle experienced various states while magnetized. She also revealed a number of interior personalities, the principal one being Angeline, a personality who looked after Estelle's health and welfare and prescribed remedies for her physical ills (Despine 1838, pp. 38ff). Despine had a difficult time sorting out Estelle's various states and personalities, and his report mentions only Angeline. But in Estelle's mother's journal, published as a note to the body of the report, a number of names and characters are mentioned—some friendly, some menacing (pp. 93–117).

As Catherine Fine points out (1988, p. 37), Despine's treatment was notable for its openness and ingenuity. Though he failed to get a clear picture of the various personalities, their roles, and how they could be dealt with, he did provide the earliest account of a case of childhood multiple personality disorder.

In 1846 Elliotson catalogued a number of cases of "double consciousness" that occurred naturally and "independent of mesmerism." He believed that "mesmerism produces no phenomenon that does not occur in nervous affections without mesmerism" (1846a, p. 157). He began with a sleepwalking case from 1706 and the cases described by Dewar and Belden, and continued with a case related to him by a Dr. Wilson (pp. 161–68). This involved a certain R. Jones, fifteen years of age, who alternated between a normal state and a state in which he was sleepy yet uncooperative and unruly and ate enormous quantities of food. In his normal state he had no knowledge of occurrences in the disordered state. He was admitted to Middlesex Hospital six times under the care of Dr. Wilson, who refused to apply conventional medical treatment such as bleeding and blistering. After this therapy Jones remained permanently in his normal state.

Elliotson also reported on the case of the German boy Sörgel (see note 4), the cases of Drs. Skae and Mayo, and three cases communicated to him personally. The first involved a seventeen-year-old girl who would suddenly enter a state in which she made unusual motions with her arms and threw objects around in the room. In this state she also played the piano with a sensitivity far beyond that of which she was capable in her normal state. In her second state she kept her eyes closed and responded to no one. These states lasted up to twelve hours. The condition cleared up after a period of six months. Elliotson next wrote about a nineteen-year-old girl who suffered from various hysterical symptoms, including the inability to keep food down. Once or twice a day her voice and mental character would radically change. Instead of being weak and childish, as in her normal condition, she was lively and amiable. This state, lasting up to two hours, ended with a short period of sleep. Upon awakening she recalled nothing of the altered state. Finally, Elliotson narrated the tale, related to him by a Dr. Lingen, of a forty-year-old woman who was subject to periods of alteration during which she would wash, clean, and straighten her household with unusual concentration and energy. In this state she was at least partially anesthetic, for she more than once scalded her leg with no awareness of the fact. Her amnesia for these events while in her normal state was complete (pp. 179–83).

Among cases of naturally occurring double consciousness, some of the most curious were those of the so-called sleeping preachers, who sermonized eloquently while in a second state of consciousness. The best known were Rachel Baker and the Reverend C. B. Sanders.[5]

Rachel Baker was born in May 1794 and lived with her family in Marcellus, New York. At age fourteen she joined the Presbyterian church, but she was uneasy with this choice and two years later became a Baptist. Rachel was very troubled by her feeling of sinfulness and was continually depressed— "sedate, reserved, and diffident, . . . little prone to talk," as Samuel L. Mitchill, who first reported the case, described her. On November 28, 1811, she experienced her first episode of sleep-preaching, in which she spoke of her inevitable doom. From that time these episodes continued, with a gathering sense of religious hope. Mitchill left a vivid account of Baker's "paroxysms":

---

5. It seems that the phenomenon of "sleeping preachers" was not rare. In the spring of 1904 alone, two cases were reported—one occurring in Wooster, Ohio, and the other in Sharon, Pennsylvania (*New York Herald*, June 12, 1904; *New York Tour*, May 14, 1904).

The fit invades her at nine o'clock in the evening or about ordinary bedtime. It commences with spasmodic agitation, heaviness of respiration, and anxiety. . . . The intermittent disorder which Miss Baker suffers, seizes her in bed or in her chair if she sits up. After a few moments of torpor or somnolency, at the usual hour she loses her consciousness and begins to speak in an audible and frequently forcible tone. . . . Thus this modest damsel falls into a devotional exercise as soon as she loses her consciousness. It would be improper to consider her asleep tho' her body and limbs are so quiet and her eyes steadily close. The exercise consists of three parts: the first an incipient or opening prayer to God . . . ; the second an address or exhortation as to a human audience present, and listening to her; and the third a closing supplication to the Supreme Being. (Mais 1814, pp. 5–6)

Mitchill described Baker's sermons as sensible and intelligent. He wrote that her words flowed rapidly and fluently. If bystanders commented or asked her questions, she listened and replied, but she remained in the altered state. The direction of her thought, however, was sometimes altered by the new idea thus introduced. Mitchill gave a detailed description of changes in pulse, temperature, and pallor during her fits, and he noted that when she finished her preaching she usually passed into a tranquil normal sleep that lasted until morning, when she would remember nothing of her preaching of the previous evening (Mais 1814, pp. 6–8).

In *Devotional Somnium; or, A Collection of Prayers and Exhortations Uttered by Miss Rachel Baker* (1815), Mitchill attempted to provide a theoretical framework for the phenomenon he had witnessed. He chose the word "somnium," drawn from Cicero, to describe a state between waking and sleep in which certain bodily and mental actions are performed without direction of the will and without recollection afterwards. Mitchill posited two kinds of somnium: symptomatic and idiopathic. Symptomatic somnium arises from specific external causes, such as indigestion, fever, debility, nightmare, an overloaded brain, sexual excitation, drugs, or alcohol. He also included in this category somnium from "old and forgotten occurrences, (*ab obsoletis*) when long-lost images are renewed to the memory, and dead friends are brought before us" (p. 31). Mitchill described idiopathic somnium as arising from internal states, such as reverie and madness, accompanied by such phenomena as talking, walking, invention (in which original poetry or musical compositions spontaneously come forth), mistaken impressions of sight or hearing, singing, and praying and preaching (pp. 29–35).

Eventually Rachel Baker was brought to New York City, where she was examined by a number of physicians. Some formed the opinion that her condition was a disease of some kind and she should look for a cure. Others concluded that she was affected by an "influx from above" and should consider herself blessed.

Mitchill cited other Americans who had exhibited the phenomenon of preaching while asleep, such as Job Cooper of Philadelphia, said to have preached in 1794 (pp. 49–55). Nor was Baker the last of the sleeping preachers. Among those to follow her was the Reverend Constantine Blackman Sanders, the "Sleeping Preacher of North Alabama," who in 1854, at the age of twenty-three, fell ill and began to be subject to "occasional convulsions" during which he was heard to sing, pray, and deliver religious exhortations in his room. This was the beginning of the development of a peculiar state that was to continue for twenty-two years in which Sanders would be "unconscious" or "asleep" and yet carry on a number of activities.

In 1876 the Reverend G. W. Mitchell, who knew Sanders well, wrote an account of the Sleeping Preacher, whom he considered to be an agent of God:

> There are manifestly various degrees in these peculiar states. When not suffering much, he is more social, more disposed to answer questions; his mental powers are more than ordinarily lucid. In this condition, he will sing and play on instruments of music. It is very interesting, pleasant and generally edifying to be with him in such a case. . . . As a general thing, in cases of increased, but not extreme suffering, his mind is more occupied with religious subjects than ordinary, and often giving expression to the most exalted views of the divine character, and the governance of God. . . . In these religious effusions he often is endowed with such an unction and pathos, that those present are moved and melted to tears. . . . In these exercises, sometimes continued for one or more hours, his remarks are, at times of a general character; at other, personal and pointed. (pp. 29–33)

Before and during his "sleeping" spells, Sanders was subject to physical problems that included headaches, cramps, and breathing difficulties. While in his altered state, he often described events occurring at a distance, found lost articles, revealed the contents of sealed letters, predicted the future, and accomplished other paranormal feats that astounded those who knew him.

Sanders normally experienced complete amnesia for all that occurred while asleep, and his sleeping state had the characteristics of a separate personality, which he called by the peculiar name of $X + Y = Z$. He referred to himself in his normal state as a completely separate person, called My Casket, never using his given name. As $X + Y = Z$ he wrote letters to My

Casket, or his waking self, addressing him as a friend and companion, encouraging him and supporting him in his religious work. $X + Y = Z$ eventually wrote a letter of this kind announcing that he would be departing from the body of Sanders and returning only after a long period of time. The departure took place in 1876, and at the time Mitchell wrote his book $X + Y = Z$ had been gone for four months.

Sanders's contemporaries had a variety of explanations for the phenomena connected with the Sleeping Preacher, "some regarding them all as humbuggery; some, as real, but the result of animal magnetism or mesmerism; some, that of somnambulism, or clairvoyance; and others the manifestations of spiritualism, or rather spiritism" (Mitchell 1876, p. 196). $X + Y = Z$, however, had his own ideas about the matter: "My peculiar developments will not be explained from a scientific standpoint, at least so long as it is assumed that my physical sufferings are the cause of my mental phenomena. The solution may be sought successfully, only from a theological and scriptural standpoint. I am no spiritist nor clairvoyant; neither am I the subject of mesmerism or animal-magnetism. But I am a 'vessel of mercy' whom the Lord hath chosen to this end" (Quoted in Mitchell 1876, p. 198).

Another remarkable case of separation of personalities was described by William James in "Notes on Automatic Writing" (1889). The case, reported by Dr. Ira Barrows of Providence, Rhode Island, involved a young woman, Anna Winsor, who on September 29, 1860, spoke of great pains in her right arm. As the pains grew, the arm suddenly fell limp at her side. The young woman looked at the arm in amazement, thinking that it belonged to someone else. She could not be convinced that it was her right arm, which she believed to be drawn back along her spine. No matter what was done to the right arm—cutting, pricking—she took no notice of it. At the same time she complained of pains in her neck and back, which she now called her right shoulder and arm.

When Barrows wrote of the case in 1865, the hallucination had been firmly in place for five years. Then Anna believed her right arm was indeed an arm, but not her own. She treated it as an intelligent thing and wanted to keep it away from her, biting it or hitting it and generally trying to get rid of it. She saw it as an interference in her life, sometimes taking things that belonged to her. She called the right arm Stump or Old Stump. At the same time Anna's *left* hand sometimes carried out very violent, self-destructive acts. It would tear her hair, rip the bedclothes, and shred her nightdress. Old Stump would protect the woman against the left hand, grabbing the vicious member and restraining it.

When things were quieter and Anna was either asleep or in a state of magnetic somnambulism, Old Stump would engage in all kinds of constructive activities. Old Stump would often write, sometimes producing poetry, sometimes messages from departed persons. The poetry included original pieces. Barrows stated that Old Stump wrote out "Hasty Pudding" by Barlow, even though Anna had never read it. Old Stump also produced a set of verses with English and Latin phrases cleverly combined, although Barrows insisted that Anna knew no Latin.

Old Stump also wrote letters, some of them quite amusing. At times it would answer questions put to it, or give directions about how to care for Anna. Old Stump never slept but was always available to help, sometimes adjusting the bedclothes of the sleeping woman, sometimes knocking on the headboard to get the attention of Anna's mother if some special need arose. When Anna was at her most delirious, Old Stump remained completely rational and helpful.

Anna remained adamant that the arm was not hers. She was not aware of its activities and usually was unconcerned about them. Barrows commented that the arm seemed to have a separate intelligence and an independent life. James noted that Old Stump sometimes wrote about Miss Winsor, but always in the third person, as Anna.

Anna Winsor died in 1873. There is no record of the final ten years of her life. Boris Sidis, commenting on this case in 1898, speculated that Old Stump was probably a "peculiarly crystallized personality formed of the sane remnants of the patient's subconscious self" (1911, p. 289).

In 1869 Samuel Jackson, professor at the Institute of Medicine at the University of Pennsylvania, reported four cases of double consciousness that he had personally treated. The first involved a man who, after a series of seizures, was subject to brief periods during which he would talk as another person and describe a fictional romantic situation in which he believed himself to be involved. The second was a young woman from Virginia with whom Jackson used acupuncture needles to remove neuralgic face pains. Her mother informed Jackson that each day the patient entered into a state of consciousness in which she conversed with imaginary people and spoke of fictitious circumstances. The third case was a depressed girl of fifteen or sixteen who had daily alterations of consciousness in which she held conversations with absent friends. Jackson was later informed by the family that she had returned to a full state of health. The fourth case involved a woman who, when four or five, had been a patient of Jackson's. She was brought to him again in her teens with an eating disorder (an inability to swallow food) and various bodily pains. She soon began to experience states of altered consciousness in which she would carry out household tasks with no subsequent memory of them. In her second consciousness she exhibited a very different character and an altered voice. Sometimes she would leave the house in her second state and return in her normal one. She always had total amnesia in her normal personality for what occurred in her altered state. These alternations of personality continued up to the time of Jackson's report.

A case of double personality that had some notoriety in the nineteenth century was that of Ansel Bourne (see Hodgson 1891; Myers 1903, pp. 309–317; Kenny 1986, pp. 63–96). In 1857 Bourne, a resident of Rhode Island, suffered an illness that left him unconscious for two days. While recovering he went for a walk, and the idea occurred to him that he would rather be struck deaf and dumb than go to church again. He was immediately rendered deaf, dumb, and blind and had to be carried home in a cart. He was converted to Christianity and, in the midst of the congregation, had his sight and hearing restored. Later he became an itinerant preacher, but because of the hardships of that occupation he took up carpentry.

Then, on the morning of January 17, 1887, Bourne left his home to visit his bank in Providence, expecting to return home in the afternoon. He drew out money and did some errands, but then disappeared from the vicinity and was not heard of again for two months. In that time Bourne had traveled to Norristown, Pennsylvania, and under the name of A. J. Brown had opened a small store. The people who made his acquaintance in Norristown considered him to be quiet and dependable. He attended the Methodist church. On the morning of March 14, he awoke after hearing what he described to be a sound like a gunshot to find himself in a strange room and a strange city. The last thing he could remember was walking down the street in Providence in January on his way to his nephew's store. Bourne's nephew was notified and came to Norristown. Having settled the accounts of "Mr. Brown's" store, the two returned to Rhode Island.

In 1890 William James, who had heard of this strange case, invited Bourne to Boston for an interview. James attempted to hypnotize Bourne by using magnetic passes and discovered that he was a good subject. While somnambulistic Bourne was plunged back into the life of John Brown, being able to remember that period very well but incapable of recalling the events of his normal life. James described the Brown personality as "nothing but a rather shrunken, dejected, and amnesic extract of Mr. Bourne himself." An attempt to unify hypnotically the two personalities failed and "Mr. Bourne's skull to-day still covers two distinct personal selves" (James 1890a, 1:392).

Félida X and Louis Vivé

From 1875 to the close of the century, cases of dual or multiple personality were reported with increasing frequency (see Goettman, Greaves, and Coons 1991). Two in particular stand out because they were observed over a long period of time and reported in great detail, and especially because of the light they cast on the relationship between multiple personality and the consciousness of artificial somnambulism.

In 1858 Eugène Azam, who was to be instrumental in importing Braid's hypnotism into France, began to observe a thirty-two-year-old woman who showed signs of double personality. In 1876 he made the first of several reports on the case (see Azam 1876a, 1876b, 1877, 1887, 1893, and Dufay and Azam 1879). He wrote that at age thirteen the woman, whom he called Félida X, had shown signs of hysterical conversions. At fourteen she was subject to spontaneous profound sleeps, after which she would wake up in a different personality—her "second state." In her primary state she was intelligent and industrious but unimaginative and melancholy, suffering from pains in various parts of her body, especially her head. In her second state, which occurred nearly every day, she seemed entirely different. She smiled a great deal, felt very gay, and was emotional and imaginative; she was also completely free of the pains that plagued her in her primary state.

In her secondary state Félida could remember everything that previously occurred in that state as well as in her primary state. But in her primary state she had no access to memories of events that took place in the secondary state. On returning from her secondary to her primary state, she would take up things precisely where she had left them.

Félida became pregnant in 1859 and in her secondary state knew very well what was going on and who the father was, and she was not frightened by her pregnancy. In her primary state, however, she was puzzled and alarmed to note that her abdomen was swelling, a change she attributed to a worsening of her hysterical condition. When the pregnancy progressed so far that in her primary state she could no longer ignore the truth, she was thrown into violent hysterical convulsions. The child, a boy, was born in 1860. Félida later married the father.

Azam was out of touch with Félida from 1859 to 1875, when he again had occasion to talk with her and observe her condition. In the earlier period of observation, Félida's second state had taken up about one-tenth of her life. By 1875 it had become dominant, with her primary state seldom in control. That seemed a blessing to her and her family, for in her primary state she was extremely depressed and in great pain. By now Félida had a family and successfully managed a grocery store, and she dreaded the interference that the onset of the primary state produced.

In attempting to understand the nature of Félida's "double life," Azam perceptively realized that there were phenomena held in common by Félida's second state and the state of artificial somnambulism. He concluded that Félida's primary state corresponded to the normal waking condition of a hypnotic subject and that her second state corresponded to the somnambulistic state. Azam pointed out that hypnotic somnambulists tended to experience limited sensory awareness and curtailed intellectual functioning, and he called this a state of "imperfect somnambulism." He also noted that some artificial somnambulists retain full use of their senses and intellectual agility. He concluded that Félida's second state was a form of this "perfect" somnambulism (called by some "vigilambulism") in which all of her senses and mental functions were active, and so she appeared to be awake (Azam 1887, pp. 155–58).

Azam was thus left with the problem of explaining how Félida's "normal" condition should be the one with defective memory, while her second, "morbid" condition had perfect memory. He proposed a theory of dual cerebral function, along with inhibited blood flow to the brain. In her second state, Azam speculated, Félida functioned from one hemisphere of the brain, a hemisphere that enjoyed perfect blood circulation and therefore perfect memory. In her primary state she functioned from the other hemisphere, one damaged through the physiological alteration that causes hysteria and suffering from inhibited blood flow, and therefore producing imperfect memory (Azam 1887, p. 182). This theory satisfied Azam's need for a physiologically based explanation for the psychological phenomena he observed.

But Azam also attempted to look more deeply into *psychological* factors that might be involved in Félida's double life. When he was observing Félida in 1859, he had just become acquainted with Braid's hypnotism and decided to see if she could be made somnambulistic. Using Braid's eye fixation method he succeeded, and in the process noted a peculiar phenomenon: whether he hypnotized Félida when she was in her primary state or in her secondary state, when she was awakened she always returned to her primary state (Azam 1876a, p. 483). Had Azam followed through with this observation he might have taken the next step to a psychological explanation for multiple personality. But because of his concentration on the physiology of the condition, this step was left to a team of investigators who were at work studying a case that showed characteristics most unusual for the time—a male who exhibited six distinct personalities.

Louis Vivé was born in 1863 to a mentally disturbed mother. At age ten he was sent to a reform school, where he was quiet and well behaved. At four-teen, still in the reform school, he was frightened by a snake while working in a field. This trauma marked the beginning of a cycle of disturbed emotional states, including hysterical symptoms such as epileptic seizures and paralysis of the legs. Then, after being sent to the asylum at Bonneval, he had a hysterical attack that caused a radical change in his mental state. He was no longer paralyzed, he was no longer well behaved, and, whereas he had previously done competent work as a tailor, he no longer could handle needle and thread. Louis now was violent and greedy, drank excessively, and stole. His memory went back to the incident with the snake, but no further. Louis escaped from Bonneval, joined the military, was caught stealing, and was returned to the asylum of Rochefort, where he had lived earlier.

At this time Louis came to the attention of two physicians, Henri Bourru and P. Burot, and from then on he was closely observed not only by Bourru and Burot but also by Drs. Camuset and Jules Voisin. Bourru and Burot first published their observations in 1885. They described Louis as a young man who had six distinct personalities,[6] each with its peculiar set of muscular contractions and anesthesias and its individual set of memories. Each of these personalities or "states" was tied to a particular period of Louis's life and held the memories for only that period. When Louis was in one of these states, he believed himself to be of an age that corresponded to that period. The one exception was his sixth personality, whose memory overlapped with four of the others. Personality 1 was violent and unruly. Personalities 2 and 3 were quiet and well educated. Personality 4 was shy, childlike in speech, and had the skill of a tailor but little education. Personality 5 was obedient, boyish, and well educated. Personality 6 was the best balanced of them all, with a decent character, moderate education, good physical strength, and a memory for nearly all the events of Louis's life.

In 1888 Bourru and Burot published a full study of Louis Vivé under the title *Variations de la personnalité*. This book ranks as the most important study of a single case of multiple personality to be published in the nineteenth century and contains significant advances in understanding the genesis and therapy of that condition. Here at last the link between multiple personality and states of artificial somnambulism was solidly established and the connection between specific personalities and specific memories was acknowledged. Control of switching from personality to personality was achieved through hypnotism. And a therapy that used artificial somnambulism to recover lost

6. In fact, they noted eight, but two of these personalities were not clearly defined (1886, pp. 73–74).

memories and deal with traumatic events was used to improve the psychological health of the subject.

In attempting to understand the data provided them by Louis Vivé, Bourru and Burot laid out the basic concepts they would apply. Following the framework devised by Ribot in his *Les maladies de la personnalité* (Ribot 1885), they defined personality as the result of two basic factors: the constitution of the body with the tendencies and sentiments it expresses, and memory. In discussing the second element, they stated: "The comparison of anterior states of consciousness with actual states is the bond that unites the past psychic life with the present psychic life. This is the fundament of personality. A consciousness that compares itself with the past person is a true person. It judges itself in time and can in this way, with a powerful force, affirm its existence" (p. 263).

From this starting point Bourru and Burot described the condition exemplified by Louis Vivé as "the alternation or entanglement of two or more personalities." They then went on to outline various types of relation between personalities housed in one body:

> Sometimes two personalities follow one after the other and are reciprocally unaware of each other. Sometimes one embraces the whole of the person's life, and the other only a part. One can even find cases where the second personality constantly encroaches on the first which, as time goes on, becomes more and more curtailed, so that one can foresee the time when it will disappear and the second alone will continue to exist. It can happen that many personalities are entangled with each other. In some cases one might use the term "double consciousness," but in the presence of the multiplicity of states of consciousness that we sometimes see, it is convenient, in our opinion to designate them by the more vague but more realistic term of *variations* of personality. (1888, pp. 9–10)

By hypnotic suggestion, Bourru and Burot succeeded in reviving dormant memories "through transporting [subjects] to a certain period of their past" (p. 13). They described a case of somnambulistic regression of a seventeen-year-old girl to the age of six. During the regression she relived in vivid detail an incident that involved peeling chestnuts and carrying out a particular prank. While regressed she spoke in her original patois and could not utter a word of French.

Bourru and Burot used this knowledge of the effect of hypnotism on memory to reach a significant conclusion about Louis Vivé. Because his different personalities were tied to definite periods of his life, they connected specific personalities to specific memories. From this they learned to control the switching between personalities by inducing specific memories. They described how Louis spontaneously switched from personality to personality (pp. 91–98), but they also discussed how they could control that switching and cause Louis to enter into whatever personality (and period of his life) they chose (pp. 98–120). They were able to use muscular excitation, contact with various metals, electrical stimulation, and drugs to bring on the personalities, and they catalogued which type of muscular, metallic, electrical, or narcotic excitation would trigger which personality.

They also discovered that artificial somnambulism could be used to bring forward personalities and the accompanying memories: "February 10 . . . After a few minutes in the hypnotic state, he goes into his habitual crisis. But for the first time he experiences a delirium with terrifying hallucinations. He sees before him the snake—the contact with which occasioned his first attack at Saint-Urbain. He lets out sharp cries. He tears his clothes. His features are disturbed. He is calmed by our assuring him that the viper has been killed" (p. 107). Jules Voisin, working hypnotically with Louis, was able to use suggestion to control what memory (and personality) would come forward. "Seeing the magnet produced an attack and he found himself transported to the hospice of Bicêtre. . . . If one suggests to the patient when in a state of somnambulism that he is being transported to Saint-Urbain or to Rochefort or to Bonneval, etc., etc., we get the same results: the patient has an attack that corresponds to his state [in those places] described above" (pp. 113–15).

In looking for an overall explanation for multiple personalities, the authors rejected as inadequate Eugène Azam's theory of dual cerebral hemispheric functioning to explain duality of consciousness and his notion that inhibition of blood flow to the brain accounted for differences of memory in the two states. The association of the two hemispheres to alternating states might work for cases such as Félida X, but, they pointed out, in the case of Louis Vivé a multiplicity of alterations of personalities must be explained. In this instance Azam's theory of duality is useless (pp. 270–90).

The explanation arrived at by Bourru and Burot was influenced by research conducted by Hippolyte Bernheim, who noted that in some cases a simple closing of the eyes could create a new state of consciousness. Any impressions gained in that new state are then associated with that state. When the eyes are opened, the state of consciousness is altered and the impressions of the closed-eyes state are extinguished, but will return when the eyes are once again closed. Bernheim considered such specially created states to be comparable to hypnotically induced states. It is important to note that for Bernheim the hypnotic state was not abnormal at all, but an artificially induced condition that mobilizes functions already available in normal life. In that framework, Bernheim asked: Is it not possible that in reality there is not one state of consciousness in human beings, nor two states, but rather any number of states that are infinitely variable (pp. 272–77)?

Bourru and Burot decided to look at multiple personality in a similar way, seeing it as a special instance of successive states of consciousness. "Our body can hold two contrary attitudes in rapid succession without ceasing to be the same body. It is also clear that even three states can succeed each other (appearing to coexist) by the same mechanism. We are not stuck on the number two. . . . Experience seems to favor a very rapid succession—equivalent to a coexistence. The two or three or four contrary states would at bottom be a succession" (pp. 289–90). To give a physiological base to this view, Bourru and Burot posited a special energy activated by psycho-motor stimuli. These stimuli served as suggestions that triggered a corresponding state of consciousness and, when sufficiently unified, a corresponding personality (pp. 294–97).

This view had important practical consequences for the therapy of multiple personality—consequences that had often been confirmed in their work with Louis Vivé. If Louis went into a crisis involving a disturbed personality, they could use hypnotic suggestion (or specific physical stimulation that served as a specific suggestion) to restore him to his normal state. They were even able to use these methods to put an end, for the most part, to his switches from personality to personality. They accomplished this by spotting which personality and which memory were active and by intervening at an opportune time to restore Louis's normal personality. They considered the importance of this technique to be that it enabled the revival and restoration of blocked memories (pp. 298–300).

With their explanation of successive multiple personalities, Bourru and Burot made a central contribution to the evolution of an effective approach to psychological healing. While these investigators were working and writing, Pierre Janet was carrying out his own experiments with hysterical patients, some of whom exhibited multiple personalities. His experiences took him in similar directions, but, as we shall see, Janet reached different conclusions about the temporal relationship between the personalities.

Chapter 15

Pierre Janet: Dissociation and Subconscious Acts

The year 1884 marked the one hundredth anniversary of the discovery of magnetic sleep. The revelation of the existence of a second consciousness in human beings had intrigued psychologists and physiologists, but the practical implications of that revelation were slow in coming. The recognition of cases of spontaneously occurring divided consciousness and multiple personality was an important factor in the development of a new form of psychological healing. But in 1884 the final step in the consolidation of the basis for effective psychotherapy had yet to be taken. The nature and mode of action of the alternate consciousness of magnetic sleep needed to be further clarified.

That clarification came in the 1880s, when four essential characteristics of the alternate consciousness were identified. The alternate consciousness was seen as *intelligent*, capable of understanding facts and events and making judgments based on reasoning; *reactive*, aware of what is happening in the environment and capable of responding to those events; *purposeful*, able to pursue its own goals and take action based on its own criteria, which may differ from those of the individual's normal consciousness; and *co-conscious*, existing simultaneously with the consciousness of daily life (even though unrecognized by that consciousness) and carrying out its own operations concurrently with those of normal consciousness.

The identification of these characteristics was the work of a number of researchers who drew upon the whole magnetic tradition. Among them, no one contributed more to the outcome than Pierre Janet.

Pierre Janet

Janet was born in Paris in 1859 to a middle-class family that had produced many professionals—scholars, lawyers, and engineers.[1] At nineteen he was admitted to the prestigious Ecole normal supérieure, which trained its students for professorships in the French lyceums and universities. In

---

1. Much of the material presented here on Janet's life is drawn from Ellenberger (1970, pp. 331–417).

1882, then age twenty-two, he was appointed professor of philosophy at the lyceum in Châteauroux. A year later he transferred to the lyceum in Le Havre, where he remained until 1889.

While in Le Havre he began work on his doctoral thesis, *L'automatisme psychologique* (1889), which he based on experiments with hysterical subjects. A physician of that city, Dr. Gilbert, acquainted Janet with the case of Léonie, who, it was said, could be hypnotized from a distance, and Dr. Powilewicz at the hospital in Le Havre gave Janet a ward in which he could carry out experiments with Léonie and other hysterical patients. Janet's experiments with nineteen hysterics—fourteen women and five men—became the basis for three extremely important articles published in the *Revue philosophique* (Janet 1886c, 1887, 1888).

In 1889 Janet moved to Paris to take up medical studies. Meanwhile he taught at the Lycée Louis-le-Grand and continued his study of psychopathology. He spent a great deal of time in Charcot's wards at the Salpêtrière Hospital, concentrating on the phenomena of hysteria. When he received his medical degree in 1893, Janet was already internationally respected as a psychological researcher.

Janet's key contributions to the development of psychological healing based on the alternate-consciousness paradigm stem from his work in Le Havre. There he elaborated the concepts of unconscious or subconscious acts, dissociation, mental disaggregation, and psychological automatisms.

Unconscious or Subconscious Acts

Janet defined an unconscious act as an action having all the characteristics of a psychological happening except that, at the moment of its execution, the person carrying it out is totally unaware of it. He listed four kinds of unconscious acts: (1) those deriving from post-hypnotic suggestion, (2) those produced by anesthesia, (3) those that occur during "distraction," and (4) spontaneous unconscious acts. In the last category are acts performed by individuals suffering from hysteria (1888, pp. 239–40).

Writing of his work with "Lucie," Janet noted that, when placed in a state of artificial somnambulism and given post-hypnotic suggestions, she would execute those suggestions perfectly in the waking state. But how could commands issued during somnambulism and forgotten nonetheless be obeyed? How could Lucie know how to fulfill suggestions of which she had no conscious awareness? Janet decided that since the waking Lucie had no awareness of the command (or of her fulfillment of the command), there must have been another consciousness involved in the execution of the order. There must have been a doubling of the "I," so that Lucie did not know what her second consciousness knew.

To confirm his suspicion, Janet decided to contact the second consciousness through automatic writing, and this strange dialogue ensued:

> "Do you hear me?"—(She responds in writing) *"No."*—"But one must hear in order to answer."—*"Yes! Absolutely!"*—"Well, how do you do this?"—*"I do not know."*—"There must be someone who hears me."—*"Yes."*—"Who is it?"—*"Someone other than L."*—"Indeed! Another person; should we call you Blanche?"—*"Yes, Blanche."*— "Well, Blanche, do you hear me?"—*"Yes."* (Janet 1886c, p. 589)

As Janet soon discovered, it was Blanche, or actually Adrienne (the name she eventually chose for herself), who carried out post-hypnotic suggestions. Adrienne also revealed that she carried memories of a traumatic event that was not known to Lucie. That traumatic memory and the need to keep it hidden from Lucie were the cause of Lucie's hysterical condition. Janet also learned that Adrienne sometimes experienced emotions—such as fear or anger—to which Lucie remained oblivious. As his work progressed, Janet decided to use hypnotic suggestion as a psychotherapeutic tool to see if he could improve Lucie's condition. To his delight, she gained ground steadily to the point that the worst of her hysterical symptoms (convulsions and severe headaches) disappeared. Janet was intrigued to note that as the hysteria diminished, Adrienne became less and less available and finally disappeared (pp. 588–92).

In his second article (1887) Janet noted that Lucie's improvement had only been temporary: eight months later the hysteria returned. Though this was unfortunate from her point of view, it did allow Janet to continue his experiments with doubled consciousness.

Reminding the reader that all Lucie's unconscious acts were conscious to Adrienne, Janet further specified that Lucie did not forget the post-hypnotic suggestions given her; in fact, she never knew them. It was Adrienne who heard them and Adrienne who executed them. After Lucie had been brought out of her state of somnambulism, and at the moment when Adrienne carried out the command, Lucie entered into a special state of consciousness, which Janet labeled "somno-vigil" or "somnambulic waking." In this special state Lucie was unaware of Adrienne's actions in carrying out the post-hypnotic suggestion (pp. 449–51).

Janet's second article considered the status of Adrienne. Was she a creation of the experimental situation, the result of a maladroit suggestion on his part? Did she have any previous existence in her own right? Through an examination of Lucie's history, her present states, and the sequence of events that led to

Adrienne's emergence in their work, he came to the conclusion that Adrienne was not created by his suggestion. On the contrary, he believed, this kind of doubling of personality was at the very heart of the hysterical condition (pp. 451–52).

To further understand the relationship between Lucie and Adrienne, Janet set up an experiment. After putting Lucie into a state of artificial somnambulism, he gave her ten blank cards and suggested to her that a specific card had a picture on it. In her suggestive state she hallucinated the picture on that particular card. Janet then shuffled the cards, and when he displayed them to her again, she again saw the picture, and on the very same card. Binet and Féré had previously pointed out that in similar experiments hypnotic subjects unconsciously noted minute details that distinguished the specified particular blank card from the rest and would use that information to identify the "picture" card. It was only after the proper blank card had been noted that the suggested hallucination was projected onto it.

There is a problem in explaining how all this can occur. The somnambulist knows nothing of this process. She does not know that she is hallucinating a picture, and she does not know that she is using subtle information to pick out the right card onto which to make the projection. If the specified blank card was noted through minute observation, who or what was doing the observing and noting? Who or what was making the judgment that this was the card onto which to make the suggested projection? In an effort to answer these questions Janet, having performed this experiment with Lucie, asked Adrienne what she saw on the "picture" card. Adrienne said she saw a spot on the upper righthand corner but nothing else—no portrait. In this way Janet determined that it was Adrienne who picked out the specific card and Lucie who experienced the hallucination (pp. 452–53).

To look more deeply into the matter, Janet decided to attempt experiments with "negative hallucinations" or "systematized anesthesias." This phenomenon consisted in blocking the ability of the somnambulist to see a person or an object that was present. Deleuze, Bertrand, Charpignon, Teste, Braid, Durand de Gros, and Liébeault had experimented with this phenomenon before Janet, but they had accomplished the feat by replacing the vanished object with an imaginary one. Janet noted that Bernheim, who invented the term "negative hallucination," was the first to produce clear cases of blocked perception that did not involve substitution.

Recent investigators had discovered some interesting characteristics of negative hallucinations. If, for instance, the suggestion was made that a certain person in the room, Mr. X, had disappeared, the somnambulist would be unable to see him even when looking right at him. If Mr. X were then to put on a hat, the somnambulist would see it floating in mid-air. If Mr. X was standing in front of a piece of furniture, the somnambulist would use her imagination to fill in the missing perception and actually believe she could see it.

In another kind of experiment, if six boxes were placed in front ,of a somnambulist and the suggestion were made that one specific box had disappeared, the somnambulist would not see that box no matter how the boxes were arranged, provided that there was some minute flaw in that particular box that could be unconsciously noted. The researchers (mainly Binet and Féré) pointed out that to have a negative hallucination of this sort, it was necessary that the subject *recognize* the box that she could not *perceive*: only then could she decide which box to avoid seeing. But since the subject did not perceive the box and did not know that the box was blocked from her perception, the recognition must have taken place on an unconscious level. The unconscious recognition was then used to guide the suggested negative hallucination (pp. 455–57).

To create a negative hallucination for Lucie, Janet took five blank cards and marked two with a cross. He gave Lucie the suggestion that when she awoke she would be unable to see the marked cards. When awakening Lucie had no memory of Janet's suggestion. Janet asked Lucie to count the cards and hand them to him one by one. She handed him the three unmarked cards, and when he insisted that she hand him the rest, she said there were no more. But when Janet turned the cards over, Lucie handed him all five.

Janet then addressed Adrienne, using a procedure he called "distraction." While Lucie was deeply engaged in conversation with a third person (totally distracted), Janet would speak to her, or more properly to Adrienne, in a firm, low tone, asking her to respond to him in writing. Lucie would be completely unaware both of Janet's command and of the writing that resulted. In this fashion, Janet asked Adrienne to indicate how the cards appeared to her. She said that two of them were marked with crosses. When Janet asked her why Lucie had not handed those cards to him, Adrienne responded that Lucie could not see them. So it was Adrienne, free of all hallucination, who saw the cards and their markings and guided the blocking of Lucie's perception (pp. 457–59).

Through further experiments, Janet confirmed a number of important facts about the relationship between Lucie and Adrienne. When given suggestions for negative hallucinations, Adrienne always saw what Lucie was unable to see. Adrienne had knowledge of life events for which Lucie had naturally induced amnesia (traumatic events of her past). Although Lucie was subject to specific hysterical symptoms related to sensation (complete anesthesia in her whole body, inability to detect physical contact or changes in temperature or pressure), Adrienne had no such symptoms (pp. 457–61). These observations led Janet to a profound insight—that psychological functions commonly called unconscious are not unconscious at all:

There are certain words used by some authors that I have not been able to understand; they are: "unconscious perception" and "unconscious reasoning." If a phenomenon is not conscious, it cannot be a psychological event—that is, a perception or a reasoning. Rather it changes and becomes something else—perhaps a simple movement. A reasoning can only occur in the head of a person if he is conscious. In the present case why is the reasoning not known by the somnambulist or the hysteric? Because a psychological phenomenon can be conscious and yet not attached by association to the group of sensations and memories that constitute the idea of the "I." When this phenomenon presents itself, it can remain isolated and totally separated, after having fulfilled its role, without being associated with any manifestation and without the memory of it begin awakened by any deed, since nothing is tied to it. Or indeed it could be associated with other facts equally separated from all consciousness and form, as it were, a second personality, as we see in the case of Adrienne. In a word, the anesthesia is a simple dissociation of phenomena, in such a way that every sensation or every idea brought to normal consciousness still subsists and can sometimes be rediscovered as part of another consciousness. (p. 462)

With this formulation, Janet opened the door to a new way of understanding hysterical and hypnotic phenomena based on his concept of dissociation.

Dissociation

Janet's 1887 article contained his first use of the word "dissociation." Around this term he built a cohesive notion of the nature of hysterical functioning, which enabled him to throw light on the intrinsic structure of the human psyche.

"Dissociation" had rarely been used in a psychological context before Janet.[2] Jacques Joseph Moreau (1845) had written about the "dissociation of

2. In 1811, Puységur described three distinctive qualities of magnetic somnambulism that to some degree foreshadow much later discussion of the nature of psychological dissociation. The first quality was isolation, by which the somnambulist has no relationship with objects or people around him, with the exception of the magnetizer, with whom he has a special rapport. The second quality was concentration, or an inner preoccupation of the somnambulist of such

ideas" in his treatise on hashish. In a section describing the sense of self, Moreau stated that it is through memory that new ideas are assimilated and made a part of our being. It is through memory, he wrote, that we live in the past, and it is through memory that we are able to have consciousness of being a self continuing in existence and renewing our existence at each moment. This important assimilative function of memory occurs through a process by which our will brings about an association of the images or ideas retained in memory. Hashish has the effect of weakening the will and breaking down this association of ideas. The resulting dissociation creates confusion and mental disorder (pp. 62–67).

Janet's formulation of dissociation went far beyond that of Moreau. Working from his experience with hysterics, he condensed his ideas into two laws of dissociation. The first law was that ideas can be conscious but not associated with that grouping of sensations and memories that make up the "I." This means that these ideas are not conscious to the normal personality. Further, they can be associated with a second personality, distinct from the normal personality, and be conscious to that second personality.

The second law of dissociation derived from the observation that every phenomenon artificially associated with or attached to the second personality was withdrawn from the awareness of the normal personality. This showed that dissociation did not involve the undoing of a unity already existing; rather it involved the *association* or assignment of a phenomenon, immediately when it occurs, to one system and not to the other. Dissociation, then, is in no way to be thought of as dis-association. Nothing is being undone; there is no breaking away or splitting off of one group from another. Rather, psychic events are grouped as they occur. Those groups will have various degrees of complexity; some of them are sufficiently complex to constitute a personality. The groups begin as and remain isolated units. If the normal consciousness does not have knowledge of an event, it never did have knowledge of it. Forgetting is not involved.

From experimentation with hysterical patients, Janet concluded that hysterical anesthesias were of the same order as artificially induced anesthesias and were not a problem of sensation but of association. Adrienne served as a good example. Certain ideas and sensations in Lucie's life were not assimilated into her normal consciousness but were assigned to or associated with the grouping that came to be called Adrienne. This grouping was sufficiently

---

intensity that he cannot be distracted from it. The third quality was magnetic mobility, that is to say, the somnambulist is subject to the thoughts and will of the magnetizer in a very direct and compelling way (Puységur 1811, pp. 42–46).

complex to warrant being called a personality. This personality was dissociated from Lucie—that is, Lucie knew nothing of Adrienne, since ideas and sensations associated to Adrienne never reached Lucie's consciousness. As a dissociated personality, Adrienne could perceive, think, respond, and act. But all of these activities remained outside Lucie's awareness.

Janet discovered that one of the most effective ways to remove hysterical symptoms was through suggestion, both waking and hypnotic. The effectiveness of suggestion was not surprising to Janet, since he believed that hysterical symptoms were simply the manifestation of a dissociated group of ideas crystallized in a personality. The hysteric was in a perpetual but unrecognized dream state, in which a second personality was able to manifest undetected in daily life. This second personality was intelligent and purposeful, with palpable good sense, capable of carrying on a line of thought simultaneous with but completely independent of the thinking taking place in the normal personality. This meant that the second personality had a continuous existence that did not disappear when the normal personality was functioning. Also, the second personality was able to initiate a line of action in accordance with its own thinking, even if it contradicted the desires of the normal personality (1888, pp. 249–54, 262).

Psychological Automatism

From the beginning Janet was careful to say that he carried out his experiments and derived his conclusions as a psychologist, not a physiologist. (He did not receive his medical degree until 1893.) Again and again he pointed out that he, not being a medical experimenter, could not judge the merits of the explanations offered by physiologists for some of the phenomena he had observed. In fact Janet believed that physiological explanations for psychological issues had severe defects and that psychological data had to be treated psychologically. This can be seen most clearly in his central notion of "psychological automatism."

In 1889 Janet published his monumental *L'automatisme psychologique* in which he attempted to outline a new scientific theory of psychological pathology. In his introduction, Janet stated that all sciences must study the most basic and rudimentary forms of the phenomena with which they are concerned. For the science of psychology the basic phenomenon is automatic activity. This elementary activity partakes of both determinism and spontaneity, both automatic action and consciousness, and for that reason he calls it "psychological automatism" (pp. 1–3).

Setting the framework for his approach, Janet stated his opposition to those, such as Prosper Despine (1880), who believed that somnambulists were totally unconscious and their actions were physiologically automatic.[3] Such a view, wrote Janet, cannot be sustained when one hears somnambulists speak, resolve problems, manifest spontaneous sympathies and antipathies, and sometimes resist the commands of their magnetizer or hypnotist. These are not the actions of an automatic puppet. Despine and many others argued that since the somnambulist in the waking state knows nothing of what happened during somnambulism, and since there is only one conscious "I" in any human being, all actions performed during somnambulism must be automatic. Janet insisted that the unity of the "I" had to be established by facts, not assumed by virtue of some metaphysical theory. In this way Janet placed the issue of the unity of consciousness at the heart of his psychological researches (pp. 22–26).

The elementary activity to be studied in psychology only appears to occur automatically and without consciousness. In fact it is performed with the participation of consciousness, *but not the ordinary consciousness of the individual.* The act is produced by a consciousness distinct from the normal consciousness of daily life and entirely unknown to normal consciousness. It seems to be automatic, but that is an illusion. To understand how one could have a second consciousness that is active but hidden was the central task of *L'automatisme psychologique.*

In his work with hysterics, Janet found it fairly easy to show that they experienced successive states of consciousness that were distinct from one another. These successive states had all the characteristics necessary to deem them successive personalities. Lucie and Adrienne, for example, were distinguishable personal units with different views, values, and abilities. Janet claimed that personality involves the grouping of psychological phenomena in a synthesis that experiences itself as an "I." Whenever one finds this synthesis and the corresponding judgment of an "I" within that synthesis, there is a personality. For Janet the role of judgment in the structuring of personality was central:

> When a certain number of psychological phenomena are united, ordinarily there is produced in the mind a new, very important reality: their unity, noted and understood, gives birth to a particular *judgment* that one calls the ideas of the "I." It is, we say, a judgment and not an association of ideas. The latter reproduces the phenomena, one after the

3. Neither did Janet agree with a view from the opposite end of the spectrum, expressed by Durand de Gros (1868), that the human individual is the seat of a polyzoism by which various parts of the organism might be considered centers of independent intelligent action (pp. 94–99).

other, it juxtaposes them automatically and from that gives us occasion
to note their unity, to judge their resemblance; but it does not of itself
constitute this connection of unity and resemblance. Judgment, on the
contrary, synthesizes different acts, establishes their unity and . . .
forms a new idea: that of personality. (Janet 1889, p. 117)

When Lucie or Adrienne speaks of herself as "I," Janet says, she reveals that
she has made this judgment and shows that she is a separate personality.

Janet provided many examples of successive personalities in hysterics. In
L'automatisme psychologique we learn that Lucie actually had three successive states or personalities, that a subject named Léonie, discussed in the
articles in *Revue philosophique,* also had three, and that a hysteric whom Janet
called Marie had four or five (p. 334).

Janet discovered that the hysteric's alternate state of consciousness, or
alternate personality, could quite easily be provoked through the induction of
somnambulism. On the basis of his experiments he went further and insisted
that somnambulistic states themselves could show all the characteristics of
personality and could be seen as successive personalities for the hypnotic
subject. He showed that the old magnetizers had repeatedly made the same
observation, noting, with surprise, that in magnetic sleep the subject would
often exhibit qualities at variance with the normal state and speak of the
waking self as another person (pp. 125, 131). Janet believed that the second
personality of somnambulism was usually formed as a result of a number of
external influences, the most potent of which was the magnetizer himself:
"The second personality who is being born undergoes the influence of ideas
and manners of his magnetizer as an infant undergoes the influence of his
parents. It takes on the habits, manners, and beliefs which have been inspired
in him, almost without knowing or intending it. As the magnetizer, so the
somnambulist, one might say. Show me a somnambulist and I will immediately know who induces the sleep state and the opinions and beliefs, scientific or otherwise, of his first master" (pp. 127–28).

Having established the reality of total psychological automatism or successive personalities in hysteria and in somnambulism, Janet moved on to partial
psychological automatism: a consciousness (or consciousnesses) existing and
acting simultaneously with the normal consciousness, but totally outside its
awareness: "Psychological automatism, instead of being complete and regu-
lating all conscious thought, can be partial and regulate a small group of
phenomena separated from the others, isolated from the total consciousness of
the individual, and continuing to develop by themselves on their own count
and in a different way" (p. 224).

Janet discovered, through specially designed experiments, that the second consciousness that operates behind the scenes in an individual is intelligent, capable of complex mental operations, able to resist suggestions, and a source of ideas that could be at variance with those of the normal consciousness (pp. 237–65). The second, hidden consciousness was also capable of producing actions, feelings, hallucinations, and impulses in the normal consciousness which it could not account for. Janet called a phenomenon of this kind an "unconscious act," which he defined as "an action having all the characteristics of a psychological act, save one: that it is always unknown by the person himself who executes it at the moment of its execution" (p. 225).

Janet quickly established that he preferred the term "subconscious" to "unconscious": "These phenomena seem to appertain to a particular consciousness below the normal consciousness of an individual. This is not an explanation beyond doubt, but the observation of a fact, as bizarre as it may seem. We can simply summarize these observations by henceforth calling these acts subconscious happenings [*faits subconscients*], having a consciousness below normal consciousness" (p. 265).

The presence of subconscious acts was a sign that a "narrowing of the field of consciousness" had occurred (pp. 190–99). Janet held that a normal consciousness should be able to take in many sensations and impressions at one time. In some circumstances, however, this "field of consciousness" is found to be greatly narrowed, so that the individual is focused in on a very small range of impressions. Some impressions received while the field consciousness is restricted do not register with the primary consciousness but are received by the individual's subconscious strata. In hysterics the field of consciousness is restricted by their illness. In hypnotic somnambulism the subject's field of consciousness is narrowed by an artificially induced state. While the field of consciousness is restricted, the person is very suggestible. Needless to say, such diminished states of awareness also make it difficult to deal with the complexities of life.

Experimentally, the reality of subconscious acts was brought home most clearly by negative hallucinations, wherein because of a post-hypnotic suggestion the waking subject is unable to perceive an object directly in the field of vision. Janet found that an idea suggested during somnambulism did not disappear after waking, even though the subject seemed to have no awareness of it. He insisted that the suggested idea was nonetheless there to be acted upon by the individual, that it "subsists and develops itself outside of and below normal consciousness" (p. 269). This was the only way to explain how the subject could have the information needed not to perceive what was there.

The existence of subconscious acts indicated that the person in whom they occurred was in a state of "mental disaggregation" (p. 282). Ideas could group themselves in various units, each with a sense of self or "I." The separation of these units accounted for the amnesia. In Janet's view, this state of affairs could exist only because of a weakness of the "mental force" that synthesizes ideas and sensations. In the absence of that synthesis, the individual exists in a state of psychological disaggregation. This is the basis of the formation of the second psychological state existing simultaneously with the normal one.

To illustrate this, Janet cited the example of automatic writing. In automatic writing a person holds a pen or pencil lightly on a piece of paper and allows movements to occur outside voluntary control, movements that might become intelligible writing. The writer might have no awareness of what is being written until he reads the finished product, or may have an awareness of the content as the writing is produced but without a sense that he is producing it or thinking the ideas being expressed. In Janet's experiments the automatic script sometimes revealed intelligence and original thought, yet the thinking processes involved were never a part of the normal consciousness: "Subconscious writing constantly uses the word 'I.' This is the manifestation of a person exactly as in the normal speech of the subject. . . . Here is a secondary personality, a secondary self. . . . Without doubt this 'secondary self' is very rudimentary at the beginning and can hardly be compared to the 'normal self,' but it develops itself in a way that is quite remarkable" (p. 317).

Just as Janet had shown the existence of multiple successive existences or consciousnesses, so he also demonstrated that there were various levels of simultaneous subconscious existences or consciousnesses:

> The unconscious life of . . . Lucie, for example, seems to be composed of three parallel streams one under the other. When the subject is awake, the three streams still exist: the first is the normal consciousness of the subject who speaks to us, the two others are groups of sensations and acts more or less associated among themselves, but absolutely unknown by the person who speaks. When the subject is put to sleep in the first somnambulism, the first stream is interrupted and the second surfaces. It shows itself in broad daylight and lets us see the memories that it has acquired in its subterranean life. If we pass to the second somnambulism, the second stream is interrupted in its turn and only the third subsists, which then forms the whole conscious life of the individual in which one does not see anesthesias or subconscious acts. It would be necessary to complicate the diagram to represent other subjects who have more numerous somnambulistic states, natural somnambulisms,

crises of hysteria, etc., but the general disposition can, I believe, remain the same. (p. 335)

Janet's notion of the layering of somnambulisms is of great psychotherapeutic importance. When the last level of somnambulism is reached, there are no subconscious acts—in other words, that level seems to be "*the state of perfect psychological health: the power to synthesize being very great, all psychological phenomena, whatever their origin, are united in the same personal perception,* and consequently the second personality does not exist. In such a state there would be no distraction, no anesthesia (systematic or general), no suggestibility and no possibility of producing a somnambulism, since one could not develop subconscious phenomena, which would not exist" (p. 336).[4]

In Janet's view, if a disturbed individual is going to be helped to perfect or near-perfect health, it is crucial to understand the subconscious structuring involved. He remarked that the revelation of various subconscious personalities, each operating in its own sphere and yet influencing the others, may not seem of great theoretical value, but it had crucial implications in therapeutic practice: "The systems of psychological phenomena that form the successive personalities of somnambulism do not disappear upon awakening, but subsist, more or less complete, below normal consciousness, which they can change and disturb in a most singular way. . . . The synthesis that forms the personal perception at each moment of life therefore shows us the original activity which previously had been the source of what we today call automatism" (p. 365).

To deal with the subconscious personalities and discover the original activity that produced them often meant uncovering memories of events in the life of the individual that could not be synthesized into the normal personality. This approach became the basis for a comprehensive psychotherapy that Janet was to practice and elaborate for decades to come.[5]

Subconscious Fixed Ideas and Psychotherapy

For Janet, psychological health consists in being able to unite diverse experiences in one grouping through an internal "synthesizing force." Weakness of the synthesizing force produces a plurality of groupings. If the experiences

4. This would explain a phenomenon noted by all magnetizers: the healthy could not be put into a state of magnetic sleep.

5. On Janet's psychotherapy, see Ellenberger 1970, pp. 331–417; Haule 1986; Van der Hart and Friedman 1989; and Laurence and Perry 1984.

that escape synthesis under the normal personality are of sufficient number and importance, the secondary groupings take the form of subconscious personalities. These subconscious personalities are fed by new experiences and gain more and more internal unity. Operating independently from the normal personality, they can affect the perceptions, emotions, and actions of the individual in such a way that the normal personality feels at odds with himself or herself, subject to phobias, compulsions, hallucinations, and other symptoms for which there is no apparent explanation.

People who suffer from hysteria are, according to Janet, the clearest examples of how this works. They have a particularly weak synthesizing force, and as events in their lives unfold, they are often unable to synthesize them or experience them as belonging to their normal personality. Instead they form secondary subconscious centers of synthesis that take the form of personalities. In his work with these individuals, Janet discovered that the unassimilated experiences were often traumatic events. Because those events were very uncomfortable, a greater synthesizing force was needed to assimilate them. That force was not available, so the events formed new subconscious personalities or were assimilated to already existing secondary personalities.

The weakness of psychic force characteristic of the hysteric is, in Janet's view, congenital. For that reason, the basic hysterical condition is incurable. Nevertheless, certain symptoms can be removed through uncovering the original traumatic events and helping the individual synthesize them under the normal personality.

But did Janet believe that his theory was relevant on a practical, psychotherapeutic level only to hysterics? What about people who are not hysterics yet suffer from psychological disturbances that, in addition to hysterical symptoms, manifest "fixed ideas, impulsions, anesthesias due to distraction, automatic writing, and finally somnambulism itself"? Janet's answer is firm: "It is not hysteria which constitutes terrain favorable to hypnotism, but it is hypnotic sensibility that constitutes favorable terrain for hysteria and other illnesses" (pp. 451–52).

Weakness of the synthesizing force, wrote Janet, is the basic characteristic that underlies all such psychological problems. This "moral weakness" explains a whole spectrum of psychological disturbances, and hysteria is merely one of its manifestations. Janet labeled this weakness "psychological misery" (p. 455). With this position, Janet departed radically from Charcot, who believed that hysteria is the basic condition that accounts for all dissociative phenomena, including artificial somnambulism.

The state of psychological misery need not be ongoing and permanent, as is the case with hysterics. It can also be temporary and passing. Physical problems, fatigue, or drugs can produce this passing state of misery, and even emotion "has an action that dissolves the mind, diminishes the synthesis and makes it miserable for a moment" (p. 457). This means that "even the most normal men are far from always existing in such a state of moral health; and as to our [hysterical] subjects, they attain that very rarely. . . . Outside of this state of perfect health, *the power of psychic synthesis is very weak, and fairly sizable numbers of psychological phenomena are allowed to escape outside personal perception. This is the state of disaggregation*" (pp. 336–37).

What, then, is the result of these temporary states of psychological misery and disaggregation?

> If during this unhappy period, the ill person has not been impressed by any abnormal sensation, if he has not been struck by any specific, dangerous idea, he will be healed with little difficulty. He will preserve little or no memory of this accidental state, and remain, during the rest of his life, perfectly free and reasonable. . . . But if, unhappily, a new distinctive and dangerous impetus is brought to bear on the mind at the moment when it is incapable of resisting, it takes root in a group of abnormal phenomena, it develops there and stays there. When the troublesome circumstances disappear, the mind will try in vain to regain its usual strength. The fixed idea, like a morbid virus, had been sown in it and developed in a place within his person that he cannot reach. It acts subconsciously, troubles the conscious mind, and provokes all the symptoms of hysteria or insanity. (p. 457)

The concept of a fixed idea as an enduring, irrational, obsessive notion was a familiar one in psychological literature.[6] But Janet gave the concept a specific meaning and developed it as a central feature in his theory of psychological pathology and psychological healing. Always having an eye to how theory might enhance practice, Janet tried to show that his concepts of psychological misery and subconscious fixed ideas were extremely practical and capable of being successfully applied to disturbed persons:

> Another characteristic of the fixed ideas that result from a nonpermanent, passing disaggregation is that they are much more difficult to reach and change. You can do anything you want with the conscious-

6. The term fixed idea (*idée fixe*) had long been used to denote a mental pathology characterized by a persistent unreasonable thought. Ellenberger has pointed out (1970, pp. 148–49) that Janet was indebted to Liébeault and Charcot for his formulation of the notion of subconscious fixed ideas. It seems reasonable to believe that Liébeault had in his turn been influenced by Braid's discussion of "fixed dominant ideas" (see chapter 8).

ness of a hysteric, because she is *presently* in the state of psychological misery that made her malleable. You cannot change an insane person in the same way because ordinarily you are dealing with him when his delirium is organized and when his intelligence has returned to a state of stable equilibrium which one cannot disarrange. It is necessary to see if one can take the individual back to the psychological state in which his delirium had its origin. (p. 459)

To illustrate this psychotherapeutic approach, Janet described in detail the cure of a peculiar symptom developed by Marie, one of his hysterical subjects (pp. 436–40). She was brought to the hospital at Le Havre at the age of nineteen suffering from a number of hysterical symptoms. One was the recurrence, at each menstrual period, of a state of somberness and violence, quite unlike her normal state. At the same time she underwent pains and nervous trembling in all her limbs. Then twenty hours after the beginning of menstruation the flow would cease, and she would begin to shiver throughout her whole body. A powerful pain would rise from the abdomen to the throat, and she would enter into the "great crises" of hysteria, with violent convulsions and "epileptoid" tremblings. This was followed by a delirium, sometimes with cries of terror in which she would speak of blood and escaping from flames, sometimes with the words and gestures of a child speaking to her mother, and she would climb onto furniture and generally disarrange the room. These symptoms would continue for forty-eight hours, at which point she would vomit blood and become quiet. After one or two days of rest, she would become generally calm and retain no memory of anything that had occurred.

To treat this condition, Janet placed Marie into a state of deep somnambulism. There she divulged memories of a set of events of which previously she had only the sketchiest notion. At thirteen she had experienced her first menstrual period. But because of some infantile misconception she believed that it was a shameful thing and had to be stopped as quickly as possible. Twenty hours after her period began, she plunged herself into a tub of frigid water. This extreme measure caused the immediate cessation of the menstrual flow, and despite the convulsive shivering that resulted, she felt better. This was followed by an illness and several days' delirium. After that she was calm, and she did not have another menstrual period for five years.

In treating Marie, Janet at first tried to remove directly from her somnambulistic consciousness the fixed idea that menstruation could be stopped by a cold bath, but he was unsuccessful. Then he tried a different approach. He used suggestion to regress Marie to age thirteen and re-created the circumstances that were present at the beginning of her first menstruation. He then assured her that her period had lasted for three days and no uncomfortable events had accompanied it. When Marie had her next menstruation after this somnambulistic session, it indeed lasted for three days, and there followed none of the symptoms of pain to which she was accustomed.

On exploring the matter further, Janet discovered that the feelings of terror were the repetition of an emotion Marie had experienced at age sixteen when she saw an old woman die a bloody death by falling down a set of stairs. Finding no similar memory connected to the image of the fire, Janet concluded that flames were connected to the same memory by a kind of association of ideas. Janet dealt with the memory of the old woman's death by taking Marie back to the instant of the accident and changing the image, creating a new picture in which the woman regained her balance and did not die. After this session the experiences of terror ceased.

After five months' treatment for these and other symptoms, Marie returned to apparent full health. Janet did not know how long the cure would last considering her hysterical weakness of constitution, but he believed that the case "demonstrated the importance of subconscious fixed ideas and the role they play in certain physical illnesses as well as moral illnesses" (p. 440).

From the story of Marie, Janet drew some useful conclusions about a condition that had puzzled and frightened people for centuries—demonic possession. The symptoms manifested by Marie could under certain circumstances be mistaken as signs that the victim was possessed by the devil or some other spirit. A few years after the publication of *L'automatisme psychologique* Janet had occasion to treat someone who was subject to just such symptoms. The case of Achille is interesting, not only because it successfully interprets a case of apparent possession as the work of a subconscious fixed idea, but also because it illustrates how Janet was able to use the principles derived from work with hysterics and apply them to a person who was, up to the time of his attack, in a state of relatively good psychological health.

Achille, a married man thirty-three years of age, was described as belonging to a "modest" but superstitious family in central France.[7] Achille did not seem to share their superstitions, and he practiced no religion. Janet stated that one might declare him quite normal, except for periodic migraine attacks.

Toward the end of the winter of 1890 Achille was away from his family for

7. The case of Achille was first described by Janet in the thesis he wrote for his medical degree (Janet 1893, pp. 252–57), elaborated in lectures given at the University of Lyon (Janet 1894) and further detailed in 1898 in his important work on the neuroses and fixed ideas (Janet 1904, 1:377–89). The account given here is drawn from these sources.

some weeks on business. When he returned, although he said he felt fine, his family found that he had changed. He was now somber and preoccupied, and he spoke very little. Over the following days he spoke less and less, and eventually was unable to speak at all, even when he wanted to. He stopped eating and seemed to be undergoing extreme inner turmoil. Doctors were unable to help him. Achille grew weak and was confined to bed. He told his wife that he did not know why this was happening to him and that he was convinced that he was about to die.

After two days of being near death, Achille suddenly sat bolt upright, eyes wide open, and let out a frightening laugh. This laugh was not normal. It was exaggerated and convulsive, twisting his mouth and shaking his whole body, and it continued for two hours without stopping. It was the type of laugh often called "satanic." Next he jumped out of his bed, and when anyone asked him a question, he responded: "Don't do anything. It is useless. Let's drink champagne, for it is the end of the world." He also yelled out, "I am being burned, I am being cut to pieces." After sleeping the night, Achille awoke convinced that he was possessed by the devil. He spoke blasphemies, and his limbs became contracted. Several times he attempted to kill himself. To the onlooker, Achille presented the classical symptoms of possession.

Achille was brought to the Salpêtrière Hospital, where he was placed under Janet's care. Janet attempted to hypnotize him but failed completely, and any attempt he made to alleviate Achille's condition was met with rebuff and abuse. Janet then used his "distraction" method to induce automatic writing. While the man raved on, Janet took his hand and began moving it in a way that suggested writing. Having gotten the hand to write letters and even sign Achille's name, Janet attempted to communicate with Achille's subconscious intelligence. In a low tone Janet asked Achille to write his name. The response came in writing: "I do not want to." "Why do you not want to?" "Because I am stronger than you are." "Who are you?" "I am the devil."

Seizing the opportunity, Janet engaged the "devil" in a dialogue with a definite purpose:

"I do not believe in your power," I said to him, "and I do not believe you can give me a proof of it." "What proof?" responded the devil. . . . "Lift the left hand of this unfortunate person without him knowing it." The left hand of Achille immediately rose. . . . Through the same procedure I had the demon carry out a host of different actions. . . . I was able to go further and do what the exorcists [of old] had not thought to do. I asked the demon, as a final proof of his power, to put Achille to sleep in an armchair, and to put him to sleep so completely that he could not resist. I had already tried in vain to hypnotize the patient by address-ing him directly, but it was useless. This time, however, profiting from the distraction and my addressing the devil, I did it very easily. Achille tried in vain to fight the sleep that was taking him over; he fell back heavily and was completely asleep. (Janet 1904, 1:389)

Once Achille was in a state of somnambulism, Janet was able to obtain the information needed to help the man. In that state he revealed events that had occurred to him on his trip away from home—events that he had previously forgotten or did not understand. The beginning of his illness had been a guilty deed he committed while away: he had been sexually unfaithful to his wife. His initial reluctance and then inability to speak had been an attempt to keep the affair secret. His inner torture and bodily suffering had been punishment for the deed. He was living in a waking dream in which he was dead and the devil rose from hell to take him. He smelled brimstone, and flames filled the room as fiends whipped the wretched man and drove nails into his eyes. From this point he believed himself possessed by the devil and subject to his control, so that he had to rant and yell blasphemies.

Janet's psychotherapeutic treatment of Achille's condition could now pro-ceed. As in the case of Marie, he had Achille recall the events, and then he modified their content in Achille's mind by means of "dissociation of ideas" and "substitution": "It is the very memory of the transgression that Achille must be made to forget. This operation is far from easy. . . . An idea, a memory, can be considered a system of images that one can destroy by separating them, by altering them once isolated, by substituting in the com-posite such and such a partial image for those that exist. . . . The memory of his transgression was transformed in every way through suggested hallucina-tions. Finally, his wife was herself evoked by hallucination at the proper moment, come to give full pardon to this most unfortunate guilty spouse" (Janet 1904, 1:404).

With this treatment Achille was restored to health, and Janet noted that eight years after the cure he was still in sound health (1:379).

The Development of an Alternate-Consciousness Psychotherapy

Through his experience with hysterics and his psychotherapeutic endeavors, Janet reached some definite conclusions about how to view mental disorders. With the publication of *L'automatisme psychologique* in 1889 he had put in place the elements of a theory and psychotherapeutic practice that would carry him through his long career. There he took a stand that placed him in clear opposition to two powerful currents of thought about mental pathology, the organic and intrusion paradigms.

Janet could not accept the view of those who claimed that psychological disturbances were adequately explained by physiology. He did not accept that defective functioning of the nervous system could account for hysteria, and he did not agree that automatic actions were merely mechanical reflexes of the brain. Further, duality of brain function did not, in his opinion, provide an adequate explanation for doubling or multiplying personalities. In a word, Janet rejected the organic paradigm for explaining disturbances of consciousness (Janet 1889, pp. 22–44, 225–36, 413–19).

Neither could Janet accept a spiritistic explanation for mental disturbances. He believed that mediumship, thought reading, divination, table turning, and all the other phenomena sometimes attributed to the interventions of spiritual beings could be adequately explained as manifestations of subconscious activity (pp. 366–419). Janet was also convinced that cases of apparent possession by spiritual beings could best be accounted for in terms of psychological dysfunction, not demonic invasion (pp. 435–442). Thus Janet discarded the intrusion paradigm for mental disturbances.

Janet's work was the culmination of a new kind of psychological healing begun by Puységur one hundred years earlier. He viewed mental dysfunction in terms of a stream of thought and of will not accessible to the ordinary awareness, a consciousness that operates independently of the ideas and intentions of normal consciousness. This second level of consciousness can produce actions, emotions, hallucinations, and physical symptoms that are inexplicable in terms of the perceived desires of the individual. Treatment involves bringing the content of this hidden level to light and destroying its power to affect the person. Janet conceived of these subterranean or subconscious influences in terms of groupings of thought and emotion that carry with them a consciousness of their own. These secondary consciousnesses are identifiable as personalities, with a self-awareness, a unity, and an ability to act in a coordinated way that is analogous to that of the normal waking personality. Through his work, Janet showed himself to be the foremost proponent and spokesman of the alternate-consciousness paradigm for explaining disturbances of consciousness. With Janet, the alternate-consciousness paradigm had come of age, acquiring a framework that would from that time lie at the heart of every psychodynamic psychotherapy.

Chapter 16

Frederic Myers: Exploring the Second Self

In the first year of the twentieth century, William James eulogized a fellow scientist, Frederic W. H. Myers (1843–1901), as "the wary critic of evidence, the skilful handler of hypothesis, the learned neurologist and omnivorous reader of biological and cosmological matter" (1901, p. 13). Myers was respected by virtually every prominent psychologist of his time, and they paid serious attention to his work. He collaborated on psychological investigations with Janet. He was acquainted with Charcot and Bernheim and witnessed experiments at the Salpêtrière and the Hôpital Civil at Nancy. He also visited Liébeault and observed his hypnotic techniques firsthand. Myers was a friend of Charles Richet, Theodore Flournoy, and William James and was acquainted with Alfred Binet and Charles Féré. Yet the man whom Flournoy called "one of the most remarkable personalities of our time in the realm of mental science" (1911, p. 48) is today almost unknown in the field of psychology. His major work, the posthumously published *Human Personality and Its Survival of Bodily Death* (1903), and his important journal articles are rarely cited. Considering the lively and fruitful interaction that took place between Myers and nearly every other psychologist of importance at the time, this is quite surprising. Myers's psychological views were derived empirically from his work with hypnotism, his experiments with automatic writing, and his investigations of other automatisms, especially mediumistic phenomena. As James noted, he was also an avid reader of the current psychological literature, and his writings are a remarkably comprehensive bibliographical source of the relevant books and articles of the day.

Philosophically Myers's views derived from a strong religious background, coupled with an acute critical sense. James understood that in his science Myers was "seeking evidence for human immortality. His contributions to psychology were incidental to that research, and would probably never have been made had he not entered on it. But they have a value for Science entirely independent of the light they shed upon that problem" (1901, p. 14).

Whereas Janet was the first investigator to use the alternate-consciousness paradigm as the basis for explaining some of the data of psychology and the framework for an effective psychotherapy, Myers was the first (and arguably the only) worker to employ it as the foundation for a comprehensive psychological system that had a place for all the data of human experience, from the commonplace to the sublime. Once again James phrased it well:

> One cannot help admiring the great originality with which Myers wove such an extraordinarily detached and discontinuous series of phenomena together, unconscious cerebration, dreams, hypnotism, hysteria, inspirations of genius, the willing game, planchette, crystal gazing, hallucinatory voices, apparitions of the dying, medium trances, demoniacal possession, clairvoyance, thought transference—even ghosts and other facts more doubtful—these things form a chaos at first sight most discouraging. No wonder that scientists can think of no other principle of unity among them than their common appeal to men's perverse propensity to superstition. Yet Myers has actually made a system of them, stringing them continuously upon a perfectly legitimate objective hypothesis, verified in some cases and extended to others by analogy. (James 1901, p. 18)

Myers's synthesis was at bottom psychological, and psychologists were among his most faithful correspondents, even when they did not see eye to eye. Janet and Myers, for example, worked together, admired and influenced each other, but had some deep disagreements. Myers believed that Janet had made a mistake in using hysterics as the principal source of his psychological data. And he criticized Janet for assuming that the ordinary consciousness of everyday life was superior to and healthier than the submerged strata of consciousness. Myers believed that, given the evidence arising from investigation of these multiple consciousnesses, there was no basis—empirical or philosophical—for such a judgment.

Myers often emphasized that the English experiments with states of consciousness were different from the French in that the French tended to deal predominantly with the ill, especially hysterics, whereas the English, not having as many hysterics at their disposal, were forced to use persons who were basically healthy. As a result the English felt compelled to account for dissociative phenomena as an aspect of normal human functioning.

Automatic writing, for instance, was a phenomenon that could be induced in many healthy subjects. Myers conducted hundreds of experiments in automatic writing with normal subjects and obtained striking dissociative phenomena (Myers 1884, 1885, 1887a, 1889b). In England, too, hypnotic experiments were commonly carried out on healthy subjects, notably by Edmund

Gurney, with results no less spectacular than those obtained from the "French" hysterics. Because of these experiences, Myers simply could not agree with Janet's view "that the phenomena of automatism and disaggregation depend on a state that is unhealthy" (1889, p. 451).

Also, Myers refused to dismiss out of hand and on a priori grounds the myriad paranormal experiences catalogued and verified by the Society for Psychical Research. He insisted that psychology had to take *all* the data into account. He believed that sober judgment informed by sound scientific experimentation and data gathering had to be the basis for deciding whether such experiences were valid. And if they could not be dismissed on the basis of that kind of judgment, then psychology had to explain them.

Myers desired above all to arrive at a framework for the scientific examination and explanation of the full range of psychological and psychic phenomena. He outlined this vision in an early article on human personality:

> The method to which I refer is that of *experimental psychology* in its strictest sense—the attempt to attack the great problems of our being not by metaphysical argument, nor by merely introspective analysis, but by a study, as detailed and exact as in any other natural science, of all such phenomena of life as have both a psychical and a physical aspect. Pre-eminently important for such a science is the study of abnormal, and, I may add, of *supernormal* states; sleep and dreams, somnambulism, trance, hysteria, automatism, alternating consciousness, epilepsy, insanity, death and dissolution. Then parallel with these spontaneous states runs another series of *induced* states; narcotism, hypnotic catalepsy, hypnotic somnambulism and the like. (Myers 1886a, pp. 1–2)

Myers believed that such an approach should be able to account for dissociative phenomena in both the healthy and the ill, as well as paranormal phenomena. This belief led to his formulation of the notion of the "subliminal self," which James considered to be the basis for regarding Myers as "the founder of a new science. . . . *What is the precise constitution of the Subliminal*—such is the problem which deserves to figure in our Science hereafter as the *problem of Myers*; and willy-nilly, inquiry must follow the path which it has opened up" (James 1903, p. 22; 1901, p. 17).

Automatic Writing and Hypnotism

In the process of investigating a wide variety of mediumistic phenomena, Myers became intrigued by the automatic nature of most of them. Like Janet, he believed that by studying automatisms exhibited in unusual phenomena, light could be thrown on the nature of human personality. Unlike Janet, Myers was not prepared to label unusual phenomena "abnormal." Though some of the individuals involved in the production of the automatic phenomena had tendencies toward hysteria, most of Myers's subjects were people living normal lives, and some were extraordinarily intelligent and successful members of British society. Among the phenomena Myers studied, hypnotism and automatic writing were particularly important to him. Both could be produced at will in susceptible subjects, and this made a systematic investigation of the automatisms involved possible. While his friend Edmund Gurney focused on hypnotic experiments, Myers concentrated on automatic writing.

In the earliest formulation of his thoughts about automatic writing, Myers stated that it might occur as a product of "unconscious cerebration," as in dreams; some "higher unconscious intelligence or faculty of my own, as in clairvoyance"; telepathic impact from other minds; or "spirits" or extrahuman intelligences (Myers 1884, p. 224). Whereas spiritualists tended to see all automatic writing as the intervention of spirits, Myers held that where the written message "fails to convey any facts which demonstrably are not known to the writer and have never been known to him, there is no need to assume that any intelligence but his own has been concerned in the message" (1885, p. 1). He believed that there were cases in which automatic writing produced information that could not have been known to the writer, but he held that such cases could generally be explained by "telepathy" (influence from the minds of other living individuals) without the intervention of spirits (1887a, pp. 210–11).

In cases in which only the writer's mind was involved, Myers believed that some kind of unconscious mentation was at work. The nature of these automatic productions indicated an intelligence capable, in some cases, of highly original thought, but not operating within the awareness of the writing subject.

Although in his first article on automatic writing, Myers used Carpenter's term "unconscious cerebration" to describe the mental action occurring "behind" automatic writing, upon further reflection he considered this concept, implying an action subsidiary to consciousness, to be inadequate. For he noted that in automatic writing the consciousness that produced the writing presented itself as coordinate with conscious action and able to force its attention on the conscious mind. Also, the consciousness producing the writing showed cleverness and originality. Myers described the case of Clelia, in which anagrams were produced through automatic writing which the writer had great difficulty deciphering: "It must be confessed . . . that in advancing this case I am already overpassing very considerably the recognized limits of unconscious cerebration. And, moreover, I do not even advance the 'Clelia' case as in my view an altogether *exceptional* one. I conceive, rather, that this kind of active duality of mentation—this kind of colloquy between a conscious and an unconscious self—is not a rare, but a fairly common phenomenon" (1885, p. 25).

Myers called this hidden intelligence a "secondary self" and variously described it as a "second focus of cerebral energy," an "inner self," and an "unconscious self." At this stage of his thinking on the matter, Myers posited "a secondary self possessing our brains, as it were, in a kind of sleeping co-partnership" (pp. 26–28). And this state of "possession" was not the result of pathology but an experience of automatism that occurs in normal individuals (pp. 47, 57).

Myers pointed out that this sense of possession could lead one to think, erroneously, that another entity or entities were involved. He mentioned cases in which automatic writing was signed by different personal names and presented personal qualities and initiatives independent of, or even contrary to, those of the writer (1887a, pp. 216–32).[1] But whereas others attributed these messages to spirits, Myers pointed out that hysterics who manifest apparent possession by spirits have a tendency to attribute independent external existence to internal divisions. He believed that a parallel could be drawn between these hysterical phenomena and the various personages who manifest themselves in the automatic writing of some healthy subjects. He saw the presence of purported spirits or "guides" in these instances of automatic writing as "showing the tendency of the individuality to split itself up into various coordinate and alternating trains of personality, each of which may seem for a time to be dominant and obsessing, while yet the habitual sense of the ordinary self may persist through all these invasions" (1887a, p. 233).

Myers pointed out the parallels between the multiple subterranean personalities of healthy writing automatists and some of Janet's hysterical subjects (pp. 236–51). He found Janet's experiments extremely intriguing but empha-

---

1. Myers (1903) often mentioned the tendency of the subliminal self to dramatize the message it conveys to the supraliminal self (2:119): "I have myself received so many cases of these dramatised utterances—as though a number of different spirits were writing in turn through some automatist's hand—that I have come to recognise the operation of some law of dreams, so to call it, as yet but obscurely understood. The alleged personalities are for the most part not only unidentified but purposely unidentifiable; they give themselves romantic or ludicrous names, and they are produced and disappear as lightly as puppets on a mimic stage. The main curiosity of such cases lies in their very persistence and complexity" (p. 130).

sized that this personality-forming tendency was a *normal* function of the human psyche:

> I hold that each of us contain the potentialities of many different arrangements of the elements of our personality, each arrangement being distinguishable from the rest by differences in the chain of memories which pertains to it. . . . I consider that dreams, with natural somnambulism, automatic writing, with so-called mediumistic trance, as well as certain intoxications, epilepsies, hysterias, and recurrent insanities, afford examples of the development of what I have called secondary mnemonic chains,—fresh personalities, more or less complete, alongside the normal state. And I would add that hypnotism is only the name given to a group of empirical methods of inducing these fresh personalities, of shifting the centres of maximum energy, and starting a new mnemonic chain. (Myers 1889a, p. 387)

Thus Myers appreciated hypnotism as a source of automatic phenomena. The material derived from his work with automatic writing was augmented and enhanced by hypnotic experiments that he and Edmund Gurney carried out.[2]

Myers believed that hypnotism afforded "as though by a painless and harmless psychical vivisection, an unequalled insight into the mysteries of man" (1886a, p. 2). The subjects of the vivisectional experiments carried out by Gurney and Myers were not, like Janet's subjects, women living under the burden of hysteria, but ordinary English men and women from every stratum of society. Evaluating the data obtained from his own experiments and those of French investigators, Myers concluded that personality was not a unitary and stable element of human consciousness, but was shifting and discontinuous. Multiple chains of memory, revolving around multiple personal centers and evincing multiple sets of character traits, indicate that human personality is richer and more complex than ordinarily thought. To explore that richness and complexity through experimental psychology was Myers's chief aim (1886a, pp. 19–20).

---

2. Myers's view of hypnotism vs. mesmerism should be noted: "I have used the term 'hypnotism' throughout this paper, but I do not concede that the hypnotic phenomena are always produced by mere monotonous stimulation or other mechanical causes. I still hold to the view of Cuvier, that there is in some cases a specific action of one organism on another, of a kind as yet unknown. This theory is generally connoted by the term 'mesmerism.' Since the days of Braid there has been a tendency to exclude it as unnecessary and even fantastic. Mr. Gurney and I (with Dr. Despine, in France) stand almost alone among recent writers in adhering to it. Our contention has steadily been that no one has as yet advanced experiments numerous or careful enough to disprove the specific influence in question, and that certain of our own experiments, of Esdaile's, &c., come very near to proving it" (Myers 1886a, p. 6n).

From an analysis of the data of hypnotism and other altered states of consciousness, Myers posed a question that struck at the foundations of how people ordinarily think of themselves. He wondered on what basis we could assume that our ordinary waking consciousness is superior to other types of consciousness, such as sleep states, states of naturally occurring somnambulism, multiple personality states, or hypnotic states. To him there seemed good evidence that states of consciousness other than the ordinary waking state were superior in very important ways. These states sometimes manifested more acute memory, higher moral values, greater control over the physical organism, and closer contact with paranormal abilities (Myers 1886a, 1887b). Moreover, in direct contradiction to Janet, he strongly objected to viewing these heightened powers as manifestations of mental degeneration: "So long as we try to explain all the phenomena of hypnotism, double consciousness, &c., as mere morbid disaggregations of the empirical personality—repartitions among several selves of powers habitually appertaining to one alone—so long, I think, shall we be condemning ourselves to a failure which will become more evident with each new batch of experiments, each fresh manifestation of the profundity and strangeness of the subliminal forces at work" (Myers 1892a, p. 301). The need to provide a coherent framework for all the data of the multiple strata of human consciousness led Myers to formulate his theory of the subliminal self.

The Subliminal Self

Myers insisted that the "stream of consciousness in which we habitually live" is by no means our only consciousness. He accorded no primacy to this ordinary waking self, which, as far as he was concerned, had only one thing to recommend it: that it "has shown itself the fittest to meet the needs of common life." He held that the other consciousnesses that exist "in some kind of coordination with [the human] organism" form part of one's total individuality, and he expressed the hope that at some point it may be possible for people to empirically experience this multiplicity and "assume these personalities under one single consciousness" (Myers 1892a, p. 301).

In attempting to devise a vocabulary for these matters, Myers decided to avoid terms like "soul" and "spirit" with their historical weight of meaning. To designate the underlying psychical unity of the human being, he chose the word "individuality." "Personality" was his choice for referring to "something more external and transitory," those "apparent characters, or chains of memory and desire, which may at any time mask at once and manifest a psychical existence deeper and more perdurable than our own. . . . Each of us is in reality an abiding psychical entity far more extensive than he knows—an individuality which can never express itself completely through any corporeal manifestation. The Self manifests through the organism; but there is always some part of the Self unmanifested" (1892a, p. 305). The part of the self that exists below the threshold of our habitual consciousness Myers called the "subliminal self."[3] He preferred this term to "unconscious," "subconscious," or "secondary" self because subliminal strata involve consciousness, the ordinary consciousness of daily life has no superiority over the subliminal, and the subliminal self harbors a multitude of consciousnesses, not just a second consciousness. What one ordinarily identifies as oneself, the self of common experience, the empirical self, the part of the individuality that is *above* the threshold, Myers called the "supraliminal self" (p. 305).

Myers contended that subliminal consciousness and subliminal memory embraced a far wider range of both physiological and psychical activity than did the supraliminal. What the supraliminal self knows and remembers is quite circumscribed, as an examination of ordinary consciousness reveals. Also, the supraliminal consciousness exercises very limited control over the physiological functions of the organism. The subliminal region, on the other hand, embraces consciousnesses that exhibit remarkable awareness (intellectual, moral, and paranormal) and phenomenal memory (for facts, previous subliminal states, and past events) as revealed in hypnotic experiments. Hypnotic investigations also show that the subliminal self has an extraordinary ability to affect functions of that body that are usually thought to be beyond the control of waking consciousness (for example, control of bleeding, formation of blisters, increasing, decreasing, or altering the sense of hearing, sight, taste, smell, or touch—including pain sensitivity). Referring to these heightened psychical and physiological powers, Myers said that "the spectrum of consciousness . . . is in the subliminal self indefinitely extended at both ends" (Myers 1892a, p. 306).

According to Myers, the supraliminal self evolves to deal with all that is related to living in the mundane world, and it tends to ignore elements of experience that are not relevant to that task: "It seems likely . . . that the greater part of the contents of our supraliminal consciousness may be determined in some such fashion as by natural selection so operating as to keep ready at hand those perceptions which are most needed for the conduct of life" (Myers 1903, 1:39). In this way, the supraliminal self is but a small facet of the

3. "I here use the word 'self' as a brief descriptive term for any chain of memory sufficiently continuous and embracing sufficient particulars to acquire what is popularly called a 'character' of its own. There will thus be one distinct supraliminal self at a time; but more than one subliminal self may exist, or may be capable of being called into existence" (Myers 1892a, p. 305n).

greater self or individuality. But because human beings are largely taken up with surviving and prospering in the mundane world, they identify themselves completely with the supraliminal self and can easily brush aside stimuli emanating from below the threshold of daily consciousness. Psychologists too tend to think of the supraliminal self as the norm of mental existence, but Myers believed that according it primacy was based on mere convenience: "I . . . regard supraliminal life merely as a *privileged case* of personality; a special phase of our personality, which is easiest for us to study, because it is simplified for us by our ready consciousness of what is going on in it; yet which is by no means necessarily either central or prepotent, could we see our whole being in comprehensive view" (Myers 1903, 1:223).

But how is the supraliminal self shielded from the subliminal? Myers posited a kind of barrier or psychic membrane between them. That membrane easily allows information to flow from the supraliminal to the subliminal, but the reverse movement is much more inhibited. The way Myers saw it, the screening of messages from the subliminal was necessary for the proper functioning of the supraliminal self. If it were flooded by awareness of the diverse activities of the subliminal consciousness, it would be incapable of dealing with the problems and challenges of everyday existence.

Although some messages do rise from the subliminal to the supraliminal, in ordinary life these often take the form of "uprushes," suddenly emerging impressions that strike the supraliminal consciousness as coming out of the blue. In this way the supraliminal self perceives these impressions as having a different quality from any element in the ordinary supraliminal life. They imply a faculty of which the person has no previous knowledge, operating in an environment of which he has been totally unaware.

Outlining the operations of the subliminal self and its effect upon supraliminal life was Myers's main task as a psychologist. To do that, he divided subliminal activities into several types, the principal ones being hypnotism, disintegrations of personality, genius, sensory automatisms, and motor automatisms.

Hypnotism

One of the ways that messages rise from the subliminal to the supraliminal level is through hypnotism. Myers saw hypnotism on the one hand as inhibiting supraliminal functioning and on the other as providing access to the subliminal self (Myers 1903, 1:163, 169).

For Myers, suggestion was central in the production of the hypnotic state. He believed that in the last analysis all suggestion was reducible to self-suggestion (Myers 1892a, pp. 350–51). Hypnotism comes about through a self-suggested "successful appeal to the subliminal consciousness." Understanding hypnotism, then, involves investigating the "obscure relationships and interdependencies of the supraliminal and the subliminal self," through which a successful appeal to the subliminal consciousness can be made (Myers 1903, 1:169).

When such an appeal is carried out and a state of hypnotism is induced, the subliminal self comes to the surface and displaces the supraliminal self to the extent it judges necessary. Myers did not find the notion of specific "degrees" or defined "stages" of hypnotism (such as Charcot's lethargy, catalepsy, and somnambulism) to be useful. He thought that Gurney's simple distinction between the "alert stage" and the "deep stage" of hypnotism was sufficient, and that all other stages were merely independent trains of memory that could be of any number and appear in any order. All intermediary hypnotic stages were "secondary or alternating personalities of a very shallow type" (Myers 1903, 1:172).

The phenomena of hypnotism, presented in such profusion and detail in the literature of animal magnetism and hypnotism, manifested the capacities and powers of the subliminal self. Alterations in personal characteristics, memory, mental ability, physical functioning—these and other phenomena show how the subliminal self functions and provide a glimpse of the inner nature of the subliminal world. Even those thousands of instances of well-attested physical cures—mesmeric and hypnotic—could now be understood as resulting from the successful mobilization of subliminal powers that control physiological functioning. Myers held that whether or not one accepts the notion of a mesmeric agent that passes between operator and subject, in the last analysis all cure is accomplished through the "vitalizing" power of the subliminal self (Myers 1903, 1:210–17).

Disintegrations of Personality

For Myers, the human being is psychologically a "colonial organism," with the various strata, supraliminal and subliminal, operating independently but in harmony. This harmonious functioning or unity is enabled by a mysterious, overreaching psychical entity that maintains a continuum for the smaller psychic entities (Myers 1903, 1:34–38). Disintegration occurs when this unity breaks down and some psychic centers begin to operate shut off from free and healthy interchange with the rest of the personality. Myers considers the first symptom of disintegration to be the *idée fixe* described by Janet—that is, "the persistence of an uncontrolled and unmodifiable group of thoughts or emotions, which from their brooding isolation,—from the very fact of deficient interchange with the general current of thought—become alien and intrusive, so that some special idea or image presses into consciousness with undue and painful frequency" (Myers 1903, 1:40).

Myers pointed out that since fixed ideas can be cured through hypnotic suggestion—that is, through the action of the subliminal self—the disorder probably first occurred in the subliminal stratum. If a person is subject, for instance, to a fixed idea of agoraphobia, it is probably because the thoughts involved in the formation of the idea have sunk below the threshold and can no longer be summoned into ordinary consciousness. Then the faulty functioning has to be laid at the doorstep of the subliminal self, which is supposed to keep available any thoughts needed for the proper functioning of daily life. Put another way, disintegrations of the personality may be seen as resulting from the excessive permeability of the psychic barrier separating the supraliminal from the subliminal. Because of that permeability, the supraliminal self is subject to powerful chaotic uprushes from the subliminal that it cannot handle.

Myers saw hysteria in the same terms. In this case it is some submerged primitive instinct that rises as a subliminal uprush into the supraliminal consciousness and operates there with a powerfully disturbing effect. This indicates an instability of the threshold of consciousness, which implies a "disorderly or diseased condition of the hypnotic stratum,—of that region of the personality which . . . is best known to us through the fact that it is reached by hypnotic suggestion" (Myers 1903, 1:42). In the region above the threshold, the supraliminal, certain faculties necessary for proper mundane functioning should be maintained; in the region below the threshold, the region affected by hypnotic suggestion, operations are dreamlike and capricious. Conscious groups of psychic phenomena that have gotten isolated from healthy interchange with the supraliminal may still be unified in the subliminal strata, but they need to be unified with the supraliminal as well.

Myers believed that this view of disintegrations of personality was in complete agreement with Janet's findings related to hysteria. Janet discovered that although emergent fixed ideas were out of harmony with the ordinary self (the supraliminal self), they were known and coordinated by the deeper centers discovered in hypnotism (the subliminal strata). Myers found additional confirmation of this concept of hysteria in the case of Anna O., described in Freud and Breuer's *Studien über Hysterie*:

Here also the first symptoms are subliminal *idées fixes,* translating themselves into somatic symptoms, whose origin is only recovered by help of the profounder memory of hypnotic trance. But with Fräulein Anna these submerged ideas, these hidden ulcers of the mind, become, so to say, *confluent*. We have a transition from *idées fixes* to a secondary personality, dominated by those ideas, and sinking into incoherent insanity. Yet even from that depth a certain resolute firmness of the patient's temper, aided by Dr. Breuer's skill in suggestion, raises her once more, and replaces her uninjured among sane and vigorous women. (Myers 1903, 1:51)

This led Myers to consider further disintegrations of personality: the spontaneously occurring phenomena of natural somnambulism and multiple personality. These too he saw as resulting from a malfunction in the hypnotic strata. This was demonstrated by the fact that hypnotism could be used to reach the alternate personalities and uncover disturbing memories tied to the multiple states (Myers 1903, 1:55–65).

Genius

From a consideration of disintegrations of personality resulting from disturbing uprushes into the supraliminal self, Myers turned to the investigation of uprushes of a very different kind: those of genius. Myers thought it as important to account for the creative and intuitive experiences of individuals as for the problematic and bizarre. He noted that throughout human history there had been references to "flashes of inspiration" and creative ideas that seemed to arise from some unknown source. Taken at face value these experiences appear to be automatisms, for they do not emanate from conscious thinking. Yet the quality of the creative flashes compels one to conclude that the source possesses not simply intelligence, but intelligence of an extremely potent and original kind. The data Myers investigated in this regard ranged from youthful "arithmetic prodigies" to some of the greatest creative geniuses of the century.

Myers pointed out that flashes of inspiration could come in various ways, most frequently through an internal sense experience, such as hearing inner voices or sounds or seeing inner sights. Moreover, the person who has the flashes feels that she is not the true author of what is experienced. Myers turned his attention to the arts, and particularly to musical inspiration:

It is like something discovered, not like something manufactured. . . . And the subjective sensations of the musician himself accord with the view of the essentially subliminal character of the gift with which he deals. In no direction is "genius" or "inspiration" more essential to true success. It is not from careful poring over the mutual relations of musical notes that the masterpieces of melody have been born. They have come as they came to Mozart, . . . in an uprush of unsummoned audition, of unpremeditated and self-revealing joy. . . . We may say that we

have reached a point where the subliminal uprush is felt by the supraliminal personality to be deeper, truer, more permanent than the product
of voluntary thought. (Myers 1892b, p. 344)

Within this framework Myers defined genius as the power to utilize more
fully than most people could a faculty possessed by all—the power to make
the results of subliminal mentation available to the supraliminal stream of
thought. He saw an inspiration of genius as an uprush of ideas originating in
the subliminal consciousness into the stream of ideas that a person is in the
process of consciously manipulating. Myers insisted, therefore, that genius
and inspirations of genius were not abnormal but, on the contrary, the fulfillment of the true nature of human beings, the realization of an evolutionary
stage hitherto only attained by a few (Myers 1903, 1:71).

Exactly how someone becomes aware of and utilizes the productions of the
subliminal self varies greatly. Myers cited the case of Robert Louis Stevenson,
who used his dreams as literary material. The dreams were very vivid, and he
discovered that they would often be produced when he needed money and so
was desperate to write something he could sell. He spoke of how the "little
people" who managed his "internal theatre" would construct a clever story
ready for the market. According to Stevenson, these "brownies" knew how to
develop a story and seemed to possess more talent than he. As the dream
progressed and the story unfolded, he, the dreamer, was as much in ignorance
of the outcome as anyone else. Even when Stevenson awoke and began to
polish the tale he had dreamed, he felt guided by his "brownies," so he
believed that, all in all, the end product was not his own (Myers 1903, 1:91).

Myers found this description to be an excellent example of the activity of
the subliminal self in a literary genius. He noted the "incommensurability"
between the conscious, logical input on the supraliminal level and the inspiration that characterizes genius. The result is that sometimes the supraliminal
self has difficulty connecting with the creative product and identifying itself as
its author. Myers saw this feeling of separation from the result, of being a
channel for some greater intelligence or power, as the hallmark of the work of
the subliminal self in creations of genius.

Since the productions of genius contain elements that are intelligent but not
brought about through ordinary consciousness, Myers considered them to be a
species of psychological automatism:

Genius represents a narrow selection among a great many cognate
phenomena;—among a great many uprushes or emergences of subliminal faculty both within and beyond the limits of the ordinary conscious
spectrum. It will be more convenient to study all these together under

the heading of sensory or of motor automatism. . . . When the subliminal mentation co-operates with and supplements the supraliminal, without changing the apparent phase of personality, we have *genius*. When
subliminal operations change the apparent phase of personality from the
state of waking in the direction of trance, we have *hypnotism*. When the
subliminal mentation forces itself up through the supraliminal, with
amalgamation, as in crystal-vision, automatic writing, &c., we have
sensory or motor automatism (Myers 1903, 1:78, 103–4).

Sensory and Motor Automatisms

Myers defined sensory automatisms as "the products of inner vision or inner
audition externalised into quasi-percepts," and motor automatisms as "messages conveyed by movement of limbs or hand or tongue, initiated by an inner
motor impulse beyond the conscious will." He saw both as attempts of "submerged tracts of our personality to present to ordinary waking thought fragments of a knowledge which no ordinary waking thought could attain" (Myers
1903, 1:222).

At the turn of the century the term commonly applied to sensory automatism was "hallucination." Hallucinations could come from a variety of
sources, morbid or normal. Myers was concerned with examining hallucinations produced by healthy individuals. Dreams, for instance, are hallucinations that occur regularly in the lives of all people. We accept them, take them
for granted, and see nothing remarkable in their recurrence, says Myers.
There are, however, hallucinations or sensory automatisms that, although
known to occur to healthy people, are nonetheless seen as extraordinary.
Among these are hypnotic hallucinations, crystal visions, telepathic and clairvoyant impressions, and apparitions.

Sensory automatisms—sensory messages transmitted from the subliminal
to the supraliminal self—may, according to Myers, originate in the automatist's own mind or may arise as a result of communication from the mind of
another. Distinguishing between these two sources may be very difficult, and
one should only conclude that an external agent is at work if no other explanation suffices. Myers noted that experimental sensory automatisms had long
been produced by means of suggestions made during artificial somnambulism, and he credited Mesmer with having devised the method for this inquiry.
The possibility of bringing about hallucinations at will in this way has a great
advantage for the scientific psychologist, said Myers, for the mechanism of
hallucination, its various forms, and the participation of consciousness in the
process can be observed closely and repeatedly. He credits Bernheim and

Gurney as the investigators who had done the most to elucidate sensory automatism by this means (Myers 1892c, pp. 442–50).

Crystal vision, produced by steadily gazing into any clear medium that gives a sense of depth (such as a crystal ball, a bowl of water, or a dark mirror), was a hallucinatory phenomenon of special interest to Myers. He recommended it as an experimental method for producing harmless and easily observable sensory automatisms. He noted that whereas some investigators had induced crystal visions through post-hypnotic suggestion, his own experiments, and those of many others, were carried out while the subject was in the normal state (Myers 1892c, p. 459).

Myers believed that the subliminal self of one individual was capable of directly communicating a sensory image to the mind of another. In his opinion, this "telepathy" (a term Myers coined) could be scientifically verified through experiment, and crystal vision offered one of the best means to carry out systematic trials of this sort. In his writings on the subliminal consciousness he detailed crystal vision experiments strongly indicating that telepathy was at work and called for coordination of future crystal vision experiments in the population at large with a view to amassing useful data (Myers 1892c, pp. 459–535).

In Myers's view, it sometimes happens that sensory automatisms are experienced spontaneously and directly in the mind, without the involvement of suggestion or the use of any external device. These spontaneous perceptions can be in the form of telepathy, clairvoyance, "telaesthesia" (perception at a distance without the use of the senses), precognition or retrocognition (direct knowledge of a future or past event), or apparition (perception of a person— living or deceased—who is not present). Myers used the term "supernormal faculties" to designate abilities that go beyond ordinary experience but are nonetheless part of the natural evolution of mankind and, like all natural phenomena, subject to fixed laws.

Sensory automatisms involve awareness by the supraliminal self of images or sensations that originate in the subliminal. How the subliminal self attains these sensory experiences is hidden from us; all we know is the end result. We can be confident that the subliminal self has the ability to acquire these impressions because careful observation of controlled experiments and testimonials of personal experiences support that view. Experiments with crystal vision, automatic writing, and telepathy confirm the existence of these phenomena. Collected data verify the reality of apparitions. As to precognition and retrocognition, Myers believed that there was solid evidence in their favor, although how they occurred was not clear (Myers 1903, 1:246, 248; 2:262–74).[4]

Motor automatisms, said Myers, involve messages conveyed from the subliminal to the supraliminal self through bodily movements. Most commonly these are movements of the hand or the tongue, since automatic writing and automatic speaking are the most frequent automatic motor phenomena.[5]

Myers noted that even though the same principles that apply to automatic writing pertain to automatic speaking, in practice automatic speaking is usually connected with spiritualistic experiences. Its occurrence during mediumistic trance had for decades been a normal part of spiritualistic practice. On the surface, it involves a kind of "possession" of the medium by a "spirit" in such a way that a more or less complete substitution of personality resulted. The medium no longer speaks from her own personality; rather, a spiritual entity, usually a discarnate human being, takes charge of the body and addresses those present. After the possessing entity had delivered its message and left, the medium ordinarily remembered nothing of what had occurred.

From his experience with automatic writing, Myers knew all too well how a secondary personality originating in the individual's subliminal self might appear to be a separately existing entity. Criteria other than the claims of the speaking voice or the conviction of the medium had to be brought to bear to determine the status of the source. On the other hand, Myers did not rule out the possibility that in some instances a separate entity might indeed take over the physical organism of an individual and write or speak by that means (Myers 1903, 2:189–91). In the cases of certain thoroughly studied mediums, there was evidence that information was produced that could have been

4. "If there is a transcendental world at all, there is a transcendental view of Past and Future fuller and further-reaching than the empirical; and in that view we may ourselves to some extent participate, either directly, as being ourselves denizens all along of the transcendental world, or indirectly, as receiving intimations from spirits from whom the shadow in which our own spirits are 'half lost' has melted away" (Myers 1903, 2:263).

5. Myers discussed messages conveyed by table turning and "spirit rapping" in this context. Admitting the possibility that such phenomena may be produced without physical contact on the part of medium or sitters, Myers was reluctant to attribute them to spirits: "I do not prejudge the question as to their real occurrence; but assuming that such disturbances of the physical order do occur, there is at least no *prima facie* need to refer them to disembodied spirits. If a table moves when no one is touching it, this is not obviously more likely to have been effected by my deceased grandfather than by myself. We cannot tell how *I* could move it; but then we cannot tell how *he* could move it either. The question must be argued on its merits in each case; and our present argument is not therefore vitiated by our postponement of this further problem" (Myers 1903, 2:92–93).

known only to the departed spirit who claimed to be communicating. Myers was inclined to believe that if possession of human beings by spirits was a reality, it probably occurred through a mechanism that was an extension of the concept of telepathy. He held telepathy between the minds of living human beings to be a fact backed by convincing scientific evidence. He also accepted telepathic communication with the departed as a well-substantiated phenomenon. He thought it logical to take one further step and hypothesize that a spirit could communicate telepathically with such intensity that it actually made its presence felt in the physical organism, especially when the individual involved was in a state of trance or partial vacation of the body: "So far as his organism is concerned, the invasion seems complete; and it indicates a power which is indeed telepathic in a true sense;—yet not quite in the sense which we originally attached to the word. We first thought of telepathy as a communication between two minds, whereas what we have here looks more like a communication between a mind and a body,—an external mind, in place of the mind which is accustomed to rule that particular body" (Myers 1903, 2:196).

Although he was prepared to admit the possibility of possession by a discarnate human spirit, Myers was not ready to accept demoniacal possession:

> A devil is not a creature whose existence is independently known to science; and the accounts of the behaviour of the invading devils seems due to mere self-suggestion. With uncivilised races, even more than among our own friends, we are bound to insist on the rule that there must be some supernormal knowledge shown before we may assume an external influence. It may of course be replied that the character shown by the "devils" was fiendish and actually *hostile* to the possessed person. Can we suppose that the tormentor was actually a fraction of the tormented? I reply that such a supposition, so far from being absurd, is supported by well-known phenomena both in insanity and in mere hysteria. (Myers 1903, 2:199)

If possession did involve spirits, Myers held, then they were "spirits who have been men like ourselves, and who are still animated by much the same motives as those which influence us" (Myers 1903, 2:200–201).

The Evolving Self

Throughout Myers's writings on the subliminal self, one glimpses a theory of human evolution that can account not only for the physical and social characteristics of the race but also for the psychological and spiritual experiences of its members. Granting due consideration to Lamarckian and Darwinian explanatory concepts, Myers believed that they had little to offer in the search for a truly comprehensive view of human beings.

In the protoplasm or primary basis of all human life there must have been the inherent ability to manifest all the faculties that eventually would show themselves in human experience—faculties that have been revealed in the activity of the subliminal and supraliminal selves. But Myers did not agree that those faculties were created through the chance combination of hereditary elements. Instead, the capricious events of human existence and the resulting protoplasmic changes over the eons merely revealed powers that were already there, raising an already existing faculty above the threshold of supraliminal consciousness. Myers realized that this conception was novel: "This view, if pushed back far enough, is no doubt inconsistent with the way evolution is generally conceived. For it denies that all human faculties need have been evoked by terrene experience. It assumes a subliminal self, with unknown faculties, originated in some unknown way, and not merely by contact with the needs which the terrene organism has had to meet" (Myers 1903, 2:358).

Myers believed that studying the activity and products of the subliminal self affords glimpses of the central source of unity that makes the individual a whole person: "Sometimes we seem to see our subliminal perceptions and faculties acting truly in unity, truly as a Self;—co-ordinated into some harmonious 'inspiration of genius,' or some profound and reasonable hypnotic self-reformation, or some far-reaching supernormal achievement of clairvoyant vision or of self-projection into a spiritual world. Whatever of subliminal personality is thus acting corresponds with the highest-level centres of supraliminal life. At such moments the *subliminal* represents (as I believe) most nearly what will become the *surviving* Self" (Myers 1903, 1:73).

In this vision Myers finds the beginnings of an answer to a question that was central for him and for many before him: Do human beings have an existence that in some form or other survives bodily death? He believed that experimental psychology provided a sure method for addressing that question, and he saw in the phenomena emanating from an alternate consciousness the most promising indications of an affirmative answer.

The Hidden Self

Myers was not alone in believing that a second layer of consciousness coexists with the normal consciousness in all human beings—healthy and ill. Psychological investigators in France, Germany, and the United States were rapidly coming to the same conclusion.

Alfred Binet

Alfred Binet (1857–1911), director of the laboratory of physiological psychology at the Sorbonne, published two important works on secondary states of consciousness. *On Double Consciousness* (1890b), published only in English, first appeared as a series of articles in *The Open Court*. In *Les altérations de la personnalité* (1892) Binet summarized previous work in the area and discussed his own ongoing experiments.

In *On Double Consciousness* Binet recounted the existing proofs of the doubling of consciousness in hysterical persons and the relationship between secondary consciousnesses and the primary consciousness. He also defended the notion of a subconscious mental life against the view that all the phenomena could be explained by the mechanical action of habit and instinct. Like Janet before him, he pointed out that the intellectual complexity of products of the subconscious and the initiative it exhibited contradicted such an explanation. Binet also agreed with Janet in seeing "retrenchment of the field of consciousness" as the basis for suggestibility in a subject. He parted company with Janet, however, in the final section, "Double Consciousness in Health," where he described experiments that he carried out on five healthy subjects. The results confirmed the presence of "rudimentary" phenomena of double consciousness, but no phenomena sufficiently developed to demonstrate the fully doubled condition.

By the time Binet wrote *Les altérations de la personnalité*, the climate had changed to the point that he could say, "It has now become trite that the majority of experiments performed upon hysterical patients give very nearly the same result with healthy persons, but less conspicuously" (1896 [American ed.], p. 217). Discussing psychological automatisms, in which thought expresses itself in action without the participation of the normal consciousness, Binet proposed that even in healthy persons there are "several mental syntheses of consciousness" (Binet 1896, p. 220).

The experiments Binet mentioned for investigating dissociative states in the healthy used the "exploring pendulum" (a hand-held pendulum indicating answers to questions), automatic writing, and unconscious communicative movements that occur in various circumstances. To these Binet added his own experiments involving the presence of two mental operations that were in conflict:

When a person divides his attention between two voluntary mental transactions, endeavouring to perform them simultaneously, each of these transactions, especially on the first trial, is performed less correctly than if it were done separately; and, in the second place, it often happens that one of these transactions tends to force its particular form or rhythm on the other. But the dominant fact, which seems to me most important, is that with some persons a division of consciousness is produced; one of the two conflicting transactions leaves the subject's consciousness and continues to act without his direction and without his clearly perceiving it. (Binet 1896, p. 241)

Binet's experiments also indicated that suggestion always produced a division of consciousness, and, further, that it could even create artificial personalities. Charles Richet (1883) had long before demonstrated the creation of artificial personalities during somnambulism. Binet described how this occurred: "Everything that is inconsistent with the suggestion [of being a specific person] gets inhibited and leaves the subject's consciousness. As has been said, alterations of personality imply phenomena of amnesia. In order that the subject may assume the fictitious personality he must begin by forgetting his true personality. The infinite number of memories that represent his past existence and constitute the basis of his normal ego are for the time being effaced, because these memories are inconsistent with the idea of the suggestion" (Binet 1896, p. 257). The division produced in the subject's consciousness causes the true personality to leave the scene of action, being relegated to a second plane, where it is "temporarily forgotten." The new personality unconsciously borrows certain elements—such as habits of gesture and speech, without which it could not express itself—from the normal personality. But the suggested personality carries on completely absorbed in its artificial role with no awareness that its identity has been manufactured.

Quoting experiments in automatic writing carried out by F. W. H. Myers, Binet proposed that this kind of artificial creation of personalities also occurred in "spiritism." And he noted that mediumistic subjects of the kind described by Myers did not exhibit changes in sensation of the kind found in hysterical patients.

Max Dessoir

Secretary of the Berlin Society for Experimental Psychology, the German counterpart of the Society for Psychical Research, Max Dessoir (1867–1947) was already internationally recognized for his *Bibliographie des modernen Hypnotismus* (1888) when he wrote a small treatise on the subconscious titled *Das Doppel-Ich* (1889). Dessoir posited two spheres of consciousness, which he called the "overconsciousness" (*Oberbewusstsein*) and the "underconsciousness" (*Unterbewusstsein*) (1896 [2d ed.], p. 13). He boldly stated that all human beings have within themselves the seeds of a second personality, which is gifted with understanding, feeling, and will and is capable of initiating actions. They may in fact form any number of subsidiary selves or personalities. In other words, every individual is potentially multiple (pp. 29–34).

Like Myers, Dessoir thought that the underconsciousness presides over the powers of clairvoyance and telepathy. He drew this inference from experiments in thought transference in which the receiving subjects were able to accurately reproduce telepathically transmitted images through automatic drawing. Since the overconsciousness of the subject was not involved, the underconsciousness must have been (Dessoir 1896, p. 31).

William James

The year after the publication of *Das Doppel-Ich,* William James (1842–1910) wrote his own response to the investigations into double consciousness being conducted on the other side of the Atlantic. In "The Hidden Self" (1890b) James pointed up the importance of the discoveries of Janet, Gurney, Bernheim, Binet, and others in regard to the "simultaneous coexistence of the different personages into which one human being may be split. . . . It seems to me a very great step to have ascertained that the secondary self, or selves, coexist with the primary one. . . . But just what these secondary selves may be, and what are their remoter relations and conditions of existence, are questions to which the answer is anything but clear" (pp. 368, 373).

James took up the nature of the self and the phenomenon of alternating selves in his masterwork, *The Principles of Psychology* (1890). The human self, he said, is naturally multiple, but it expresses this multiplicity in terms of separated functions. His section on alternating personalities was simply descriptive, with no attempt to fathom the psychical structure that could account for the phenomenon.

In his Lowell Lectures, delivered in 1896, James took up the subject more thoroughly. He spoke of a secondary intelligence that could attend to its own affairs without interfering with ordinary active consciousness. In some circumstances it shows itself in waking life, sometimes in dramatic ways. These manifestations of the secondary intelligence included automatic writing, hysteria, multiple personality, demoniacal possession, and witchcraft. James believed that these phenomena were best explained in terms of the action of an intelligent subconscious mind, and he saw in the psychology of the subconscious the potential for a therapy that could be effective in healing some of the greatest human psychological ills (Taylor 1983).

As James struggled with the mysteries of the subconscious he continually referred to the work of Janet and Myers. In Janet's writings he perceived the first revelation of coexisting consciousnesses in one individual. In Myers's exposition of the subliminal self he found what he believed to be the most comprehensive and enlightening attempt yet made to fathom the nature of the hidden self.

Boris Sidis

The Russian-born American Boris Sidis (1867–1923), a student and later an associate of William James, developed a psychology that bore his distinctive stamp. Through experimentation with suggestibility in both normal and abnormal subjects, he came to the conclusion that in everyone two streams of consciousness coexist and constitute two selves: the waking self and the subwaking self. Sidis first formulated these views in *The Psychology of Suggestion* (1911 [1898]).

According to Sidis, the waking self is the ruling self, "a person having the power to investigate his own nature, to discover faults, to create ideals, to strive after them, to struggle for them, and by continuous, strenuous efforts of will to attain higher and higher states of personality" (Sidis 1911, p. 296). The subwaking self is a consciousness that has an awareness broader than that of the waking self, being aware of the life of the waking self, while the waking self knows nothing of subwaking life. The subwaking self is first of all stupid, totally lacking in critical sense. It is also highly suggestible, willing to follow commands in an extremely literal way. Furthermore, the subwaking self is devoid of morality, apparently ready to carry out any act, no matter how destructive, without scruple. Being both suggestible and amoral, the subwaking self is susceptible to the emotional forces that operate in crowds and mobs. It is, at bottom, without personality, individuality, willpower, and goals—a "brutal self."

Nevertheless, the subwaking self can make some progress toward self-awareness and becoming a self-conscious personality. Indeed, it can become sufficiently individualized to lead a submerged life independent of the waking self. It can, as a matter of fact, rise to the surface and assume control of the organism. For instance, it can take possession of some part of the body formerly under the control of the waking self and make it anesthetic, as in hysteria. Or it may take over the whole waking life of the individual, alternating control with the ordinary self, as in the case of multiple personality. In this way Sidis saw all instances of the emergence of dissociated ideas or systems as forms of possession of the organism by the subwaking self. He considered these intrusions by the subwaking self to be a sort of plundering of the riches of the waking self. For no matter how personalized it may become, the subwaking self remains a brute, essentially lacking the psychological resources of the waking self (pp. 91–187).

Morton Prince

Morton Prince (1854–1929), professor of neurology at Tufts Medical School and founding editor of the *Journal of Abnormal Psychology,* was the last major contributor to a system of psychological healing based on the alternate-consciousness paradigm. Drawing largely on his work with automatism and multiple personality,[6] he formulated a notion of the subconscious that provided much-needed clarity about the problem of the simultaneous existence of multiple centers of consciousness.

Like Myers, Prince used automatic writing as a principal source of data on the activity of the subconscious. He too thought that any attempt to explain such phenomena purely in terms of physiological processes was doomed to failure because of the complex nature of some automatic productions. He noted that automatic writing consisted not merely of words, phrases, or paragraphs that were repetitions of content previously shaped by the individual's conscious mind but were frequently elaborate compositions of great complexity and originality. Moreover, the products of automatic writing manifest themselves as the creation of a personality quite unknown to the main personality of the writer but showing all the characteristics usually ascribed to individual human personality. Such phenomena, said Prince, suggest a subconscious intelligence (Prince 1907, pp. 67–80).

In his rather extensive psychological writings (see Campbell et al. 1932, pp. 420–27) Prince developed a comprehensive framework for understanding dissociation based on a careful definition of terms. He pointed out that "subconscious" and "unconscious" had been often used as synonyms although they referred to different classes of facts. He suggested that "subconscious" be replaced by the term "co-conscious" and that "unconscious" be reserved for those basically physiological processes that were devoid of the attributes of consciousness.

Unconscious processes, according to Prince, have to do with the registration, storage, and retrieval of memories, which happen on a purely neurological level. But co-consciousness involves a dissociated consciousness that coexists with one's normal consciousness. Co-conscious ideas have been

6. Prince's most famous multiple personality case was that of "Miss Beauchamp," first reported in the *Proceedings of the Society for Psychical Research* (Prince 1901) and then published in detail in his well-known *Dissociation of a Personality* (Prince 1905).

misnamed unconscious (for example, by Freud) because the personal consciousness is not aware of them. But this designation, Prince contends, is
inaccurate and confusing. Co-conscious ideas include both states that our
normal consciousness is not aware of—because they are not the focus of our
attention—and also pathologically split-off and independently active ideas or
systems of ideas, such as occur in hysteria, multiple personalities, and automatic writing. Prince preferred the term "co-conscious" to Janet's "subconscious," first, because it expressed the simultaneous activity of an alternate consciousness and, second, because the coactive ideas might not be
outside the awareness of the personal consciousness but still might be recognized as a distinct consciousness existing alongside it.

Through his redefinition of terms, Prince made the simultaneous activity of
two or more systems of consciousness in one individual the key element in
dissociation. In this way he formulated the notion of a continuum of dissociation, ranging from automatisms in healthy persons to multiple personality, in
which dissociated systems have taken on a stable personalized form.

Chapter 17

The Mesmeric Legacy to Freud

While psychological theory and psychotherapeutic practice based on the alternate-consciousness paradigm was just reaching a point of maturity, another psychodynamic approach was about to come into existence. The birth of psychoanalysis was a milestone in the evolution of human self-knowledge. It brought with it a novel formulation of child development and a model of a dynamic unconscious that would permanently alter psychological thinking. But psychoanalysis emerged on the scene with such impact that the psychological approaches based on the alternate-consciousness paradigm, just coming into their own, were largely eclipsed. Nonetheless, psychoanalysis was itself the beneficiary of the alternate-consciousness tradition and could not have come into being without it. That it so quickly overshadowed that tradition is one of the intriguing ironies of the history of psychology.

Josef Breuer

The Viennese physician Josef Breuer (1842–1925) can be situated in the stream of psychological thinking that arose out of the tradition of animal magnetism, and he was in his own right one of the pioneers in the evolution of the alternate-consciousness paradigm. At the same time, his psychotherapeutic work with his colleague Sigmund Freud (1856–1939) initiated a series of events that was to give a striking new direction to the development of psychological healing.

Breuer did not write specifically about his exposure to mesmerism and hypnotism, but it has been pointed out (Hirschmüller 1978, pp. 92–95) that he had ample opportunity to observe and practice the technique. It is very likely that a well-read physician like Breuer would have been acquainted with Richet's articles on artificial somnambulism and Charcot's view on hypnotism. It is known that in 1868 Breuer observed experiments in animal magnetism by Moriz Benedikt, and it is likely that he was also aware of later hypnotic trials carried out by Benedikt. Breuer commented on an 1873 paper on hypnotism by Johann Czermak and in 1874 wrote one of his own on the use of hypnotism on animals. This background proved to be specially useful when, in 1880, Breuer undertook the treatment of a young woman suffering from a puzzling but intriguing psychological condition.

The Case of Anna O.

Bertha Pappenheim was born in Vienna in 1859 to an Orthodox Jewish family. Nevertheless, she attended a Catholic private school and received a broad education that included foreign languages, music, and literature. In 1880 Bertha's father fell seriously ill with peripleuritis, probably with a background of tuberculosis. He needed constant care at home, and Bertha took on the responsibility of looking after him at night. On July 17, a surgeon was called in to drain the abscess. That evening, awaiting the surgeon's arrival at her father's bedside, Bertha suddenly became paralyzed in her right leg. This was the first of a series of symptoms that would virtually incapacitate the young woman.[1]

Breuer, who was probably serving as the Pappenheim family physician, was consulted some months after the onset of Bertha's illness. What he saw was a young woman who, in addition to the paralysis, suffered from headache, muscular contractions, anesthesias, ocular disturbances, and speech anomalies. She manifested two distinct personalities: a "normal," melancholy personality and an "abnormal," crude, agitated personality that was subject to hallucinations. She experienced discontinuity in consciousness so that she was sometimes not aware of things she had just done. She spoke of this as "lost" time.

Bertha's symptoms were recognized as typically hysterical, but Breuer did not choose to use the conventional treatment methods of the time, such as "antihysterical" drugs, electrotherapy, hydrotherapy, movement therapy, or dietetic regimes. Instead of approaching her condition as a physiological problem requiring physiological countermeasures, Breuer chose to view it as fundamentally psychological in origin. He took the time to observe her mental states and try to improve her condition through psychological means.

---

1. There are two reports describing Bertha Pappenheim's illness. One is the well-known essay by Breuer, "Fräulein Anna O.," *Studien über Hysterie,* 1895. The other is the little-known transcript of the case history written by Breuer in 1882 and discovered by Henri Ellenberger at the Bellevue Sanatorium in Kreuzlingen near Konstanz. It was Freud's biographer Ernest Jones who realized that the "Anna O." of the 1895 report was Bertha Pappenheim, a connection particularly difficult to make because Breuer had so well disguised the conditions of the case to preserve the patient's anonymity. In 1972 Hirschmüller came across more documents at the Bellevue Sanatorium, including letters written by Breuer, Bertha, her mother, and her cousin.

Breuer noted that every day at dusk Bertha fell into a state of self-hypnosis and he capitalized on that fact, visiting her every evening and beginning what she called her "talking cure." When in her secondary state during the day, Bertha would mention single words that hinted at the nature of her fantasy life. While in the hypnotic state in the evening, she could be induced to tell a story built around one of those words. On awakening, she would feel comforted and peaceful. Breuer took this as an indication that some kind of relief was produced by the release of psychic tension through the stories. This method of treatment was continued for some months, and by the beginning of April her symptoms had just about disappeared.

Then on April 5 her father died, and Bertha was thrust back into a disturbed state. She became hostile to her family, would eat only when Breuer fed her, and ceased to understand German. At one point Breuer had to go away for a few days, and he returned to find that her state had worsened even further. She was now subject to disturbing hallucinations and made a number of suicide attempts. Breuer attempted to get Bertha to continue the evening "stories," and her condition again improved somewhat.

Now Breuer began a more intensive treatment, seeing her for longer periods and lengthening the time during which Bertha told her stories. The stories shifted in content, containing not mere poetic fantasy but material relating to her hallucinations and to things that had irritated her during the past few days—and, most important, elements of the history of her illness. Breuer now made the significant discovery that if Bertha related the actual incident that occasioned the initial formation of a symptom, that symptom would disappear no matter how long it had been since its onset. In this way he removed symptoms one by one—her eye spasms, contracture of the right leg, inhibition of drinking, and others. From this point his treatment centered around this newly discovered therapeutic framework.

The course of the treatment was not smooth, however, and in the fall of 1881 Breuer considered sending Bertha to the Bellevue Sanatorium at Kreuzlingen. In 1882 he continued his work with the symptoms, now adding the task of revealing previously concealed symptoms of the incubation of Bertha's illness in 1880. He had to trace each occurrence of a symptom back sequentially in time to discover the nucleus of the problem. At this time Breuer began seeing Bertha in the morning as well as the evening and for the first time started artificially inducing her hypnotic state. Finally the original traumatic events at the core of her illness were unveiled.

By 1895 Breuer, along with his fellow physician Sigmund Freud, was able to put into systematic form the elements of the therapy he had evolved through his work with Bertha.[2] He saw in Bertha an example of what he and Freud termed a "traumatic hysteria" or "psychically acquired hysteria" that could be treated through psychological means (Breuer and Freud 1964, pp. 5, 12–13). In her treatment, he first of all prevented the buildup of day-to-day anxiety by processing traumatic daily events through the storytelling—this was the "talking cure." He attempted to deal with the morbid symptoms by having the patient recall the occasion on which they first occurred, releasing the affect that had been trapped with the traumatic memory—this was the "abreaction of pathological affects." "Abreaction," a term chosen by Breuer and Freud, led to a "catharsis" and freedom from the symptoms. From this Breuer concluded that the patient had "ideas inadmissible to consciousness" that were avoided through the symptom formation (p. 222). His psychotherapy would bring the operative force of the idea to an end by allowing the patient to release its "strangulated affect" through speech, in that way introducing the inadmissible idea into consciousness (p. 17).

Breuer and Freud described the process involved in this sequence of events in their *Studien über Hysterie* (1895). Reading that account one would think that Bertha was completely cured by the therapy. In fact, the remission of her symptoms was temporary, and she soon was admitted to Bellevue. Ellenberger mentions that Jung learned from Freud that Bertha was never cured. He also points out that it is ironic that Bertha Pappenheim's unsuccessful treatment became the prototype and model for cathartic cure (1970, pp. 484–85).

Hypnoid States

Breuer noted that before the onset of her illness, Bertha had a tendency to daydreaming (which she called her "private theatre"), which laid the foundation for the dissociation that characterized her illness. He stated that reveries of this kind were normal and did not necessarily imply a pathological "splitting of consciousness." Under the stress of her father's illness, however, a more complete dissociation set in, producing a second state of consciousness that alternated with Bertha's primary or normal state. Eventually she spent more time in her second state than in the normal (Breuer and Freud 1964, pp. 42–43).

When Bertha was in her self-induced hypnotic states she regained memories of significant events that had played a part in the formation of her illness. When in her normal state she had no recall of these events. The problem was that material from her second state, disturbing memories and the emotions

2. See Breuer, "Fräulein Anna O.," and Freud and Breuer, "On the psychical mechanism of hysterical phenomena: Preliminary communication," in Breuer and Freud 1964.

attached to them, intruded into her normal state indirectly, in the form of symptoms. While in her primary state, the products of her secondary state acted as a stimulus "in the unconscious" (Breuer and Freud 1895, pp. 45, 47).[3]

It was Breuer's contention that, at least with "traumatic hysteria," treatment should concentrate on the psychological conditions that are present. Bertha showed Breuer the way through her spontaneous autohypnotic state, in which she was able to get in touch with the relevant memories and emotions. In this way the buildup of psychic tension that expressed itself in symptoms could be relieved through the talking cure, and the symptoms themselves could be removed by reverting to their point of origin.

Breuer called Bertha's self-induced hypnosis-like states "hypnoid states." He believed that such individuals as Bertha have an innate disposition to go into such states. He also believed that hypnoid states can be generated by buildup of affect or emotion that cannot be integrated into normal consciousness, be discharged, or be "worn away" through a normal expression of feeling. This buildup will sometimes produce a hypnoid state in which the affect exists in isolation from ordinary consciousness (Breuer and Freud 1895, pp. 215–21).

In this way Breuer agreed with Janet that the memory of a psychic trauma can exist within a person like a foreign body affecting waking life. Like Janet, he spoke of "unconscious ideas," but in this matter there were differences. He noted that in many patients there is a large complex of ideas that are admissible to consciousness, and a smaller complex of ideas that are not, so that the "field of ideational psychical activity" is broader than consciousness and is divided into a conscious and an unconscious part. Correspondingly, one's ideas are divided into those that are admissible and those that are inadmissible to consciousness. Breuer called this division the "splitting of the mind."

Janet and Binet had described a "splitting of consciousness" involving the presence of two portions of psychical activity that alternate with each other. Breuer pointed out that Bertha experienced this kind of double consciousness when her hypnoid state alternated with her normal state. But he also called attention to a split-off portion that was "thrust into darkness" and, like the Titans imprisoned in the crater of Etna, can shake the earth but can never emerge into the light of day (Breuer and Freud 1964, pp. 217, 227–29).

How does this take place? When a traumatic event occurs, the person is

---

3. There was also evidence of a third state, a clear-sighted and calm observer that looked on from a corner in her mind even during the most disturbed periods (Breuer and Freud 1964, p. 46). Bertha said that this "observer brain" saw things clearly and gave her information about obscure matters (Hirschmüller 1978, p. 283).

thrown into a hypnoid state. For some time afterwards, whenever the traumatic event is recalled, the person reenters the state of hypnoid fright. After a time the intensity of the state diminishes, and the hypnoid state no longer alternates with normal consciousness but exists side by side with it. At this point the hypnoid state is continually present below the surface, and the somatic symptoms that previously arose during the alternating hypnoid states acquire a permanent existence. This process occurred with Bertha when the alternation between the waking state and the hypnoid state was replaced by a coexistence of the normal and hypnoid complexes of ideas (Breuer and Freud 1964, pp. 229, 235).

Breuer noted that a kind of duplication of psychical functioning is present in all human beings—both the healthy and the ill. For instance, a "double ego" is operative when a healthy person goes into a state of reverie or creative abstraction. Then the mind is temporarily divided into two streams that reunite when the abstracted state is finished. But with a pathological splitting of the mind the two streams cannot reunite, because the ideas of the second stream are inadmissible to the normal consciousness.

It is clear from Breuer's treatment of Bertha that he believed that the original traumatic memories could be accessed and the hysterical symptom removed through hypnotic re-creation of the hypnoid state. For that reason there is some ambiguity in Breuer's statement that where a second complex of ideas has been split off it is no longer a splitting of consciousness but a splitting of the mind, involving memories that are "thrust in darkness" and made totally inaccessible (Breuer and Freud 1964, p. 225). The split-off train of ideas becomes accessible during spontaneous hypnoid states and during artificially induced hypnosis. In these conditions the unconscious complex regains its identity as a self-conscious state. Given this state of affairs, it is difficult to see on what basis Breuer distinguished between Bertha (and similar cases) and Janet's hysterics (such as Lucie), who exhibited copresent self-conscious states. Like Bertha, Lucie had an organized complex that produced physical symptoms "in the background" when Lucie was in her normal state. Like Bertha's second personality, Lucie's could be brought forward by hypnotism. Breuer, like Janet, made use of the continuing existence of the hypnoid personality to reveal what was going on below the subject's conscious awareness, and he assumed that once the secondary complex of ideas began to coexist with the primary, it would no longer appear in consciousness—except, it seems, when called forth in hypnotic states.

This apparent ambiguity was not clarified in Breuer's writings or in those of Freud, in *Studien über Hysterie*. However, Freud later made his ideas on the matter clear and in so doing separated himself from the notion of a second consciousness operating outside the awareness of normal consciousness.

Freud and the Nature of Consciousness

In 1882 Breuer told the young Freud about his treatment of Bertha Pappenheim. Freud did not actually make use of Breuer's method in his work until 1889. Freud and Breuer seemed to agree in the main on issues involving the cathartic method, but they differed with regard to hypnoid states. Writing on his theory of hysteria, Breuer noted that Freud believed that a splitting of the mind could be produced by the deliberate deflection of consciousness from distressing ideas, but Breuer himself held that the assistance of hypnoid states was necessary if a true splitting of the mind was to occur (Breuer and Freud 1964, pp. 235–36).

Freud took issue with Breuer's description of cases of "hypnoid hysteria" (such as Bertha Pappenheim) in which an idea becomes pathogenic because it has been attached to a hypnoid state and remains outside the ego. Freud said that he had never encountered a genuine hypnoid hysteria. His explanation was that an incompatible idea was repressed from consciousness from a motive of defense. The repressed idea persists as a weak memory trace, while the affect, which has been separated from the idea, manifests as a somatic symptom. Freud called this condition a "defense hysteria" (Breuer and Freud 1895, pp. 285–87).

Later, in describing his work with "Dora," Freud completely rejected Breuer's view that hypnoid states had an important role in the generation of hysteria. He noted that the term "hypnoid state" came entirely from Breuer and stated that he regarded the use of the term and the concept as "superfluous and misleading" (Freud 1964b, p. 27n). In this manner Freud drew away from the tradition of double consciousness that had for decades related split-off segments of the mind to magnetic or hypnotic somnambulism. Breuer was still firmly planted in that historical stream, seeing spontaneously induced hypnoid states as the splitting mechanism and artificially induced hypnotism as the most effective tool for psychotherapy (see Bliss 1986). Freud looked in a different direction for the splitting of the mind and in the process made the use of artificial somnambulism in psychotherapy unnecessary.

Nowhere did this divergence from the magnetic-hypnotic tradition appear more clearly than in Freud's controversy with Janet about the nature of unconscious ideas. Freud and Janet agreed in dividing human mental activity into two spheres on the basis of its availability to normal consciousness. They agreed in pinpointing the source of emotional disturbance in mental processes that operate outside of normal consciousness. They also concurred that the remedy for such disturbances involved bringing those hidden elements into ordinary awareness.

Freud and Janet differed, however, in the extent to which they believed that the concept of hidden mental processes can be applied to healthy people and in their views on the precise nature of consciousness. Janet was reluctant to attribute subconscious processes to normal, healthy individuals, since he saw the dissociated elements that constitute the subconscious as basically patho-logical. Freud, on the other hand, like Myers, Dessoir, Prince, and others, believed that everyone was subject to the hidden mental processes of the unconscious. For him, unconscious mentation was a fundamental part of human psychological life.

With regard to the nature of consciousness, Janet believed that a person can have a number of centers of consciousness operating subconsciously. He had no problem in accepting the notion of multiple streams of conscious mental activity operating simultaneously. He described these separately functioning streams as one might describe independent minds of different people. That is why Morton Prince's notion of co-consciousness was a natural elaboration of Janet's original concept.

Freud viewed it quite differently. He saw consciousness as unique—each person can have but one. He described consciousness as a "sense-organ for the perception of psychical qualities" (Freud 1964a, p. 615). Freud believed that not all that is psychical or mental takes place within a consciousness, asserting that mental processes are in themselves unconscious and that only portions of mental life are conscious. Because he could not accept the identity of the conscious and the mental, he rejected the definition of psychology as the study of the contents of consciousness and insisted that there is unconscious think-ing and willing (Freud 1964e, pp. 21–22 and 1964g, p. 31).

Freud emphasized that his "unconscious" and Janet's "subconscious" were not the same. "We shall . . . be right in rejecting the term 'subconscious' as incorrect and misleading. The well-known cases of *'double conscience'* (splitting of consciousness) prove nothing against our view. We may most aptly describe them as cases of a splitting of mental activities into two groups, and say that the same consciousness turns to one or other of these groups alternately" (Freud 1964d, pp. 170–71).

In denying the existence of a second consciousness, Freud was certainly contradicting Janet. But Janet himself seemed reluctant to admit that Freud's views were radically divergent from his own. In fact, he asserted on occasion that Freud had simply taken over his system and given it a new terminology. What Freud did beyond that, Janet said, was to transform, without sufficient basis, a series of clinical observations into an elaborate system of medical philosophy (see Janet 1914, pp. 1–35, 153–87; Ellenberger 1970, p. 344).

In 1925 Freud spelled out his position vis à vis Janet even more clearly:

> Psycho-analysis regarded everything mental as being in the first instance unconscious; the further quality of "consciousness" might also be present, or again it might be absent. This of course provoked a denial from the philosophers for whom "conscious" and "mental" were identical and who protested that they could not conceive of such an absurdity as the "unconscious mental." . . . It could be pointed out, incidentally, that this was only treating one's own mental life as one had always treated other people's. One did not hesitate to ascribe mental processes to other people, although one had not immediate consciousness of them and could only infer them from their words and actions. But what held good for other people must be applicable to oneself. Anyone who tried to push the argument further and to conclude from it that one's own hidden processes belonged actually to a second *consciousness* would be faced with the concept of a consciousness of which one knew nothing, of an "unconscious consciousness"—and this would scarcely be preferable to the assumption of an "unconscious mental." (Freud 1964g, pp. 31–32)

Freud's view of dissociated psychical elements is now clear. Dissociated systems are simply separate groups of mental but unconscious elements. Human consciousness is unitary and, like a searchlight, shines now on one group of elements, now on another. When the beam of consciousness illuminates one group, it enters into our awareness and manifests in consciousness. In themselves dissociated systems are mental, but not conscious. There is no doubling of consciousness, no second consciousness.

To those in the alternate-consciousness tradition, the complexity of the mind allows for the simultaneous existence of multiple centers of consciousness, each with its separate stream of thought. The multiple centers are in differing degrees and in varying directions aware of each other and can interact. They are like many minds in one body. Consciousness is not unitary, and any unity that exists or may be achieved is beyond that available to ordinary consciousness.

Freud seldom tackled the problem of dissociation head on. In a rare attempt to explain the intrinsic nature of automatisms and dissociative phenomena, he wrote:

If [the ego's object-identifications] obtain the upper hand and become too numerous, unduly powerful, and incompatible with one another, a pathological outcome will not be far off. It may come to a disruption of the ego in consequence of the different identifications becoming cut off from one another by resistances; perhaps the secret of the cases of what is described as "multiple personality" is that the different identifications seize hold of consciousness in turn. Even when things do not go so far as this, there remains the question of conflicts between the various identifications into which the ego comes apart, conflicts which cannot after all be described as entirely pathological. (Freud 1964f, pp. 30–31)

It was precisely in the area of automatisms and dissociative phenomena that psychoanalysis seemed to have its greatest difficulty. By separating himself from Breuer's concept of hypnoid states—and thereby from the whole tradition of multiple streams of consciousness within the psyche—Freud made the explanation of the phenomena so commonly reported in magnetic and hypnotic literature obscure. Whereas Binet, Dessoir, James, Sidis, Prince, and especially Myers were able to embrace more and more dissociative phenomena—both morbid and healthy—in their systems, Freud gave them very little attention. Freud's system could not easily handle these phenomena, and the psychoanalytic tradition that has evolved since then has done little to remedy that defect.[4]

4. This theoretical weakness in psychoanalysis was ably pointed out by Bernard Hart (1910). More recently Eugene Bliss (1986) has reemphasized this flaw and noted that Breuer's unique contribution to understanding psychopathology has been obscured by Freud's rejection of hypnoid states (pp. 54–58).

Postscript

The Alternate-Consciousness Paradigm from Freud to the Present Day

The last decades of the twentieth century have brought a renewed interest in dissociative phenomena, and as this happens the alternate-consciousness paradigm is again coming to the fore. This is in part due to recent developments in the study of multiple personality disorder.

The Rediscovery of Multiple Personality Disorder

Although some of Freud's contemporaries pointed out that psychoanalysis could not deal well with dissociative conditions and particularly multiple personality (Hart 1926), that weakness did not seem to trouble psychoanalysts. Perhaps this should not be surprising, since in the decades when psychoanalysis was coming to predominate psychotherapeutic practice, multiple personality and associated disorders were rarely diagnosed.

Goettman, Greaves, and Coons (1992) note a sharp decline in reported instances of multiple personality in the fifty years following 1920. Whether this drop was due to missed diagnoses or an actual decrease in instances is a matter of debate. In any case, the contrast in general awareness of dissociation in those decades compared to that in the period 1880–1920 is striking. Whereas many psychological writers at the turn of the century showed a fair amount of sophistication in their understanding of multiple personality, by 1950 that awareness had largely been lost.[1]

Then in 1952 Chris Sizemore, a young housewife, sought psychiatric help for headaches and depression. She was treated by two psychiatrists, Corbett Thigpen and Hervey Cleckley, and in the course of her therapy was discovered to have more than one distinct personality. At that time Thigpen and Cleckley believed her condition to be so rare that they estimated that there were probably only a half-dozen persons afflicted by it in the world (Sizemore 1991), and they introduced their first clinical report of the case

1. For the history of multiple personality, see Taylor and Martin (1944) and especially Greaves (1980). A full-length historical treatment is yet to be written.

grandly: "The psychiatric manifestation called multiple personality has been extensively discussed. So too have the unicorn and the centaur" (Thigpen and Cleckley 1954).

The case of Chris Sizemore, popularized in the book and movie *The Three Faces of Eve* (1957), stimulated a renewal of interest in dissociative phenomena. But that version of her story was incomplete. As it turned out, the famous three personalities ("Eve White," "Eve Black," and "Jane") that surfaced during Sizemore's two and a half years with Thigpen and Cleckley were not her only personalities. After vainly seeking help from six other psychiatrists, Sizemore worked successfully with Drs. Tony Tsitos and Tibor Ham to eventually uncover twenty-two personalities. Today she is a single personality.[2] But the popularized version of her story had such impact that "Eve" remains the prototypical multiple personality in the public mind.

In 1973 the work of the Kentucky psychiatrist Cornelia Wilbur with a young woman suffering from multiple personality was published under the title *Sybil* (Schreiber 1973). "Subjected to cruel sexual and physical abuse throughout childhood, Sybil had developed her personalities as a means of coping with the traumata." This book too was made into a movie that had notable impact.

In the following year, the psychiatrist Ralph Allison published his first articles on the treatment of multiple personality (Allison 1974a, 1974b) and became involved in educating members of the American Psychiatric Association about the condition. The association asked Cornelia Wilbur to form a panel to make a presentation on multiple personality at the annual meeting in 1977. Allison was invited to chair the panel. Because of the strong interest generated by this presentation, the program committee of the association asked Allison to put on a two-day workshop on multiple personality at the 1978 meeting. Again the response was lively, so he repeated the workshop the next two years. Then the Pennsylvania psychiatrist Rick Kluft took on the organization of the workshop, which continued as an annual APA event (Allison 1992).

The diagnosis of multiple personality disorder was first included in the APA's *Diagnostic and Statistical Manual of Mental Disorder* in the third edition (*DSM-III*, 1980).[3] The acceptance of multiple personality disorder as a legitimate psychiatric diagnosis was an important achievement. As Ross

2. For the full story of Chris Sizemore and her recovery from multiple personality see Sizemore 1977 and 1989.

3. The revised version of the manual (*DSM-IIIR*, 1987) further refined the description of salient features of multiple personality disorder.

has pointed out (1989, p. 77), it was undoubtedly this official recognition of multiple personality that led to a sudden rise in diagnosed cases in the 1980s.

In the early 1980s, clinical publications relating to cases of multiple personality disorder suddenly mushroomed. Articles by Coons (1980), Braun (1980, 1983a, 1983b), Kluft (1982, 1983), and Putnam (1982), for example, highlighted diagnosis and treatment, and in 1984 whole issues of several periodicals were devoted to the examination of multiple personality disorder (*American Journal of Clinical Hypnosis*, October 1984; *International Journal of Clinical and Experimental Hypnosis*, April 1984; *Psychiatric Clinics of North America*, March 1984; *Psychiatric Annals*, January 1984). That same year the International Society for the Study of Multiple Personality and Dissociation held its first conference in Chicago. Eventually that society produced the first journal to be devoted exclusively to clinical and experimental research on dissociative disorders, titled *Dissociation: Progress in the Dissociative Disorders* (1988–).

As more mental health workers have become involved in treating multiple personality, knowledge of various aspects has progressed. The relation of multiple personality to childhood sexual and physical abuse,[4] etiology,[5] the number and origin of personalities,[6] epidemiology (Ross 1991b), physiological effects (Coons 1988), and the relationship to post-traumatic stress disorder (Branscomb 1991) are issues that continue to be clarified through clinical work and research. Many of these findings have been illustrated in significant case histories, such as *The Minds of Billy Milligan* (Keyes 1981), *Prism: Andrea's World* (Bliss and Bliss 1985), *When Rabbit Howls* (Chase 1987), and *Through Divided Minds* (Mayer 1988). At the same time very good manuals of treatment have become available (Kluft 1985, Braun 1986, Putnam 1989, Ross 1989).

---

4. It is generally accepted today that multiple personality disorder almost always arises as a result of ongoing traumatic abuse in childhood. It seems that Putnam et al. (1986) gives the highest estimate, showing a background of abuse in 97 percent in a population of one hundred cases of multiple personality disorder interviewed, with sexual abuse occurring in most instances.

5. Kluft (1984) presents a four-factor model of the etiology of multiple personality disorder, involving (1) excellent ability to dissociate, (2) presence of severe childhood trauma, (3) dependence of the form and structure of the multiplicity on temperament and nonabuse experience, (4) the ongoing nature of the abuse and the failure to counterbalance it with love and care.

6. Studies of the number of personalities in a multiple give varying results. Today it is generally acknowledged that to have 15–30 personalities per individual is quite common. There are reports of polyfragmented multiples with personalities in the hundreds.

A New Look at Dissociation

This renewed interest in multiple personality disorder brought with it a reappraisal of dissociation. From this point of view, Ellenberger's *The Discovery of the Unconscious* (1970) could not have been written at a more opportune time. It aroused curiosity in a segment of medical history that had been almost completely ignored. Ellenberger's history of the ancestry and evolution of dynamic psychiatry made it clear that any thorough investigation of dissociation had to study mesmerism and its offshoots and to take into account how much had already been discovered.

The appearance of Ernest Hilgard's *Divided Consciousness* (1977) took the discussion of dissociation to a new level. Harking back to the work of Janet and other investigators of the late 1800s, Hilgard pointed out that their findings had been neglected not because they had been scientifically superseded but also because of factors arising from social history. Using data from his own hypnotic research, he formulated a revised understanding of dissociation, into which he incorporated modern findings about information processing, divided attention, and brain function. Hilgard's "neodissociation" theory has had a notable impact on contemporary thinking about the divided mind.

That impact can be discerned in Peter McKellar's *Mindsplit: The Psychology of Multiple Personality and the Dissociated Self*. In this valuable but little-known book, McKellar used the findings of his own hypnotic experiments and a study of the case literature on dissociation to construct a framework for understanding both normal and abnormal dissociative phenomena.

In the late 1970s John Watkins developed his own dissociative view of psychological functioning (Watkins 1976, 1977, Watkins and Watkins 1979). His "ego-state therapy" held that the ego is not a unity but a confederation of states or part persons that operate with relative degrees of autonomy. His theory evolved into a comprehensive psychotherapeutic approach that was presented in his book *The Therapeutic Self* (1978).

In 1982 John Beahrs published *Unity and Multiplicity,* which took the experience of hypnosis as the starting point for a new evaluation of dissociative disorders. Beahrs attempted to deal with a number of thorny theoretical issues, including the problem of the basic unity of the individual who manifests discrete identities. His notion of three selves (the conscious or executive self, the self composed of an aggregate of lesser personalities, and the conscious experience of one's basic biological existence) served as the basis for an effective treatment of dissociative disorders.

Allison's *Mind in Many Pieces* (1980) was another effort to combine a new theory of dissociation with a practical psychotherapy. This work was the first modern attempt to incorporate possession phenomena into dissociation theory and therapy. Adam Crabtree's *Multiple Man* (1985a) united a variety of psychological experiences (including multiple personality, possession, past-life memory, paranormal phenomena, ecstasy, and creative inspiration) within a comprehensive dissociative framework that considered both clinical and theoretical perspectives.

The theoretical understanding of dissociation was further advanced by Bennett Braun's BASK model of dissociative experiences (Braun 1989a, 1989b). Braun identified four aspects of dissociative experience—behavior, affect, sensation, and knowledge—and allowed for their independent manifestation and recall. This model, based on a solid phenomenological approach, made sense of the confusing variety of memories experienced by dissociated individuals.

Colin Ross (1991a) introduced an important new dimension to the theoretical understanding of dissociation with his concept of a "dissociated executive self." For Ross, multiplicity is a normal organizational principle of the human psyche, and the executive self or ego is just one of many parts that make up the whole human being. He pointed out that in the western industrialized world the executive self has suppressed all other part selves. A cultural dissociation barrier has been erected that effectively removes from consideration those parts of the self that deal with experiences that are unacceptable to western culture. These rejected experiences fall into three main categories: paranormal experiences, deep intuitive consciousness, and programs responsible for running the physical organism. Because of the dissociation barrier, the executive self—what we ordinarily call "I"—is disconnected from these vital experiences and must relegate them to second-class status or risk feeling at odds with what is culturally accepted as real.

In his book on multiple personality (1986) Eugene Bliss brought valuable historical insight to bear on the role of psychoanalysis in obscuring the issues of dissociation and elaborated a theory of dissociation and hypnoid states. In 1986 Jacques Quen edited a collection of historical essays (*Split Minds/Split Brains*) about the historical roots of dissociated phenomena. At the same time essays on the dissociation theory of Pierre Janet began to appear (Perry and Laurence 1984, Haule 1986, Van der Hart and Friedman 1989), signifying the recognition that modern experience of dissociation can be enriched by historical perspectives.

In his 1991 book *First Person Plural: Multiple Personality and the Philosophy of Mind* Stephen Braude for the first time brought a searching philosophical analysis to bear on overt concepts and covert assumptions in the language that have been used, in the past and present, in talking about dissociation and multiple personality. Braude provides a valuable analytical perspective that is not likely to be found in the writings of those who deal mainly with the daily struggles of clinical treatment of dissociation.

The alternate-consciousness paradigm supposes a division in consciousness with a realm of mental activity not available to ordinary awareness. That realm can contain many streams of consciousness that are all active concurrently. In recent years multiple personality and other dissociative disorders have once more emerged into general awareness. Those who try to understand these disorders have found themselves ineluctably drawn to the alternate-consciousness framework of thought. Though the alternate-consciousness paradigm originally arose as the byproduct of a technique of physical healing, today that paradigm is reasserting itself as the result of modern needs of psychological healing.

References

Allison, Ralph B. 1974a. A new treatment approach for multiple personalities. *Amer. J. Clinical Hypnosis* 17:15–32.

———. 1974b. A guide to parents: How to raise your daughter to have multiple personalities. *Family Therapy* 1:83–88.

———. 1980. *Mind in Many Pieces: The Making of a Very Special Doctor.* New York: Rawson, Wade.

———. 1992. Personal communication. August 4, 1992.

Almignana, abbé. [1853]. *Du somnambulisme, des tables tournantes et des mediums. Considérés dans leurs rapports avec la théologie et la physique.* Paris: Dentu.

Alvarado, Carlos S. 1990. The history of MPD: A brief guide to the literature. *ISSMP&D News* 8:8–9.

Amadou, Robert, ed. 1971. *Franz-Anton Mesmer. Le magnétisme animal. Oeuvres publiées par Robert Amadou avec des commentaires et des notes de Frank Pattie et Jean Vinchon.* Paris: Payot.

Animal Magnetism. Report of Dr. Franklin and Other Commissioners, Charged by the King of France with the Examination of the Animal Magnetism as Practised at Paris. Translated from the French. With an Historical Outline of the "Science," an Abstract of the Report on Magnetic Experiments, Made by a Committee of the Royal Academy of Medicine, in 1831, and Remarks on Col. Stone's Pamphlet. 1837. Philadelphia: H. Perkins.

The Animal Magnetizer; or, History, Phenomena and Curative Effects of Animal Magnetism; with Instructions for Conducting the Magnetic Operation. 1841. Philadelphia: J. Kau.

Annals of Animal Magnetism. 1838. 1 volume.

Archiv für den thierischen Magnetismus. 1817–1824. Vols. 1–12.

Archives du magnétisme animal. 1820–1823. Vols. 1–8.

Artelt, Walter. 1965. *Der Mesmerimus in Berlin.* Wiesbaden: Akademie der Wissenschaften und der Literatur.

Ashburner, John. 1867. *Notes and Studies in the Philosophy of Animal Magnetism and Spiritualism. With Observations upon Catarrh, Bronchitis, Rheumatism, Gout, Scrofula, and Cognate Diseases.* London: H. Baillière.

Audry, J. 1924. Le mesmérisme et le somnambulisme à Lyon avant la révolution. Académie des sciences, belles lettres et arts de Lyon. Classe des sciences et lettres. *Mémoires* 18:57–101.

Avis aux chrétiens sur les tables tournantes et parlantes, par un ecclésiastique. 1853. Paris: Devarenne and Perisse.

Azam, Eugène. 1860. Note sur le sommeil nerveux ou hypnotisme. *Archives générales de médecine*, January 1860:5–24.

———. 1876a. Amnésie périodique ou dédoublement de la vie. *Revue scientifique* 16:481–89.

———. 1876b. Le dédoublement de la personnalité: Suite de l'histoire de Félida X***. *Revue scientifique* 18:265–69.

———. 1877. Le dédoublement de la personnalité et l'amnésie périodique. *Revue scientifique* 20:577–81.

———. 1887. *Hypnotisme, double conscience, et altérations de la personnalité*. Paris: J. B. Baillière.

———. 1893. *Hypnotisme et double conscience. Origine de leur étude et divers travaux sur des sujets analogues*. Paris: Félix Alcan.

Babinet, Jacques. 1856. *Etudes et lectures sur les sciences d'observation et leurs applications pratiques*. Vol. 2. Paris: Mallet-Bachelier.

Bachelier d'Agès, P. J. [1800]. *De la nature, de l'homme, et des moyens de le rendre plus heureux*. Paris: F. Buisson, Pichard, Desenne, and Petit.

Bailly, Jean Sylvain. 1784a. *Rapport des commissaires chargés par le roi de l'examen du magnétisme animal*. Paris: Imprimerie Royale.

———. 1784b. *Rapport secret présenté au ministre et signé par la commission précédente*. N.p. [This report was seen only by the king of France, Louis XVI. For the first public printing, see Bailly 1800.]

———. 1800. Rapport secret sur le mesmérisme. *Le conservateur . . . de N. François (de Neufchateau)* 1:146–55.

Ballou, Adin. 1853. *An Exposition of Views Respecting the Principal Facts, Causes and Peculiarities Involved in Spirit Manifestations: Together with Interesting Phenomenal Statements and Communications*. Boston: Bela Marsh.

Barberin, Chevalier de. 1786. *Système raisonné du magnétisme universel. D'aprés les principes de M. Mesmer, ouvrage auquel on a joint l'explication des procédés du magnétisme animal accommodés au cures de différentes maladies, tant par M. Mesmer que par, M. le Chevalier de Barberin et par M. de Puységur, relativement au somnambulisme ainsi qu'une notice de la constitution des Sociétés dites de l'harmonie . . . Par la Société de l'harmonie d'Ostende*. Ostende: N.p.

Baréty, A. 1882. *Des Propriétés physiques d'une force particulière du corps humain (force neurique rayonnante) connue vulgairement sous le nom de magnétisme animal*. Paris: Octave Doin and Jacques Lechevalier.

———. 1887. *Le magnétisme animal étudié sous le nom de force neurique rayonnante et circulante, dans ses propriétés physiques, physiologiques et thérapeutiques*. Paris: Octave Doin and J. Lechevalier.

Barret, W. F., Edmund Gurney, Frederic W. H. Myers, et al. 1883a. First report of the committee on mesmerism. *Proceedings of the Society for Psychical Research* 1:217–29.

———. 1883b. Second report of the committee on mesmerism. *Proceedings of the Society for Psychical Research* 1:251–62.

————. 1883c. Appendix to the Report on Mesmerism. *Proceedings of the Society for Psychical Research* 1:284–90.

Barrow, Logie. 1986. *Independent Spirits: Spiritualism and English Plebeians, 1850–1910.* New York: Routledge.

Barrucand, Dominique. 1967. *Histoire de l'hypnose en France.* Paris: Presses Universitaires de France.

Barth, George H. 1850. *The Mesmerist's Manual of Phenomena and Practice; with Directions for Applying Mesmerism to the Cure of Diseases, and the Methods of Producing Mesmeric Phenomena. Intended for Domestic Use and the Instruction of Beginners.* London: H. Baillière.

————. 1853. *What Is Mesmerism? The Question Answered by a Mesmeric Practitioner, or, Mesmerism Not Miracle: An Attempt to Show That Mesmeric Phenomena and Mesmeric Cures Are Not Supernatural; to Which Is Appended Useful Remarks and Hints for Sufferers Who Are Trying Mesmerism for a Cure.* London: H. Baillière.

Bautain, Louis Eugène Marie. 1853. *Avis aux chrétiens sur les tables tournantes et parlantes.* Paris: Devarenne.

Beahrs, John O. 1982. *Unity and Multiplicity: Multilevel Consciousness of Self in Hypnosis, Psychiatric Disorder, and Mental Health.* New York: Brunner/Mazel.

Beaunis, Henri. 1886a. *Le somnambulisme provoqué: Etudes physiologiques et psychologiques.* Paris: J. B. Baillière.

————. 1886b. *Recherches expérimentales sur les conditions de l'activité cérébrale et sur la physiologie des nerfs.* Paris: J. B. Baillière.

Beecher, Charles. *A Review of the "Spiritual Manifestations."* New York: G. P. Putnam, 1853.

Belden, Lemuel W. 1834. *An account of Jane C. Rider, the Springfield Somnambulist: The Substance of Which Was Delivered as a Lecture Before the Springfield Lyceum, Jan. 22, 1834.* Springfield, Mass.: G. and C. Merriam.

Bellanger. 1854. *Le magnétisme, vérités et chimères de cette science occulte. Un drame dans le somnambulisme, épisode historique, les tables tournantes, etc., etc.* Paris: Guilhermet.

Bénézet, E. 1854. *Des tables tournantes et du panthéisme.* Paris: N.p.

Bennett, John Hughes. 1851. *The Mesmeric Mania of 1851, with a Physiological Explanation of the Phenomena Produced. A Lecture.* Edinburgh: Sutherland and Knox.

Benz, Ernest. 1976. *Franz Anton Mesmer und seine Ausstrahlung in Europa und Amerika.* Munich: Wilhelm Fink.

Bergasse, Nicolas. 1781. *Lettre d'un médecin de la Faculté de Paris à un médecin du College de Londres; ouvrage dans lequel on prouve contre M. Mesmer que le magnétisme animal n'existe pas.* The Hague: N.p.

————. 1784a. *Considérations sur le magnétisme animal, ou Sur la théorie du monde et des êtres organisés, d'aprés les principes de M. Mesmer.* The Hague: N.p.

————. 1784b. *Dialogue entre un docteur de toutes les universités et académies du*

monde connu, notamment de la faculté de médecine fondée à Paris dans la rue de la
Bucherie, l'an de notre salut 1472 et un homme de bon sens, ancien malade du
docteur. Paris: Gastellier.

[————.] 1784c. Théorie du monde et des êtres organisés suivant les principes de M
. . . Paris: N.p.

————. 1785a. Confession d'un médecin académicien et commissaire d'un rapport
sur le magnétisme animal, avec les remontrances et avis de son directeur. N.p.: N.p.

————. 1785b. Observations de M. Bergasse sur un écrit du docteur Mesmer, avant
pour titre: Lettre de L'inventeur du magnétisme animal à L'auteur des Réflexions
préliminaires. London: N.p.

Bérillon, Edgar. 1884. Hypnotisme expérimental. La dualité cérébrale et l'indépen-
dance fonctionnelle des deux hémisphères cérébraux. Paris: A. Delahay and E.
Lecrosnier.

Berna, Didier Jules. 1835. Expériences et considérations à l'appui du magnétisme
animal, thèse présentée et soutenue à la faculté de Paris. Paris: N.p.

————. 1838. Magnétisme animal. Examen et réfutation du rapport fait par M. E. F.
Dubois (d'Amiens) à l'Académie royale de médecine, le 8 août, 1837. Paris: Just.
Rouvier and E. Le Bouvier.

Bernheim, Hippolyte. 1884. De la suggestion dans l'état hypnotique et dans l'état de
veille. Paris: Octave Doin.

————. 1886. De la suggestion et de ses applications à la thérapeutique. Paris:
Octave Doin.

————. 1891. Hypnotisme, suggestion, psychothérapie; études nouvelles. Paris: Octave Doin.

————. 1897. L'hypnotisme et la suggestion dans leur rapports avec la médecine
légale. Nancy: A. Crépin-Leblond.

————. 1911. De la suggestion. Paris: Aubin Michel.

————. 1920. Automatisme et suggestion. Paris: Félix Alcan.

————. 1980. Bernheim's New Studies in Hypnotism. Translated by Richard S.
Sandor. New York: International Universities Press.

Berry, Thomas E. 1985. Spiritualism in Tsarist Society and Literature. Baltimore:
Edgar Allan Poe Society.

Bersot, Ernest. Mesmer, le magnétisme animal, les tables tournantes et les esprits. 5th
ed. Paris: Librairie Hachette, 1884.

Bertrand, Alexandre. 1823. Traité du somnambulisme, et des différentes modifications
qu'il presente. Paris: J. G. Dentu.

————. 1826. Du magnétisme animal en France, et des jugements qu'en ont portés les
sociétés savantes, avec le texte des divers rapports faits en 1784 par le commissaires
de l'Académie des Sciences, de la Faculté et de la Société Royale de Médecine, et
une analyse des dernières séances de l'Académie Royale de Médecine et du rapport
de M. Husson; suivi de considérations sur l'apparition de l'extase, dans les trait-
ments magnétiques. Paris: J. B. Baillière.

Billot, G. P. 1839. Recherches psychologique sur la cause de phénomènes extraordi-

*naires observés ches les modernes voyans improprement dits somnambules magnéti-
ques, ou Correspondence sur le magnétisme vital entre un solitaire et M.
Deleuze, bibliothécaire du Muséum à Paris*, 2 vols. Paris: Albanel et Martin.

Binet, Alfred. 1890. *On Double Consciousness: Experimental Psychological Studies.*
Chicago: Open Court.

——. 1892. *Les altérations de la personnalité.* Paris: J. B. Baillière.

——. 1896. *Alterations of Personality.* New York: D. Appleton.

——. 1897. Plural states of being. In *Appletons' Popular Science Monthly* 50:539–43.

Binet, Alfred, and Charles Féré. 1887. *Le magnétisme animal.* Paris: Ancienne Librairie Germer Baillière et Cie.

——. 1890. *Animal Magnetism.* New York: D. Appleton.

Birt, William Radcliff. 1853. *Table-Moving Popularly Explained; with an Inquiry into Reichenbach's Theory of Od Force; Also an Investigation into the Spiritual Manifestations Known as Spirit-Rapping.* London: N.p.

Blakeman, Rufus. 1849. *A Philosophical Essay on Credulity and Superstition: And Also on Animal Fascination, or Charming.* New York and New Haven: D. Appleton and S. Babcock.

Bliss, Eugene L. 1986. *Multiple Personality, Allied Disorders, and Hypnosis.* New York: Oxford University Press.

Bliss, Jonathan, and Eugene Bliss. 1985. *Prism: Andrea's World.* New York: Stein and Day.

Bloch, George, ed. and trans. 1980. *Mesmerism: A Translation of the Original Scientific and Medical Writings of F. A. Mesmer.* Los Altos, Calif.: William Kaufmann.

Bonnefoy, Jean Baptiste. 1784. *Analyse raisonée des rapports des commissaires chargés par le roi de l'examen du magnétisme animal.* Lyon and Paris: Prault.

Boring, Edwin G. 1957. *A History of Experimental Psychology.* 2d ed. New York: Appleton-Century-Crofts.

Bourru, H., and P. Burot. 1885. Observations et documents de la multiplicité des états de conscience chez un hystéro-épileptique. *Revue philosophique* 20:411–16.

——. 1886. Sur les variations de la personnalité. *Revue philosophique* 21:73–74.

——. 1888. *Variations de la personnalité.* Paris: J. B. Baillière.

Braid, James. 1842a. Animal magnetism. *Medical Times,* March 12, 1842, p. 283.

——. 1842b. Animal magnetism. *Medical Times,* March 26, 1842, p. 308.

——. 1842c. *Satanic Agency and Mesmerism Reviewed, in a Letter to the Rev. H. Mc.Neile, A. M. of Liverpool, in Reply to a Sermon Preached by Him in St. Jude's Church, Liverpool, on Sunday, April 10th, 1842.* Manchester: Sims and Dinham, Galt and Anderson.

——. 1842d. Neuro-hypnotism. *Medical Times,* July 9, 1842, p. 239.

——. 1843. *Neurypnology or the Rationale of Nervous Sleep Considered in Relation with Animal Magnetism. Illustrated by Numerous Cases of Its Successful Application in the Relief and Cure of Disease.* London: John Churchill.

————. 1844. Observations on mesmeric and hypnotic phenomena. *Medical Times* 10:31–32, 47–49.

————. 1844–1845. Magic, mesmerism, hypnotism, etc., etc., historically and physiologically considered. *Medical Times* 11:203–4, 224–27, 270–72, 296–99, 399–400, 439–41.

————. 1851. *Electro-Biological Phenomena Considered Physiologically and Psychologically*. Edinburgh: Sutherland and Knox.

————. 1852. *Magic, Witchcraft, Animal Magnetism, Hypnotism, and Electrobiology; Being a Digest of the Latest Views of the Author on These Subjects*. London: John Churchill.

————. 1853a. Hypnotic therapeutics, illustrated by cases. *Monthly J. Medical Science* 17:14–47.

————. 1853b. *Hypnotic Therapeutics, Illustrated by Cases. With an Appendix on Table-Moving and Spirit-Rapping*. Reprinted from *Monthly Journal of Medical Science* 17:14–47. [London:] Monthly Journal of Medical Science.

————. 1855. *The Physiology of Fascination, and the Critics Criticised*. Manchester: Grant.

————. 1899. *Braid on Hypnotism. Neurypnology; or The Rationale of Nervous Sleep Considered in Relation to Animal Magnetism or Mesmerism and Illustrated by Numerous Cases of Its Successful Application in the Relief and Cure of Disease. A New Edition, Edited with an Introduction Biographical and Bibliographical Embodying the Author's Later Views and Further Evidence on the Subject. By Arthur Waite*. London: G. Redway.

Bramwell, J. Milne. 1903. *Hypnotism: Its History, Practice and Theory*. London: Grant Richards.

Brandis, J. D. 1818. *Über psychische Heilmittel und Magnetismus*. Copenhagen: Gyldendal.

Brandon, Ruth. 1983. *The Spiritualists: The Passion for the Occult in the Nineteenth and Twentieth Centuries*. London: Weidenfeld and Nicolson.

Branscomb, Louisa. 1991. Dissociation in combat-related post-traumatic stress disorder. *Dissociation* 4:13–20.

Braude, Anne. 1989. *Radical Spirits: Spiritualism and Women's Rights in Nineteenth-Century America*. Boston: Beacon Press.

Braude, Stephen. 1991. *First Person Plural: Multiple Personality and the Philosophy of Mind*. London: Routledge.

Braun, Bennett G. 1980. Hypnosis for multiple personalities, in *Clinical Hypnosis in Medicine*, edited by H. Wain. Chicago: Year Book Medical.

————. 1983a. Psychophysiologic phenomena in multiple personality and hypnosis. *Amer. J. Clinical Hypnosis* 26:124–37.

————. 1983b. Neurophysiological changes in multiple personality due to integration: A preliminary report. *Amer. J. Clinical Hypnosis* 26:82–94.

————. 1989a. The BASK (behavior, affect, sensation, knowledge) model of dissociation. *Dissociation* 1, no. 1:4–23.

————. 1989b. The BASK model of dissociation: Part II—treatment. *Dissociation* 1, no. 2: 16–23.

Braun, Bennett G., ed. 1986. *Treatment of Multiple Personality Disorder*. Washington, D.C.: American Psychiatric Press.

Breuer, Josef, and Sigmund Freud. 1964. *Studies in Hysteria (1895)*. Vol. 2 of *The Standard Edition of the Complete Psychological Works of Sigmund Freud*, edited by James Strachey. London: Hogarth Press and the Institute of Psycho-analysis.

Brierre de Boismont, Alexandre Jacques. 1832. *Des hallucinations, ou Histoire raisonée des apparitions, des visions, des songes, de l'extase, des rêves, du magnétisme et du somnambulisme*. Paris: G. Baillière.

————. 1853. *Hallucinations; or, The Rational History of Apparitions, Visions, Dreams, Ecstasy, Magnetism, and Somnambulism*. Philadelphia: Lindsay and Blakiston.

Brigham, Amariah. 1835. *Observations on the Influence of Religion upon the Health and Physical Welfare of Mankind*. Boston: Marsh, Capen & Lyon.

[Britten, Emma Hardinge.] 1870. *Modern American Spiritualism: A Twenty Year's Record of the Communion Between Earth and the World of Spirits*. 2 vols. New York: The Author.

————. 1884. *Nineteenth Century Miracles; or, Spirits and Their Work in Every Country of the Earth. A Complete Historical Compendium of the Great Movement Known as "Modern Spiritualism."* New York: William Britten.

Broca, Paul. 1859. *Sur l'anesthésie chirurgicale hypnotique. Note présentée à l'Académie des Sciences, le 5 décembre 1859. Suivie d'une lettre adressée au rédacteur en chef du Moniteur des Sciences Médicale*. Paris: A. Henry Noblet.

Brouardel, P. 1879. Accusation de viol accompli pendant le sommeil hypnotique. *Annales d'hygiène publique et de médecine légale* 1:39–57.

Brown, John. 1843. *Mesmerism; Its Pretensions & Effects upon Society Considered*. Boston: N.p.

Brown, Slater. 1972. *The Heyday of Spiritualism*. New York: Pocket Books.

Brunn, Walter L. von. 1954. Die Anfänge der hypnotischen Anästhesie. *Deutsche Medizinische Wochenschrift* 8:336–40.

Buchanan, Joseph Rodes. 1854. *Outlines of Lectures on the Neurological System of Anthropology, as Discovered, Demonstrated and Taught in 1841 and 1842*. Cincinnati: Office of Buchanan's Journal of Man.

Buranelli, Vincent. 1975. *The Wizard from Vienna*. New York: Coward, McCann & Geoghegan.

Burdin, C., and F. Dubois. 1841. *Histoire académique du magnétisme animal, accompagnée de notes et de remarques critiques sur tout les observations et expériences faites jusqu'à ce jour*. Paris: J. B. Baillière.

Burq, Victor. 1853. *Métallotherapie; traitement des maladies nerveuses, paralysies, rhumatisme chronique, etc. . . . .* Paris: Germer Baillière.

Bush, George. 1847. *Mesmer and Swedenborg; or, The Relation of the Developments of Mesmerism to the Doctrines and Disclosures of Swedenborg*. New York: John Allen.

Cahagnet, Louis Alphonse. 1848. *Arcanes de la vie future dévoilés, ou, l'existence, la forme, les occupations de l'âme après sa séparation du corps sont prouvées . . .* Vol. 1. Paris: The Author.

———. 1850a. *Arcanes de la vie future dévoilés . . .* Vol. 2. Paris: Germer Baillière.

———. 1850b. *Sanctuaire du spiritualisme. Étude de l'âme humaine, et de ses rapports avec l'univers, d'après le somnambulisme et l'extase*. Paris: Germer Baillière.

———. 1850c. *The Celestial Telegraph; or, Secrets of the Life to Come Revealed Through Magnetism . . .* London: George Peirce.

———. 1854a. *Arcanes de la vie future dévoilés . . .* Vol. 3. Paris: Germer Baillière.

———. 1854b. *Magie magnétique, ou Traité historique et pratique de fascinations, miroirs cabalistiques, apports, suspensions, pactes, talismans, possessions, envoûtements, sortilèges, etc.* Paris: Germer Baillière.

Caldwell, Charles. 1842. *Facts in Mesmerism and Thoughts on Its Causes and Uses*. Louisville, Ky.: Prentice and Weissinger.

Campbell, C. MacFie, H. S. Langfeld, William McDougall, A. A. Roback, and E. W. Taylor. 1932. *Problems of Personality: Studies Presented to Dr. Morton Prince, Pioneer in American Psychopathology*. London: Kegan Paul, Trench, Trubner.

Capron, E. W. 1855. *Modern Spiritualism, Its Facts and Fanaticisms, Its Consistencies and Contradictions*. Boston: Bela Marsh.

Carlson, Eric T. 1960. Charles Poyen brings mesmerism to America. *J. History of Medicine and Allied Sciences* 15:121–32.

———. 1981. The history of multiple personality in the United States: I. The Beginnings. *Amer. J. Psychiatry* 138:666–68.

———. 1982. The history of multiple personality in the United States: II. Mary Reynolds and her subsequent reputation. Paper presented at the 55th annual meeting of the American Association for the History of Medicine, Bethesda, Maryland, April 30, 1982. Privately published.

———. 1984. The history of multiple personality in the United States: Mary Reynolds and her subsequent reputation. *Bull. Hist. Medicine* 58:72–82.

Carlson, Eric T., and Meribeth M. and Simpson. 1970. Perkinism vs. mesmerism. *J. Hist. Behavioral Science* 6:16–24.

Carlson, Eric T., Jeffrey L. Wollock, and Patricia S. Noel, eds. 1981. *Benjamin Rush's Lectures on the Mind*. Philadelphia: American Philosophical Society.

Carpenter, William. 1853. Electrobiology and Mesmerism. *Quarterly Review* 93: 501–57.

———. 1876. *Principles of Mental Physiology, with Their Applications to the Training and Discipline of the Mind, and the Study of Its Morbid Conditions*. 4th ed. London: Henry S. King.

Carus, Carl Gustav. 1857. *Über Lebensmagnetismus und über die magischen Wirkungen überhaupt*. Leipzig: F. A. Brockhaus.

Caullet de Veaumorel, ed. 1785. *Aphorismes de M. Mesmer, dictés à l'assemblée de ses élèves, & dans lesquels on trouve ses principles, sa théorie & les moyens de magnétiser. . . .* Paris: M. Quinquet.

Cerullo, John J. 1982. *The Secularization of the Soul.* Philadelphia: Institute for the Study of Human Issues.

Charcot, Jean Martin. 1881. *Contribution à l'étude de l'hypnotisme chez les hystériques.* Paris: Progès médical.

————. 1882. Sur les divers états nerveux déterminés par l'hypnotisation chez les hystériques. *Comptes-rendus hebdomadaires des séances de l'Académie des Sciences* 94:403–5.

Charcot, Jean Martin, J. B. Luys, and A. Dumontpallier. 1877. *Rapport fait à la Société de Biologie sur la métalloscopie du Docteur Burq au nom d'une commission composée de MM. Charcot, Luys, et Dumontpallier, Rapporteur.* Paris: Union Médicale.

Chardel, Casimir Marie. 1818. *Mémoire sur le magnétisme animal, présenté à l'Académie de Berlin, en 1818.* Paris: Bandoin frères.

————. 1826. *Esquisse de la nature humaine expliquée par le magnétisme animal précédée d'un aperçu du système général de l'univers, et contenant l'explication du somnambulisme magnétique et de tous les phénomènes du magnétisme animal.* Paris: Dentu et Delaunay.

————. 1844. [1st ed. 1831.] *Essai de psychologie physiologique.* Paris: Germer Baillière.

Charpignon, Jules. 1841. *Physiologie, médecine et métaphysique du magnétisme.* Orleans: Pesty. Paris: Germer Baillière.

Chase, Truddi. 1987. *When Rabbit Howls.* New York: Dutton.

Der Cheiroelektromagnetismus oder die Selbstbewegung und das Tanzen der Tische (Tischrücken). 1853. Berlin: Lassar.

Chertok, Leon, and Raymond De Saussure. 1979. *The Therapeutic Revolution from Mesmer to Freud.* New York: Brunner/Mazel.

Chevallier, Pierre. 1974. *Histoire de la franc-maçonnerie française. La Maçonnerie: Ecole de l'Egalité, 1725–1799.* Vol. 1. [Paris:] Fayard.

Chevenix, Richard. 1829. On mesmerism, improperly denominated animal magnetism. *London Medical and Physical Journal,* March, June, August, October.

Chevreul, Michel Eugène. 1854. *De la baguette divinatoire: Du pendule dit explorateur et des tables tournantes au point de vue de l'histoire, de la critique et de la méthode expérimentale.* Paris: Mallet-Bachelier.

[Cloquet.] 1784. *Détail des cures opérées à Buzancy, près Soissons par le magnétisme animal.* Soissons: N.p.

Close, Rev. F. 1853. *The Testers Tested: Or Table-Moving . . . Not Diabolical.* London.

Coggeshall, William Turner. 1851. *Signs of the Times, Comprising a History of the Spirit Rappings in Cincinnati and Other Places, with Notes of Clairvoyant Revealments.* Cincinnati: The Author.

Cohnfeld, Adalbert Salomo. 1853. *Die Wunder-Erscheinungen des Vitalismus (Tisch-drehen, Tischklopfen, Tischsprechen u.) nebst ihrer rationellen Erklärung in Briefen an eine Dame.* Bremen: E. Schünemann.

Collyer, Robert Hanham. 1843. *Psychology, or The Embodiment of Thought; with an Analysis of Phreno-magnetism, "Neurology," and Mental Hallucination, Including Rules to Govern and Produce the Magnetic State.* New York and Boston: Sun Office and Redding.

Colquhoun, John Campbell. 1833. *Report of the Experiments on Animal Magnetism, Made by a Committee of the Medical Section of the French Royal Academy of Sciences: Read at the Meetings of the 21st and 28th of June, 1831. Translated and Now for the First Time Published; with an Historical and Explanatory Introduction, and an Appendix.* Edinburgh: Robert Cadell.

―――. 1836. *Isis Revelata; an Inquiry into the Origin, Progress & Present State of Animal Magnetism.* 2 vols. Edinburgh: Maclachlan & Stewart.

―――. 1851. *An History of Magic, Witchcraft, and Animal Magnetism.* 2 vols. London: Longman, Brown, Green & Longmans.

Confessions of a Magnetizer, Being an Exposé of Animal Magnetism. 1845. Boston: Gleason's Pub. Hall.

Confessions of a Truth Seeker. A Narrative of Personal Investigations into the Facts and Philosophy of Spirit-Intercourse. 1859. London: William Horsell.

Coons, Philip M. 1980. Multiple personality: Diagnostic considerations. *J. Clinical Psychiatry* 41:330–36.

―――. 1988. Psychophysiologic aspects of multiple personality disorder: A review. *Dissociation* 1:47–53.

Cooter, Roger. 1985. The history of mesmerism in Britain: Poverty and promise. In *Franz Anton Mesmer und die Geschichte des Mesmerismus,* edited by Heinz Schott. Stuttgart: Franz Steiner.

Court de Gébelin. 1783. *Lettre de l'auteur de Monde primitif à messieurs ses souscripteurs sur le magnétisme animal.* Paris: Valleyre l'aîné.

Cowan, Charles. 1861. *Thoughts on Satanic Influence, or Modern Spiritualism Considered.* Reading: Lovejoy.

Cox, Edward William. 1871. *Spiritualism Answered by Science.* London: Longman.

―――. 1873–1874. *What Am I? Popular Introduction to the Study of Psychology.* 2 vols. London: Longmans.

Crabtree, Adam. 1985a. *Multiple Man: Explorations in Possession and Multiple Personality.* Toronto: Collins.

―――. 1985b. Mesmerism, divided consciousness, and multiple personality. In *Franz Anton Mesmer und die Geschichte des Mesmerismus,* edited by Heinz Schott. Stuttgart: Franz Steiner.

―――. 1986. Explanations of dissociation in the first half of the twentieth century. In *Split Minds and Split Brains,* edited by Jacques Quen. New York: New York University Press.

————. 1988. *Animal Magnetism, Early Hypnotism, and Psychical Research, 1766 to 1925: An Annotated Bibliography.* White Plains, N.Y.: Kraus International Publications.

————. 1992. Dissociation and memory: A two-hundred year perspective. *Dissociation* 5:150–54.

Crookes, William. 1871. *Psychic Force and Modern Spiritualism: A Reply to the "Quarterly Review" and Other Critics.* London: Longmans, Green.

————. 1874. *Researches in the Phenomena of Spiritualism.* Reprinted from the *Quarterly Journal of Science.* London: J. Burns.

Dalgado, D. G. 1906a. La Doctrine du Dr Braid sur l'hypnotisme comparée avec celle de l'abbé de Faria sur le sommeil lucide. *Revue de l'hypnotisme* 21.

————. 1906b. *Mémoire sur la vie de l'abbé de Faria.* Paris: Henri Jove.

[Dampierre, Antoine Esmonin, marquis de.] 1784. *Réflexions impartiales sur le magnétisme animal, faites après la publication du rapport des commissaires chargés par le roi de l'examen de cette découverte.* Geneva and Paris: Barthélemy Chirol and Périsse le jeune.

La danse des tables devoilée, expériences de magnétisme animal, manière de fair tourner une bague, un chapeau, une montre, une table, et même jusqu'eux têtes des expérientateurs et celles des spectateurs. 1853. Paris: N.p.

Darnton, Robert. 1968. *Mesmerism and the End of the Enlightenment in France.* Cambridge, Mass.: Harvard University Press.

Davies, John D. 1955. *Phrenology: Fad and Science: A 19th-Century American Crusade.* New Haven: Yale University Press.

Deleuze, Joseph Philippe. 1813. *Histoire critique du magnétisme animal.* 2 vols. Paris: Mame.

————. 1817. *Réponse aux objections contre le magnétisme.* Paris: Dentu.

————. 1819. *Défense du magnétisme animal contre les attaques dont il est l'objet dans le Dictionnaire des Sciences Médicales.* Paris: Belin Le Prieur.

————. 1821. *Observations adressés aux médecins qui désireraient établir un traitement magnétique.* Paris: Belin Le Prieur.

————. 1825. *Instruction pratique sur le magnétisme animal, suivie d'un lettre écrite à l'auteur par un médecin étranger.* Paris: Dentu.

————. 1826. *Lettre à messieurs les membres de l'Académie de médecine.* Paris: Béchet jeune.

————. 1828. *De l'état actuel du magnétisme.* [Paris:] N.p.

————. 1834. *Mémoire sur la faculté de prévision: suivi des notes et pièces justificatives recueillis par M. Mialle.* Paris: Crochard.

————. 1850. *Practical Instruction in Animal Magnetism.* Revised American edition of *Instruction pratique sur le magnétisme animal.* New York: D. Appleton.

De Mainauduc, John Benoit. 1798. *The Lectures of J. B. de Mainauduc, M.D. Part the First.* London: Printed for the Executrix.

Demarquay, Jean Nicolas, and M. A. Giraud-Teulon. 1860. *Recherches sur l'hypnotisme ou sommeil nerveux, comprenant une séries d'expériences institutées à la maison municipale de santé.* Paris: J. B. Baillière.

Dendy, Walter Cooper. 1845. *The Philosophy of Mystery.* New York: Harper & Brothers.

Denton, William, and Elizabeth Denton. 1863–1874. *Soul of Things; or, Psychometric Researches and Discoveries,* 3 vols. Boston: Walker, Wise.

D'Eslon, Charles. 1780. *Observations sur le magnétisme animal.* London and Paris: Didot.

———. 1782. *Lettre de M. d'Eslon, docteur régent de la Faculté de Paris, et médecine ordinaire de Monseigneur le comte d'Artois, à M. Philip, docteur en médecine, doyen de la Faculté.* The Hague: N.p.

———. 1784. *Observations sur les deux rapports de MM les commissaires nommés par sa majesté pour l'examen du magnétisme animal.* Philadelphia and Paris: Clousier.

Despine, Charles Humbert Antoine. 1838. *Observations de médecine pratique. Faites aux Bains d'Aix-en-Savoie.* Anneci: Burdet.

Despine, Prosper. 1868. *Psychologie naturelle. Etude sur les facultés intellectuelles et morales dans leur état normal et dans leurs manifestations anomales chez les aliénés et chez les criminels.* 3 vols. Paris: F. Savy.

———. 1880. *Etude scientifique sur le somnambulisme, sur les phénomènes qu'il présente et sur son action thérapeutique dans certaines maladies nerveuses, du role important qu'il joue dans l'épilepsie, dans l'hystérie et dans les névroses dites extraordinaires.* Paris: F. Savy.

Dessoir, Max. 1888. *Bibliographie des modernen Hypnotismus.* Berlin: Carl Duncker.

———. 1889. *Das Doppel-Ich.* Berlin: Gesellschaft für Experimental-Psychologie zu Berlin.

———. 1896. [1st ed. 1889.] *Das Doppel-Ich,* 2d enlarged ed. Leipzig: Ernst Günther.

Devotional Somnium; or, A Collection of Prayers and Exhortations, Uttered by Miss Rachel Baker, in the City of New York, in the Winter of 1815, during her Astracted and Unconscious State. . . . 1815. New York: S. Marks.

Dewar, H. 1823. Report on a communication from Dr. Dyce of Aberdeen, to the Royal Society of Edinburgh, "On Uterine Irritation, and Its Effects on the Female Constitution." *Transactions of the Royal Society of Edinburgh* 9:365–79.

Dibdin, Rev. R. W. 1871. *Table Turning. A Lecture by the Rev. R. W. Dibdin, M. A., Delivered in the Music Hall, Store Street, on Tuesday Evening, November the 8th, 1853.* London: N.p.

Dickerson, K. D. D. 1843. *The Philosophy of Mesmerism, or Animal Magnetism. Being a Compilation of Facts Ascertained by Experience, and Drawn from the Writings of the Most Celebrated Magnetizers in Europe and America.* . . . Concord, N.H.: Merrill, Silsby.

Dingwall, Eric J. 1967. *Abnormal Hypnotic Phenomena: A Survey of Nineteenth-Century Cases*. 4 vols. New York: Barnes & Noble.

Dods, John Bovee. 1843. *Six Lectures on the Philosophy of Mesmerism, Delivered in the Marlboro Chapel, Boston. Reported by a Hearer*. Boston: W. A. Hall.

————. 1850. *The Philosophy of Electrical Psychology: In a Course of Twelve Lectures*. New York: Fowler & Wells.

————. 1854. *Spirit Manifestations Examined and Explained. Judge Edmonds Refuted; or, An Exposition of the Involuntary Powers and Instincts of the Human Mind*. New York: De Witt & Davenport.

Doppet, François Amédée. 1784. *Traité théorique et pratique du magnétisme animal*. Turin: Jean Michel Briolo.

Douglas, James S. 1842. *Animal Magnetism, or Mesmerism; Being a Brief Account of the Manner of Practising Animal Magnetism; the Phenomena of that State; Its Applications in Disease, and the Precautions to be Observed in Employing It, Made So Plain That Anyone May Practice It, Experiment upon It and Test Its Effects for Himself*. New York: J. & D. Atwood.

Doyle, Arthur Conan. 1926. *The History of Spiritualism*, 2 vols. London: George H. Doran.

Drake, Daniel. 1844. *Analytical Report of a Series of Experiments in Mesmeric Somniloquism, Performed by an Association of Gentlemen: With Speculations on the Production of Its Phenomena*. Louisville, Ky.: F. W. Prescott.

Du Commun, Joseph. 1829. *Three Lectures on Animal Magnetism, as Delivered in New York, at the Hall of Science, on the 26th of July, 2d and 9th of August*. New York: Berard & Mondon.

Dufay. 1888. Deux cas de somnambulisme provoqué à distance. *Revue scientifique* 25: 240–43.

Dufay, and Eugène Azam. 1879. La double personnalité. *Revue scientifique* 23:843–46.

Dumontpallier, A. 1879. Métalloscopie, métallothérapie. *Journal de thérapie* 6:613–16.

Dumontpallier, A., and P. Magnin. 1882a. Expériences sur la métalloscopie, l'hypnotisme et la force neurique. *Comptes rendus et mémoires des sciences de la Société de Biologie* 3:359–64.

————. 1882b. Note sur les conditions que mettent en évidence le phénomène désigné sous le nom d'hyper-excitabilité neuromusculaire dans les différentes périodes de l'hypnotisme. *Comptes rendus et mémoires des sciences de la Société de Biologie* 4:147–50.

Du Potet de Sennevoy, Jules Denis. 1826. [1st ed. 1821.] *Expériences publique sur le magnétisme animal faites à l'Hôtel Dieu de Paris*. Paris: Bechet, Dentu, and the Author.

————. 1834. *Cours de magnétisme animal*. Paris: The Author.

————. 1838. *An Introduction to the Study of Animal Magnetism*. London: Sunders and Otley.

————. 1856. *Traité complet du magnétisme animal*, 3d edition. Paris: Germer Baillière.

————. 1907. [1st ed. 1852.] *La magie dévoilée, ou, Principes de Science occulte*. 4th ed. Paris: Vigot frères.

Durand, Joseph Pierre. [A. J. P. Philips.] 1855. *Electro-dynamisme vital ou les relations physiologiques de l'esprit et de la matière, démontrées par des expériences entièrement nouvelles et par l'histoire raisonnée du système nerveux*. Paris: J. B. Baillière.

————. 1860. *Cours théorique et pratique de braidisme ou hypnotisme nerveux: considéré dans ses rapports avec la psychologie, la physiologie et la pathologie et dans ses applications à la médecine, à la chirurgie, à la physiologie expérimentale, à la médecine légale et à l'éducation, par le docteur J. P. Philips*. Paris: J. B. Baillière.

————. 1868. Polyzoisme, ou pluralité animale dans l'homme. Appendix to *La philosophie physiologique et médicale à l'Académie de médecine*. Paris: Germer Baillière.

Durant, Charles Ferson. 1837. *Exposition, or a New Theory of Animal Magnetism with a Key to the Mysteries: Demonstrated by Experiments with the Most Celebrated Somnambulists in America: Also, Strictures on "Col. Wm. L. Stone's Letter of Doctor A. Brigham."* New York: Wiley & Putnam.

Edmunston, William E. 1986. *The Induction of Hypnosis*. New York: John Wiley.

Ellenberger, Henri F. 1966. The pathogenic secret and its therapeutics. *J. Hist. Behavioral Sciences* 2:29–42.

————. 1970. *The Discovery of the Unconscious: The History and Evolution of Dynamic Psychiatry*. New York: Basic Books.

Elliotson, John. 1840. *Human Physiology*. 5th ed. London: Longman, Orme, Brown, Green and Longmans.

————. 1843. *Numerous Cases of Surgical Operations without Pain in the Mesmeric State; with Remarks upon the Opposition of Many Members of the Royal Medical and Chirurgical Society and Others to the Reception of the Inestimable Blessings of Mesmerism*. Philadelphia: Lea and Blanchard.

————. 1846a. Instances of double states of consciousness independent of mesmerism. *Zoist* 4:157–87.

————. 1846b. *The Harveian Oration, Delivered before the Royal College of Physicians, London, June 27th, 1846, by John Elliotson, M.D. Cantab. F.R.S., Fellow the College. An English Version and Notes*. London: H. Baillière.

Emmons, S. B. 1859. *The Spirit Land*. Philadelphia: G. G. Evans.

Ennemoser, Joseph. 1819. *Der Magnetimus nach der allseitiger Beziehung seines Wesens, gesichtlichen Entwickelung von allen Zeiten und bei allen Volkern wissenschaftlich dargestellt*. Leipzig: F. A. Brockhaus.

————. 1842. *Der Magnetismus im Verhältnisse zur Natur und Religion*. Stuttgart and Tübingen: J. G. Cotta.

————. 1852. *Anleitung zur Mesmerischen Praxis*. Stuttgart and Tübingen: J. G. Cotta.

————. 1854. *The History of Magic*. 2 vols. Translated by William Howitt. London: H. G. Bohn. Originally published as *Geschichte der Magie* (1844).

[Eprémesnil, Jean Jacques Duval d'.] 1785. *Sommes versées entre les mains de monsieur Mesmer pour acquérir le droit de publier sa découverte*. Paris: N.p.

Erman, Wilhelm. 1925. *Der tierische Magnetismus in Preussen vor und nach den Freiheitskriegen*. Munich and Berlin: R. Oldenbourg.

Eschenmayer, Carl Adolph von. 1816. *Versuch die scheinbare Magie des thierischen Magnetismus aus physiologischen und psychischen Gesetzen zu erklären*. Stuttgart and Tübingen: J. G. Cotta.

Esdaile, James. 1846. *Mesmerism in India, and Its Practical Application in Surgery and Medicine*. London: Longman, Brown, Green and Longmans.

————. 1852a. *Natural and Mesmeric Clairvoyance, with the Practical Application of Mesmerism in Surgery and Medicine*. London: H. Baillière.

————. 1852b. *The Introduction of Mesmerism, as an Anaesthetic and Curative Agent, into the Hospitals of India*. Perth: Dewar.

Examen raisonné des prodiges récents d'Europe de l'Amerique notamment des tables tournantes et répondantes, par un philosophe. 1853. Paris: J. Vermot.

Fahnestock, William Baker. 1869. *Artificial Somnambulism. Hitherto Called Mesmerism; or, Animal Magnetism. Containing a Brief Historical Survey of Mesmer's Operations, and the Examination of the Same by the French Commissioners. Phreno-Somnambulism; or, the Exposition of Phreno-Magnetism and Neurology*. . . . Philadelphia: Barclay.

Faraday, Michael. 1853a. Table Turning. Letter to the editor. *Times* (London), June 30, 1853.

————. 1853b. Professor Faraday on Table-Moving. *Athenaeum* (London), no. 1340 (July 2, 1853): 801–3.

Faria, José Custodio de. 1819. *De la Cause du sommeil lucide, ou étude de la nature de l'homme. Tome premier*. Paris: Mme. Horiac.

————. 1906. *De la cause du sommeil lucide, ou étude de la nature de l'homme. Préface et introduction par le Dr D. G. Dalgado*. Paris: Henri Jouve.

Fechner, Gustav Theodor. 1851. *Zend-Avesta oder über die Dinge des Himmels und des Jenseits, von Standpunkt der Naturbetrachtung*. 3 vols. Leipzig: L. Voss.

————. 1861. *Über die Seelenfrage. Ein Gang durch die sichtbare Welt, um die unsichtbare zu finden*. Leipzig: C. F. Amelang.

Feuerbach, Paul Johann Anselm Ritter von. 1846. John George Sörgel, the idiot murderer. In *Narratives of Remarkable Criminal Trials*, translated by Lady Duff Gordon. New York: Harper & Brothers.

Figuier, Louis. 1860. *Histoire du marveilleux dans les temps modernes*. 4 vols. 2d ed. Paris: L. Hachette.

Fine, C. G. 1988. The work of Antoine Despine: The first scientific report on the diagnosis and treatment of a child with multiple personality disorder. *Amer. J. Clinical Hypnosis* 31:33–39.

Fischer, Friedrich. 1839. *Der Somnambulismus.* 3 vols. Basel: Schweighauser.

Flammarion, Camille. 1862–1863. *Les habitants de l'autre monde. Etudes d'outre-tombe.* 2 parts. Paris: Ledoyen.

———. 1866. *Des forces naturelles inconnues, à propos des phénomènes produits par les frères Davenport et par les médiums en général. Etude critique par Hermès.* Paris: Didier, Fred. Henry, Dentu.

Flournoy, Théodore. 1903. Review of *Human Personality and Its Survival of Bodily Death,* by Frederic Myers. *Proceedings of the Society for Psychical Research* 18:42–52.

———. 1911. *Spiritism and Psychology.* Translated and abridged by Hereward Carrington. New York: Harper and Brothers.

Foissac, Pierre. 1825. *Mémoire sur le magnétisme animal adressé à MM. les membres de l'Académie des sciences et de l'Académie de médecine.* Paris: Didot le jeune.

———. 1826. *Second mémoire sur le magnétisme animal. Observations particulières sur une somnambule présentée à la commission nommée par l'Académie royale de médecine pour l'examen du magnétisme animal.* Paris: N.p.

———. 1833. *Rapports et discussions de l'Académie royale de médecin sur le magnétisme animal, recueillis par un sténographe, et publiés, avec des notes explicatives.* Paris: J. B. Baillière.

Forbes, John. 1845. *Mesmerism True—Mesmerism False: A Critical Examination of the Facts, Claims, and Pretentions of Animal Magnetism. With an Appendix Containing a Report of Two Exhibitions by Alexis.* London: Churchill.

Fournel, Jean François. 1785. *Essai sur les probabilités du somnambulisme magnétique: Pour servir à l'histoire du magnétisme animal.* Paris: Gastelier.

Foveau de Courmelles, François. [1891.] *Hypnotism.* Translated by Laura Ensor. Philadelphia: David McKay.

Frankau, Gilbert. 1948. *Mesmerism, by Doctor Mesmer.* London: MacDonald.

Franklin, Benjamin, et al. 1837. *Animal Magnetism. Report of Dr. Franklin and Other Commissioners Charged by the King of France with the Examination of the Animal Magnetism as Practiced at Paris.* Philadelphia: Perkins.

Frapart, N. N. 1839. *Lettres sur le magnétisme et le somnambulisme, à l'occasion de mademoiselle Pigeaire.* Paris: Dentu.

Freud, Sigmund. 1964a. *The Interpretation of Dreams* (1900). In *The Standard Edition of the Complete Psychological Works of Sigmund Freud,* edited by James Strachey. Vol. 5. London: Hogarth Press.

———. 1964b. *Fragment of an analysis of a case of hysteria* (1905). In *S.E.* Vol. 7.

———. 1964c. *The Future Prospects of Psycho-analytic Therapy* (1910). In *S.E.* Vol. 11.

———. 1964d. *The Unconscious* (1915). In *S.E.* Vol. 14.

———. 1964e. *Introductory Lectures on Psycho-analysis.* (1916–1917). In *S.E.* Vol. 15.

—————. 1964f. *The Ego and the Id* (1923). In *S.E.* Vol. 19.

—————. 1964g. *An Autobiographical Study* (1925). In *S.E.* Vol. 20.

Frisz. [1853.] *Les tables et les têtes qui tournantes, ou, la fièvre de rotation en 1853. Cent et un croquis.* Paris: Librairie pittoresque.

Fuller, Robert C. 1982. *Mesmerism and the American Cure of Souls.* Philadelphia: University of Pennsylvania Press.

Gallert de Montjoye, Christophe Félix Louis. 1784. *Lettre sur le magnétisme animal, où l'on examine la conformité des opinions des peuples anciens & modernes, des savants & notamment de M. Bailly avec celles de M. Mesmer; et où l'on compare ces memes opinions au rapport des commissaires chargés par le roi de l'examen du magnétisme animal adressé à Monsieur Bailly de l'Académie des Sciences etc.* Paris and Philadelphia: Pierre J. Duplain.

Gasparin, Agénor Etienne, comte de. 1854. *Des tables tournantes, du surnaturel en général et des esprits.* 2 vols. Paris: Dentu.

—————. 1857. *Science vs. Modern Spiritualism: A Treatise on Turning Tables, the Supernatural in General, and Spirits.* 2 vols. Translated by E. W. Robert. New York: Kiggins & Kellogg.

Gauld, Alan. 1968. *The Founders of Psychical Research.* New York: Schocken Books.

—————. 1992a. *A History of Hypnotism.* Cambridge: Cambridge University Press.

—————. 1992b. Hypnosis, somnambulism, and double consciousness. *Contemporary Hypnosis* 9:69–76.

Gauthier, Aubin. 1842. *Histoire du somnambulisme chez tous les peuples, sous les noms divers d'extases, songes, oracles et visions; examen des doctrines théoriques et philosophiques. . . .* 2 vols. Paris: Félix Malteste, Dentu, and G. Baillière.

Gentil, Joseph Adolphe. 1854a. *Manuel élémentaire de l'aspirant magnétiseur.* Paris: E. Dentu.

—————. 1854b. *L'ame de la terre et les tables parlantes, ou sauvons le genre humain; ouvrage examiné au point de vue magnétique de l'influence des besoines sur le moral.* Paris: The Author.

Gies, William J., ed. 1948. *Horace Wells, Dentist. Father of Surgical Anesthesia. Proceedings of Centenary Commemorations of Wells' Discovery in 1844 and Lists of Wells Memorabilia Including Bibliographies, Memorials and Testimonials.* Hartford, Conn.: American Dental Association.

Gilles de la Tourette, George Albert. 1887. *L'hypnotisme et les états analogues au point de vue médico-légale.* Paris: E. Plon, Nourrit et cie.

Gillson, Edward. [1853.] *Table Talking: Disclosures of Satanic Wonders and Prophetic Signs.* London.

—————. [185-?]. *Where Is the Responsibility? A Letter to the Rev. F. Close, M.A., in Reply to His Pamphlets "The Testers Tested."* London: Binns & Goodwin.

Girardin. 1784. *Observations adressés à Mrs. les commissaires chargés par le Roi de l'examen du magnétisme animal; sur la manière dont ils y ont procédé, & sur leur rapport. Par un médecin de province.* London and Paris: Royez.

Gley, E. 1886. A propos d'une observation de sommeil provoqué à distance. *Revue philosophique* 21:425–28.

Gmelin, Eberhard. 1791. *Materialien für die Anthropologie*, Vol. 1. Tübingen: Cotta.

Godfrey, Nathaniel Steadman. 1853a. *Table-Moving Tested and Proved to Be the Result of Satanic Agency*. London: Seeleys.

———. 1853b. *Table-Turning, the Devil's Modern Master-piece. Being the Result of a Course of Experiment*. London and Leeds: Seeleys.

Goettman, Carole, George Greaves, and Philip Coons. 1992. *Multiple Personality and Dissociation: 1791–1990: A Complete Bibliography*. Atlanta: Privately published.

Goldsmith, Margaret. 1934. *Franz Anton Mesmer: The History of an Idea*. London: Arthur Barker.

Gottschalk, Louis. 1950. *Lafayette between the American and the French Revolution (1783–1789)*. Chicago: University of Chicago Press.

Gougenot des Mousseaux, Henri Roger. 1860. *La magie au dix-neuvième siècle, ses agents, ses vérités, ses mensonces*. Paris: Henri Plon and E. Dentu.

[Goupy.] 1853. *Quaere et invenies*. Paris: Ledoyen.

Goupy. 1860. *Explications des tables parlantes, des médiums, des esprits et du somnambulisme par divers systèmes de cosmologie, suivie de la voyante de Prevorst*. Paris: G. Baillière.

Gratton-Guinness, Ivor, ed. 1982. *Psychical Research: A Guide to Its History, Principles and Practices. In Celebration of 100 Years of the Society for Psychical Research*. Wellingborough, Northamptonshire: Aquarian Press.

Gravitz, Melvin A. 1988. Early Uses of Hypnosis as Surgical Anesthesia. *Amer. J. Clinical Hypnosis* 30:201–8.

Gravitz, Melvin A., and Manuel I. Gerton. 1984. Origins of the term Hypnotism prior to Braid. *Amer. J. Clinical Hypnosis* 27:107–10.

Greaves, George. 1980. Multiple personality 165 years after Mary Reynolds. *Journal of Nervous and Mental Disease* 168:577–96.

Gregory, Samuel. 1843. *Mesmerism, or Animal Magnetism, and Its Uses; with Particular Directions for Employing It in Removing Pains and Curing Diseases, in Producing Insensibility to Pain in Surgical and Dental Operations; and in the Examination of Internal Diseases, with Cases of Operations, Examinations and Cures*. Boston: Redding.

Gregory, William. 1851. *Letters to a Candid Inquirer, on Animal Magnetism*. London: Taylor, Walton, and Maberly.

Gridley, Josiah A. 1854. *Astounding Facts from the Spirit World, Witnessed at the House of J. A. Gridley, Southampton, Mass., by a Circle of Friends, Embracing the Extremes of Good and Evil*. Southampton, Mass.: The Author.

Grimes, James Stanley. 1839. *A New System of Phrenology*. Buffalo: O. G. Steele.

———. 1845. *Etherology; or the Philosophy of Mesmerism and Phrenology: Including a New Philosophy of Sleep and Consciousness, with a Review of the Pretensions of Neurology and Phreno-Magnetism*. Boston: Saxton Peirce. New York: Saxton and Miles.

————. 1850. *Etherology, and the Phreno-Philosophy of Mesmerism and Magic Eloquence: Including a New Philosophy of Sleep and of Consciousness, with a Review of the Pretensions of Phreno-Magnetism, Electro-biology, &c.* Rev. ed. Boston and Cambridge: James Munro.

————. 1857. *The Mysteries of Human Nature Explained by a New System of Nervous Physiology: To Which Is Added, a Review of the Errors of Spiritualism, and Instructions for Developing or Resisting the Influence by Which Subjects and Mediums Are Made.* Buffalo: R. M. Wanzer.

Guillard. 1853. *Table qui danse et table qui répond, expériences à la portée de tout le monde.* Paris: Garnier frères.

Gurney, Edmund. 1884a. The stages of hypnotism. *Proceedings of the Society for Psychical Research* 2:61–72.

————. 1884b. The problem of hypnotism. *Proceedings of the Society for Psychical Research* 2:265–92.

————. 1886a. Peculiarities of certain post-hypnotic states. *Proceedings of the Society for Psychical Research* 4:268–323.

————. 1886b. Stages of hypnotic memory. *Proceedings of the Society for Psychical Research* 4:515–31.

————. 1887. Further problems of hypnotism. *Mind* 47:212–332.

————. 1888a. Recent experiments in hypnotism. *Proceedings of the Society for Psychical Research* 5:3–17.

————. 1888b. Hypnotism and telepathy. *Proceedings of the Society for Psychical Research* 5:216–59.

Gurney, Edmund, and F. W. H. Myers. 1885. Some higher aspects of mesmerism. *Proceedings of the Society for Psychical Research* 3:401–23.

Haddock, Joseph W. 1849. *Somnolism and Psycheism, Otherwise Vital Magnetism, or Mesmerism: Considered Physiologically and Philosophically. With an Appendix Containing Notes of Mesmeric and Psychical Experience.* London: Hudson.

Hall, Charles Radcliffe. 1845. *Mesmerism, Its Rise, Progress and Mysteries in All Ages and Countries.* New York: Burgess, Stringer.

Hall, Spenser T. 1845. *Mesmeric Experiences.* London: H. Baillière.

Hare, Robert. 1855. *Experimental Investigation of the Spirit Manifestations, Demonstrating the Existence of Spirits and Their Communion with Mortals.* New York: Partridge & Brittan.

Hart, Bernard. 1910. The conception of the subconscious. *J. Abnormal Psychology* 4:351–71.

————. 1926. The conception of dissociation. *Brit. J. Med. Psychology* 6:241–63.

Harte, Richard. 1902. *Hypnotism and the Doctors.* Vols. 1 and 2. London: L. N. Fowler. New York: Fowler & Wells.

Harvey, C. H. 1853. *Millenial [sic] Dawn, or Spiritual Manifestations Tested, and Their Reality Established by Reason, Scripture, and Facts, and Their Object and End Disclosed.* Carbondale: S. S. Benedict.

Haule, John R. 1986. Pierre Janet and dissociation: The first transference theory and its origins in hypnosis. *Amer. J. Clinical Hypnosis* 29:86–94.

Haynes, Renée. 1982. *The Society for Psychical Research, 1882–1982: A History.* London: MacDonald.

Hénin de Cuvillers, Etienne Félix, baron d'. 1821. *Le magnétisme animal retrouvé dans l'antiquité, ou dissertation historique, étymologique et mythologique, sur Esculape, Hippocrate et Galien; sur Apis, Sérapis ou Osiris, et sur Isis; suivie de recherches sur l'origine de l'alchimie.* 2d ed. Paris: Barrois l'ainé, Treuttel et Vurtz, Belin le Prieur, Bataille et Bousquet.

―――. 1822. *Exposition critique du système et de la doctrine mystique des magnétistes.* Paris: Barrois, Belin le Prieur, Treuttel et Vurtz and Delannay.

Héricourt, J. 1886. Un cas de somnambulisme à distance. *Revue philosophique* 21:200–203.

Hering, Charles E. 1853. *Das Tischrücken.* Gotha: N.p.

Hilgard, Ernest. 1977. *Divided Consciousness: Multiple Controls in Human Thought and Action.* New York: John Wiley.

Hirsch, Helmut. 1943. Mesmerism and Revolutionary America. *American-German Review* 9:11–14.

Hirschmüller, Albrecht. 1978. *The Life and Work of Josef Breuer: Physiology and Psychoanalysis.* New York: New York University Press.

Hodgson, R. 1891. A case of double consciousness. *Proceedings of the Society for Psychical Research* 7:221–55.

Hudson, Thomson Jay. 1895. [1st ed. 1893.] *The Law of Psychic Phenomena: A Working Hypothesis for the Systematic Study of Hypnotism, Spiritism, Mental Therapeutics, Etc.* 7th ed. Chicago: A. C. McClurg.

―――. 1903. *The Law of Mental Medicine: The Correlation of the Facts of Psychology and Histology in Their Relation to Mental Therapeutics.* Chicago: A. C. McClurg.

Hufeland, Friedrich. 1811. *Über Sympathie.* Weimar: Landes-Industrie-Comptiors.

Husson, Henri Marie. 1831. *Rapport sur les expériences magnétiques faites par la commission de l'Académie royale de médecine, lu dans les séances des 21 et 28 Juin, par M. Husson, rapporteur.* [Paris:] N.p.

―――. 1836. *Report on the Magnetical Experiments Made by the Commission of the Royal Academy of Medicine, of Paris, Read in the Meetings of June 21 and 28, 1831, by Mr. Husson, the Reporter, Translated from the French, and Preceded with an Introduction, by Charles Poyen St. Sauveur.* Boston: D. K. Hitchcock.

Ince, R. B. 1920. *Franz Anton Mesmer: His Life and Teachings.* London: William Rider.

Inchebald. 1789. *Animal Magnetism. A Farce of Three Acts. As Performing at the Theatres Royal of London and Dublin.* Dublin: N.p.

Inglis, Brian. 1977. *Natural and Supernatural: A History of the Paranormal from Earliest Times to 1914.* London: Hodder and Stoughton.

————. 1984. *Science and Parascience: A History of the Paranormal, 1914–1939*. London: Hodder and Stoughton.

————. 1989. *Trance: A Natural History of Altered States of Mind*. London: Grafton Books.

Isaacs, Ernest Joseph. 1975. *A History of Nineteenth-Century American Spiritualism as a Religious and Social Movement*. Ph.D. diss., University of Wisconsin.

Jackson, J. W. 1853–1854. Table-moving, rappings, and spiritual manifestations. *The Zoist* 11:412–31, 12:1–17.

Jackson, S. 1869. On consciousness and cases of so-called double consciousness. *Amer. J. Medical Sciences* 57:17–24.

James, William. 1889. Notes on automatic writing. *Proceedings of the (Old) American Society for Psychical Research* 1:548–64.

————. 1890a. *The Principles of Psychology*. 2 vols. New York: Henry Holt.

————. 1890b. The hidden self. *Scribner's Magazine* 7:361–73.

————. 1901. Frederic Myers's service to psychology. *Proceedings of the Society for Psychical Research* 17:13–23.

————. 1903. Review of *Human Personality and Its Survival of Bodily Death*, by Frederic Myers. *Proceedings of the Society for Psychical Research* 18:22–33.

Janet, Pierre. 1886a. Note sur quelques phénomènes de somnambulisme. *Revue philosophique* 21:190–98.

————. 1886b. Deuxième note sur le sommeil provoqué à distance et la suggestion mentale pendant l'état somnambulique. *Revue philosophique* 22:212–23.

————. 1886c. Les actes inconscients et le dédoublement de la personnalité pendant le somnambulisme provoqué. *Revue philosophique* 22:577–92.

————. 1887. L'anesthésie systématisée et la dissociation des phénomènes psychologiques. *Revue philosophique* 23:449–72.

————. 1888. Les actes inconscients et la mémoire pendant le somnambulisme. *Revue philosophique* 25:238–79.

————. 1889. *L'automatisme psychologique: Essai de psychologie expérimentale sur les formes inférieures de l'activité humaine*. Paris: Félix Alcan.

————. 1893. *Contribution à l'étude des accidents mentaux chez les hystériques*. Paris: Reuff.

————. 1894. Un cas de possession et l'exorcisme moderne. *Bulletin de l'université de Lyon* 8:41–57.

————. 1904. [1st ed. 1898.] *Névroses et idées fixes*. 2 vols, 2d ed. Paris: Félix Alcan.

Jenkins, Elizabeth. 1982. *The Shadow and the Light: A Defense of Daniel Dunglas Home, the Medium*. London: Hamish Hamilton.

Jensen, Ann, and Mary Lou Watkins. 1967. *Franz Anton Mesmer: Physician Extraordinaire*. New York: Garrett.

Jervey, Edward D. 1976. LaRoy Sunderland: "Prince of the sons of Mesmer." *J. Popular Science,* 9:1010–26.

Johnson, Charles P. 1844. *A Treatise on Animal Magnetism*. New York: Burgess & Stringer.

Joire, Paul. 1892. *Précis théorique & pratique de neuro-hypnologie. Etudes sur l'hypnotisme et les differents phénomènes nerveux physiologiques et pathologiques qui s'y rattachent. Physiologie, pathologie, thérapeutique, médecine légale.* Paris: A. Maloine.

Jones, Henry. 1846. *Animal Magnetism Repudiated as Sorcery; Not . . . Science . . . With an Appendix on Magnetic Phenomena by William H. Beecher, D.D.* New York: J. S. Redfield.

Judah, J. Stillson. 1967. *The History and Philosophy of the Metaphysical Movements in America.* Philadelphia: 1967.

Jung-Stilling, Johann Heinrich. 1808. *Theorie der Geister-Kunde, in einer Natur-, Vernunft- und Bibelmässingen Beantwortung der Frage Was von Ahnungen, Geschichten und Geistererscheinungen geglaubt und nicht geglaubt werden müsse.* Nuremberg: Raw.

————. 1809. *Apologie der Theorie der Geisterkunde.* Nuremberg: Raw.

————. 1854. *Theory of Pneumatology; In Reply to the Question, What Ought to Be Believed or Disbelieved Concerning Presentiments, Visions, and Apparitions According to Nature, Reason, and Scripture.* Edited by George Bush. New York: J. S. Redfield.

[Jussieu, Antoine Laurent de]. 1784. *Rapport de l'un des commissaires chargés par le Roi, de l'examen du magnétisme animal.* Paris: Veuve Harissart.

Kaplan, Fred. 1974. "The Mesmeric Mania": The Victorians and animal magnetism. *J. History of Ideas* 35:691–702.

————. 1975. *Dickens and Mesmerism.* Princeton: Princeton University Press.

Kaplan, Fred, ed. 1982. *John Elliotson on Mesmerism.* New York: Da Capo Press.

Kaplan, Leo. 1917. *Hypnotismus, Animismus und Psychoanalyse. Historischkritische Versuche.* Leipzig and Vienna: Ranz Deuticke.

Kenny, Michael G. 1981. Multiple personality and spirit possession. *Psychiatry* 44:337–58.

————. 1986. *The Passion of Ansel Bourne: Multiple Personality in American Culture.* Washington, D.C.: Smithsonian Institution Press.

Kerner, Justinus. 1824. *Geschichte zweyer Somnambulen. Nebst einigen andern Denkwurdigkeiten aus dem Gebiete der magischen Heilkunde und der Psychologie.* Karlsruhe: G. Braun.

————. 1829. *Die Seherin von Prevorst: Eröffnungen über das innere Leben des Menschen und über das Hereinragen einer Geisterwelt in die unsere,* 2 vols. Stuttgart and Tübingen: J. G. Cotta.

————. 1836. *Nachricht von dem Vorkommen des Besessenseyns eines dämonischmagnetischen Leidens und seiner schon im Alterthum bekannten Heilung durch magisch-magnetisches Einwirken, in einem Sendschreiben an den Herrn Obermedicinalrath Dr. Schelling in Stuttgart.* Stuttgart and Augsburg: J. G. Cotta.

————. 1853. *Die Somnambulen Tische. Zur Geschichte und Erklärung dieser Erscheinung.* Stuttgart: Ebner & Seubert.

————. 1856. *Franz Anton Mesmer aus Schwaben, Entdecker des thierischen Magne-*

tismus. Erinnerungen an denselben, nebst Nachrichten von den letzten Jahren seines Lebens zu Meersburg am Bodensee. Frankfurt am Main: Literarische Anstalt.

———. [1890?] *Die Seherin von Prevorst. Eröffnungen über das innere Leben des Menschen und über das Hereinragen einer Geisterwelt in die unsere . . . mit einer biographische Einleitung von Dr. Carl du Prel.* Leipzig: Philipp Reclam jun.

Kerr, Howard. 1972. *Mediums, Spirit Rappers, and Roaring Radicals: Spiritualism in American Literature, 1850–1900.* Urbana: University of Illinois Press.

Keyes, Daniel. 1981. *The Minds of Billy Milligan.* New York: Random House.

Keys, Thomas E. 1963. *The History of Surgical Anesthesia.* 2d ed. New York: Dover.

A Key to the Science of Electrical Psychology. All Its Secrets Explained, with Full and Comprehensive Instructions in the Mode of Operation and Its Application to Disease, with Some Useful and Highly Interesting Experiments. Every Person and Operator. By a Professor of the Science. 1850. N.p.: N.p.

Kiesewetter, Carl. 1893. *Franz Anton Mesmer's Leben und Lehre. Nebst einer Vorgeschichte des Mesmerismus, Hypnotismus und Somnambulismus.* Leipzig: Max Spohr.

King, John. 1837. *An Essay of Instruction, on Animal Magnetism; Translated from the French of the Marquis de Puysegur, together with Various Extracts upon the Subject, and Notes.* New York: J. C. Kelley.

Kluft, Richard P. 1982. Varieties of hypnotic interventions in the treatment of multiple personality. *Amer. J. Clinical Hypnosis* 24:230–40.

———. 1983. Hypnotherapeutic crisis intervention in multiple personality. *Amer. J. Clinical Hypnosis* 26:73–83.

———. 1984. Treatment of multiple personality disorder. *Psychiatric Clinics of North America* 7:9–29.

Kluft, Richard P., ed. 1985. *Childhood Antecedents of Multiple Personality.* Washington, D.C.: American Psychiatric Press.

Koch, Charles. [1853.] *Table-Moving and Table-Talking, Reduced to Natural Causes: With Especial Reference to the Rev. E. Gillson's Recent Pamphlet.* Bath: Simms & Son.

Kopell, Bert S. 1968. Pierre Janet's description of hypnotic sleep provoked from a distance. *J. Hist. Behavioral Sciences* 4:119–23.

Kravis, Nathan Mark. 1988. James Braid's psychophysiology: A turning point in the history of dynamic psychiatry, in *Amer. J. Psychiatry* 145:1191–1206.

Ladame, P. L. 1882. La névrose hypnotique devant la médecine légale; du viol pendant le sommeil hypnotique; rapport médico-légale. *Annales d'hygiène publique et de médecine légale* 7:518–34.

———. 1887. L'hypnotisme et la médecine légale. *Archives de l'anthropologie criminelle et des sciences pénales* 2:293–335, 520–59.

———. 1888. Observation de somnambulisme hystérique avec dédoublement de la personnalité gueri par la suggestion. *Revue de l'hypnotisme expérimentale et thérapeutique* 2:257–62.

Ladret, Albert. 1976. *Le grand siècle de la Franc-Maçonnerie. La Franc-Maçonnerie lyonnaise au XVIIIe siècle*. Paris: Dervy.

Lafontaine, Charles. 1866. *Mémoires d'un magnétiseur*. 2 vols. Paris: G. Baillière.

Lafont-Gouzi, Gabriel Grégoire. 1839. *Traité du magnétisme animal considéré sous le rapport de l'hygiène, de la médecine légale et de la thérapeutique*. Toulouse: Senac and The Author.

[Lang, William.] 1843. *Mesmerism; Its History, Phenomena, and Practice: With Reports of Cases Developed in Scotland*. Edinburgh: Fraser.

Lapassade, Georges, and Philippe Pédelahore, eds. 1986. *Armand Marie-Jacques de Chastenet, Marquis de Puységur. Mémoires pour servir à l'histoire et à l'établissement du magnétisme animal*. N.p.: Editions Privat.

Laurence, Jean Roch, and Campbell Perry. 1988. *Hypnosis, Will, and Memory: A Psycho-Legal History*. New York: Guilford Press.

Lawton, George. 1932. *The Drama of Life after Death: A Study of the Spiritualist Religion*. New York: Henry Holt.

Laycock, Thomas. 1876. Reflex, automatic and unconscious cerebration: A history and a criticism. *J. Mental Science* 21:477–98; 22:1–17.

Lecture on Mysterious Knockings, Mesmerism, &c., with a Brief History of the Old Stone Mill, and a Prediction of Its Fall, Delivered before the A N ti Quarian Society of Pappagassett . . . by Benjamin Franklin Macy D.F., D.D.F., A.S.S., Professor of Hyperflutinated Philosophy. 1851. Newport, R.I.: N.p.

Lee, Edwin. 1835. *Animal Magnetism and Homeopathy; Being the Appendix to Observations on the Principal Medical Institutions and Practice of France, Italy, and Germany*. London: Churchill.

———. 1838. *Animal Magnetism and Homeopathy*. 2d enlarged ed. London: Whittaker.

———. 1843. *Report upon the Phenomena of Clairvoyance or Lucid Somnambulism, (from Personal Observation). With Additional Remarks*. London: J. Churchill.

———. 1849. *Animal Magnetism, and the Associated Phenomena, Somnambulism, Clairvoyance, &c., an Expository Lecture Delivered at the Town Hall, Brighton, in Aid of the Dispensary Building Fund: With Additional Remarks*. London: J. Churchill.

———. 1866. *Animal Magnetism and Magnetic Lucid Somnambulism. With Observations and Illustrative Instances of Analogous Phenomena Occurring Spontaneously; and an Appendix of Corroborative and Correlative Observations and Facts*. London: Longmans and Green.

Leger, Theodore. 1846. *Animal Magnetism; or Psychodunamy*. New York: D. Appleton.

Lettre à l'intendant de Soissons. 1800. In *Le conservateur, ou recueil de morceaux inedits d'histoire, de politique, de littérature et de philosophie, tirés des portefeuilles de N. François (de Neufchateau), de l'Institut National*. Vol. 1. Paris: L'imprimerie de Crapelet.

Lhermitte, Jean. 1963. *Diabolical Possession, True and False*. London: Burns & Oates.

Liébeault, Ambroise A. 1866. *Du sommeil et des états analogues considérés surtout au point de vue de l'action du moral sur le physique*. Paris: Victor Masson et fils. Nancy: Nicolas Grosjean.

———. 1873. *Ebauche de psychologie*. Paris: G. Masson.

———. 1883. *Etude sur le Zoomagnétisme*. Paris: G. Masson.

———. 1889. *Le sommeil provoqué et les états analogues*. Paris: Octave Doin.

———. 1891. *Thérapeutique suggestive, son mécanisme, propriétés diverses du sommeil provoqué et des états analogues*. Paris: Octave Doin.

Liégeois, Jules. 1884. *De la suggestion hypnotique dans ses rapports avec le droit civil et le droit criminel*. Paris: A. Picard.

———. 1889. *De la suggestion et du somnambulisme dans leurs rapports avec la jurisprudence et la médecine légale*. Paris: O. Doin.

Loisel, A. 1845a. *Observation concernant une jeune fille de dix-sept ans amputée d'une jambe à Cherbourg le 2 octobre 1845, pendant le sommeil magnétique*. Cherbourg: Beaufort et Lecauf.

———. 1845b. *Recueil d'opérations chirurgicales pratiquées sur des sujets magnétisés*. Cherbourg: Beaufort et Lecauf.

———. 1846. *Insensibilité produite au moyen du sommeil magnétique. Nouvelle opération chirurgicale faite à Cherbourg*. Cherbourg: N.p.

Lombard, A. 1819. *Les dangers du magnétisme animal, et l'importance d'en arrêter la propagation vulgaire*. Paris: Dentu and Ant. Bailleul.

London Dialectical Society. 1871. *Report on Spiritualism of the Committee of the London Dialectical Society, Together with the Evidence, Oral and Written, and a Selection from the Correspondance*. London: Longmans, Green, Reader, & Dyer.

Lopez, Claude-Anne. 1966. *Mon Cher Papa: Franklin and the Ladies of Paris*. New Haven: Yale University Press.

Lopez, Claude-Anne, and Eugenia W. Herbert. 1975. *The Private Franklin: The Man and His Family*. New York: Norton.

[Lutzelbourg, comte de.] 1786. *Cures faites par M. le Cte de L . . . sindic de la Société de Bienfaisance établie à Strasbourg . . . avec des notes sur les crises magnétiques appellées improprement somnambulisme*. [Strasbourg:] Lorenz & Schouler.

Luys, Jules Bernard. 1887. *Les émotions chez les sujets en état d'hypnotisme. Etudes de psychologie expérimentale faites à l'aide de substances médicamenteuses ou toxiques impressionnant à distance les réseaux nerveux periphériques*. Paris: Baillière.

———. 1888. Sur l'état de fascination déterminé chez l'homme à l'aide de surfaces brillantes en rotation (action somnifère des miroirs àlouettes). *Comptes rendus et mémoires de sciences de la Société de Biologie* 107:449.

McGuire, Gregory. 1984. The collective subconscious: Psychical research in French psychology (1880–1920). Paper presented at the symposium Controversies in Psychology During France's Belle Epoque, 92nd Annual Meeting of the American Psychological Association, August 25–28, 1984, Toronto, Canada.

McKellar, Peter. 1979. *Mindsplit: The Psychology of Multiple Personality and the Dissociated Self*. London: J. M. Dent.

MacWalter, J. G. 1854. *The Modern Mystery: or, Table-Tapping, Its History, Philosophy, and General Attributes*. London: John Farquhar Shaw.

Magnétisme, insensibilité absolute produite au moyen du sommeil magnétique. Trois nouvelles opérations chirurgicales pratiquées à Cherbourg, le 4 juin 1847 en présence de plus de 60 temoins. 1847. Cherbourg: Beaufort et Lecauf.

Mahan, Asa. 1855. *Modern Mysteries Explained and Exposed. In Four Parts. I. Clairvoyant Revelations of A. J. Davis. II. Phenomena of Spiritualism Explained and Exposed. III. Evidence that the Bible Is Given by Inspiration of the Spirit of God, as Compared with Evidence That These Manifestations Are from the Spirits of Men. IV. Clairvoyant Revelations of Emanuel Swedenborg*. Boston: J. P. Jewett.

Mais, Charles. 1814. *The Surprising Case of Rachel Baker, Who Prays and Preaches in Her Sleep: With Specimens of Her Extraordinary Performances Taken Down Accurately in Short Hand at the Time. . . .* New York: S. Marks.

Maitland, S. R. 1849. *Illustrations and Enquiries Relating to Mesmerism*. Part 1. London: William Stephenson.

———. 1855. *Superstition and Science: An Essay*. London: Rivingtons.

Martin, John. 1790. *Animal Magnetism Examined: In a Letter to a Country Gentleman*. London: Stockdale.

Martineau, Harriet. 1845. *Letters on Mesmerism*. New York: Harper & Brothers.

Maskelyne, John Nevil. 1875. *Modern Spiritualism. A Short Account of Its Rise and Progress, with Some Exposures of So-Called Spirit Media*. London: Frederick Warne.

[Mattison, Hiram.] 1854. *The Rappers; or, The Mysteries, Fallacies, and Absurdities of Spirit-Rapping, Table-Tipping, and Entrancement*. New York: H. Long.

Mauskopf, Seymour H., and Michael R. McVaugh. 1980. *The Elusive Science: Origins of Experimental Psychical Research*. Baltimore: Johns Hopkins University Press.

Mayer, Robert. 1988. *Through Divided Minds: Probing the Mysteries of Multiple Personalities—A Doctor's Story*. New York: Doubleday.

Mayo, Herbert. 1851. *Letters on the Truths Contained in Popular Superstitions, with an Account of Mesmerism*, 2d ed. Edinburgh: William Blackwood.

Mesmer, Franz Anton. 1766. *Dissertatio physico-medica de planetarum influxu*. Vienna: Ghelen.

———. 1776. *Schreiben über die Magnetkur von Herrn Dr. A. Mesmer*. N.p.: N.p.

———. 1779. *Mémoire sur la découverte du magnétisme animal*. Geneva and Paris: Didot le jeune.

———. 1784a. *Apologie de M. Mesmer; ou, Réponse à la brochure intitulée: Mémoire pour servir à l'histoire de la jonglerie dans lequel on démontre les phénomènes du mesmérisme*. [Paris:] N.p.

————. 1784b. *Lettre de M. Mesmer à M. Le Cte de C*** (31 aout 1784)*. N.p.: Imprimerie royale.

————. 1784c. *Lettres de M. Mesmer à messieurs les auteurs du Journal de Paris, et à M. Franklin*. N.p.: N.p.

————. 1784d. *Lettres de M. Mesmer, à M. Vicq-d'Azyr, et à messieurs les auteurs du Journal de Paris*. Brussels: N.p.

[————.] 1784e. *Lettre d'un médecin de Paris à un médecin de province*. [Paris:] N.p.

————. 1785. *Lettre de l'auteur de la découverte du magnétisme animal à l'auteur des Réflexions préliminaires. Pour servir de réponse à un imprimé ayant pour titre: Sommes verseés entre les mains de M. Mesmer pour acquérir le droit de publier sa découverte*. [Paris:] N.p.

————. 1799. *Mémoire de F. A. Mesmer, docteur en médecine, sur ses découvertes*. Paris: Fuchs.

————. 1812. *Allgemeine Erläuterungen über den Magnetismus und den Somnambulismus. Als vorläufige Einleitung in das Natursystem. Aus dem Askläpieion abgedruckt*. Halle and Berlin: Hallischen Waisenhauses.

————. 1814. *Mesmerismus. Oder System der Wechselwirkungen, Theorie und Anwendung des thierischen Magnetismus als die allgemeine Heilkunde zur Erhaltung des Menschen. Herausgegeben von Dr. Karl Christian Wolfart*. Berlin: Nikola.

Mesnet, Ernest. 1874. *De l'automatisme, de la mémoire et du souvenir dans le somnambulisme pathologique, considérations médico-légales*. Paris: Félix Malteste.

————. 1894. *Outrages à la pudeur. Violences sur les organes sexuels de la femme dans le somnambulisme provoqué et la fascination. Etude médico-légale*. Paris: Rueff.

Meyer, Th. J. A. G. 1839. *Natur-Analogien, oder die vornehmsten Erscheinungen des animalischen Magnetismus in ihrem Zusammenhange mit den Ergebnissen der gesammten Naturwissenschaften, mit besonderer Hinsicht auf die Standpunkte und Bedurfnisse heutiger Theologie*. Hamburg and Gotha: Friedrich und Andreas Perthes.

Mialle, Simon. 1826. *Exposé par ordre alphabétique des cures opérées en France par le magnétisme animal, depuis Mesmer jusqu'à nos jours*. 2 vols. Paris: J. G. Dentu.

Milt, Bernhard. 1953. *Franz Anton Mesmer und seine Beziehungen zur Schweiz: Magie und Heilkunde zu Lavaters Zeit*. Zurich: Leemann.

Mirville, Jules Eudes, marquis de. 1853. *Pneumatologie. Des esprits et de leurs manifestations fluidiques*. Vol. 1. Paris: H. Vrayet de Surcy.

Mitchell, G. W. 1876. *X+Y=Z; or The Sleeping Preacher of North Alabama. Containing an Account of Most Wonderful Mysterious Mental Phenomena, Fully Authenticated by Living Witnesses*. New York: W. C. Smith (printed for the author).

Mitchill, Samuel Latham. 1816. A double consciousness, or a duality of person in the same individual. *Medical Repository* 3:185–86.

Moll, Albert. 1889. *Der Hypnotismus*. Berlin: H. Kornfeld.

————. 1892. *Der Rapport in der Hypnose. Untersuchungen über den thierischen Magnetismus.* Leipzig: Ambr. Abel.

Moniz, Egas. 1925. *O Padre Faria na historia do hypnotismo.* Lisbon: Faculdade de Medicina de Lisboa.

Montegre, Antoine François Weninde. 1816. *Note sur le magnétisme animal et sur les dangers que font courir les magnétiseurs à leur patients.* [Paris:] Gazette de Santé.

Montferrier, Alexandre André Sarrazin de [A. de Lausanne]. 1819. *Des principes et des procédés du magnétisme animal, et de leurs rapports avec les lois de la physique et de la physiologie,* 2 vols. Paris: J. G. Dentu.

Montgomery, John Warwick, ed. 1976. *Demons Possession: A Medical, Historical, Anthropological and Theological Symposium.* Minneapolis: Bethany Fellowship.

Moore, R. Laurence. 1977. *In Search of White Crows: Spiritualism, Parapsychology, and American Culture.* New York: Oxford University Press.

Moreau de Tours, Jacques Joseph. 1845. *Du hachisch et de l'alienation mentale.* Paris: Fortin, Masson.

Morgan, R. C. [1854.] *An Inquiry into Table-Miracles, Their Cause, Character, and Consequence; Illustrated by Recent Manifestations of Spirit-Writing and Spirit-Music.* London: Binns & Goodwin.

[Morin, Alcide.] 1854. *Comment l'esprit vient aux tables, par un homme qui n'a pas perdu l'esprit.* Paris: Librairie Nouvelle.

Morin, André S. 1860. *Du magnétisme et des sciences occultes.* Paris: G. Baillière.

Morley, Charles. *Elements of Animal Magnetism, or Pneumatology.* New York: Turner & Hughes.

Moser, Fanny. 1935. *Der Okkultismus: Täuschungen und Tatsachen,* 2 vols. Zurich: Orell Fussli.

[Mouilleseaux, de.] 1787. *Appel au public sur le magnétisme animal, ou Projet d'un journal pour le seul avantage du public, et dont il servait le coopérateur.* Strasbourg: N.p.

Moutin, Lucien. 1887. *Le nouvel hypnotisme.* Paris: Perrin.

Myers, Frederic W. H. 1884. On a telepathic explanation of some so-called spiritualistic phenomena. *Proceedings of the Society for Psychical Research* 2:217–37.

————. 1885. Automatic writing. *Proceedings of the Society for Psychical Research* 3:1–63.

————. 1886a. Human personality in the light of hypnotic suggestion. *Proceedings of the Society for Psychical Research* 4:1–24.

————. 1886b. On telepathic hypnotism, and its relation to other forms of hypnotic suggestion. *Proceedings of the Society for Psychical Research* 4:127–88.

————. 1887a. Automatic writing. III. *Proceedings of the Society for Psychical Research* 4:209–61.

————. 1887b. Multiplex personality. *Proceedings of the Society for Psychical Research* 4:496–514.

————. 1889a. French experiments in strata of personality. *Proceedings of the Society for Psychical Research* 5:374–98.

————. 1889b. Automatic writing—IV—The daemon of Socrates. *Proceedings of the Society for Psychical Research* 5:522–48.

————. 1892a. The subliminal consciousness. *Proceedings of the Society for Psychical Research* 7:298–355.

————. 1892b. The subliminal consciousness. *Proceedings of the Society for Psychical Research* 8:333–404.

————. 1892c. The subliminal consciousness. *Proceedings of the Society for Psychical Research* 8:436–535.

————. 1903. *Human Personality and Its Survival of Bodily Death.* 2 vols. London: Longmans, Green.

Nelson, Geoffrey K. 1969. *Spiritualism and Society.* London: Routledge & Kegan Paul.

Nevius, J. L. 1893. *Demon Possession and Allied Themes.* Chicago: F. H. Revell.

Newman, John B. 1848. *Fascination, or The Philosophy of Charming, Illustrating the Principles of Life in Connection with Spirit and Matter.* New York: Fowler and Wells.

Newnham, William. 1830. *Essay on Superstition; Being an Inquiry into the Effects of Physical Influence on the Mind, in the Production of Dreams, Visions, Ghosts, and Other Supernatural Appearances.* London: J. Hatchard and Son.

————. 1845. *Human Magnetism; Its Claims to Dispassionate Inquiry. Being an Attempt to Show the Utility of Its Application for the Relief of Human Suffering.* New York: Wiley and Putnam.

Nicholls. [1853.] *Table Moving, Its Causes and Phenomena: With Directions How to Experiment.* London: John Wesley.

Noizet, François Joseph. 1854. *Mémoire sur le somnambulisme et le magnétisme animal adressé en 1820 à l'Académie Royale de Berlin.* Paris: Plon Frères.

Ochorowitz, Julian. 1887. *De la suggestion mentale.* Paris: Octave Doin.

————. 1897. *Magnetismus und Hypnotismus.* Leipzig: Oswald Mutze.

Oesterreich, Traugott K. 1974. *Possession and Exorcism among Primitive Races, in Antiquity, the Middle Ages, and Modern Times.* New York: Causeway Books.

Oppenheim, Janet. 1985. *The Other World: Spiritualism and Psychical Research in England, 1850–1914.* Cambridge: Cambridge University Press.

Orelut, Pierre. 1784. *Détail des cures opérées à Lyon par le magnétisme animal, selon les principes de M. Mesmer. Précédé d'une lettre à M. Mesmer.* Lyon: Faucheux.

Page, Charles G. 1853. *Psychomancy. Spirit-Rappings and Table-Tippings Exposed.* New York: D. Appleton.

Passavant, Johann Carl. 1837. [1st ed. 1821.] *Untersuchungen über den Lebensmagnetismus und das Hellsehen.* 2d ed. Frankfurt: Heinrich Ludwig Brönner.

Pattie, Frank. 1956. Mesmer's medical dissertation and its debt to Mead's *Imperio Solis ac Lunae. J. History of Medicine* 11.

————. 1967. A brief history of hypnotism. In *Handbook of Clinical and Experimental Hypnosis,* edited by Jesse E. Gordon. New York: Macmillan.

[Paulet, Jean Jacques.] 1784a. *Mesmer justifié.* Constance and Paris: N.p.

[————.] 1784b. *L'antimagnétisme, ou Origine, progrès, décadence, renouvellement et réfutation du magnétisme animal.* London: N.p.

[Pearson, John.] 1790. *A Plain and Rational Account of the Nature and Effects of Animal Magnetism.* London: W. and J. Stratford.

Perry, Campbell, and Jean-Roch Laurence. 1984. Mental processing outside of awareness: The contributions of Freud and Janet. In *The Unconscious Reconsidered,* edited by K. Bowers and D. Meichenbaum. New York: John Wiley.

Perty, Maximilian. 1861. *Die mystischen Erscheinungen der menschlichen Natur.* Leipzig and Heidelberg: C. F. Winter.

————. 1863. *Die Realität magischer Kräfte und Wirkungen des Menschen gegen die Widersacher vertheidigt.* Leipzig and Heidelberg: C. F. Winter.

————. 1869. *Blicke in das verborgene Leben des Menschengeistes.* Leipzig and Heidelberg: C. F. Winter.

————. 1877. *Der jetzige Spiritualismus und verwandte Erfahrungen der Vergangenheit und Gegenwart.* Leipzig and Heidelberg: C. F. Winter.

Petetin, Jacques Henri Desiré. 1787. *Mémoire sur la découverte des phénomènes que présentent la catalepsie et le somnambulisme, symptomes de l'affection hystérique essentielle, avec des recherches sur la cause physique des ces phénomènes. Première partie. Mémoire sur la découverte des phénomènes de l'affection hystérique essentielle, et sur la méthode curative de cette maladie. Second partie.* [Lyon?:] N.p.

————. 1802. *Nouveau mécanisme de l'électricité fondé sur les lois de l'équilibre et du mouvement, démontré par des expériences qui renversent le système de l'électricité positive et négative et qui etablissent ses rapports avec le mécanisme caché de l'aimant et l'heureuse influence du fluid électrique dans les affections nerveuses.* Lyon: Bruyset ainé.

————. 1805. *Electricité animal, prouvée par la découverte des phénomènes physiques et moraux de la catalepsie hystérique, et de ses variétés; et par les bons effets de l'électricité artificielle dans le traitement de ces maladies.* Lyon: Bruyset et Buynand.

————. 1808. *Electricité animal.* 2d ed. Paris: Brunot-Labbe and Gautier et Bretin. Lyon: Reymann.

The Philosophy of Animal Magnetism Together with the System of Manipulating Adopted to Produce Ecstasy and Somnambulism—The Effects and Rationale. By a Gentleman of Philadelphia. 1837. Philadelphia: Merrihew and Gunn.

Pigeaire, Jules. 1839. *Puissance de l'électricité animal, ou Du magnétisme vital et de ses rapports avec la physique, la physiologie et la médecine.* Paris: Dentu and G. Baillière.

Pivnicki, D. 1969. The beginnings of psychotherapy. *J. Hist. Behavioral Sciences* 5:238–47.

Plumer, William S. 1860. Mary Reynolds: A case of double consciousness. *Harper's New Monthly Magazine* 20:807–12.

Podmore, Frank. 1902. *Modern Spiritualism, a History and a Criticism*. 2 vols. London: Methuen.

———. 1909. *Mesmerism and Christian Science: A Short History of Mental Healing*. Philadelphia: George W. Jacobs.

———. 1910. *The Newer Spiritualism*. London: Fisher Unwin.

Poissonnier, Pierre Isaac, Calude Antoine Caille, Pierre Jean Claude Mauduyt de Varenne, and Charles Louis François Andry. 1784. *Rapport des commissaires de la Société Royale de Médecine nommés par le Roi pour fair l'examen du magnétisme animal, imprimié par ordre du Roi*. Paris: Imprimerie royale.

Powers, Grant. 1828. *Essay upon the Influence of the Imagination on the Nervous System, Contributing to a False Hope in Religion*. Andover: Flagg and Gould.

Poyen St. Sauveur, Charles. 1837a. *Progress of Animal Magnetism in New England. Being a Collection of Experiments, Reports and Certificates, from the Most Respectable Sources. Preceded by a Dissertation on the Proofs of Animal Magnetism*. Boston: Weeks, Jordan.

———. 1837b. *A Letter to Col. Wm. L. Stone, of New York, on the Facts Related in His Letter to Dr. Brigham, and a Plain Refutation of Durant's Exposition of Animal Magnetism, &c, by Charles Poyen. With Remarks on the Manner in Which the Claims of Animal Magnetism Should Be Met and Discussed. By a Member of the Massachusetts Bench*. Boston: Weeks, Jordan.

Practical Instructions in Table-Moving, with Physical Demonstrations. By a Physician. 6th ed. 1853. London: H. Baillière.

Prichard, John. 1853a. *A Few Sober Words of Table-Talk About Table-Spirits, and the Rev. N. S. Godfrey's Incantations*. Leamington: G. C. Liebenrood.

———. 1853b. *A Few Sober Words of Table-Talk about Table-Spirits, and the Rev. N. S. Godfrey's Incantations*. 2d ed. Leamington: G. C. Liebenrood.

Prince, Morton. 1901. The development and genealogy of the Misses Beauchamp: A preliminary report on a case of multiple personality. *Proceedings of the Society for Psychical Research* 15:466–83.

———. 1907. A symposium on the subconscious. *J. Abnormal Psychology* 2:67–80.

———. 1908. [1st ed. 1905.] *The Dissociation of a Personality: A Biographical Study in Abnormal Psychology*. 2d ed. London: Longmans, Green.

Putnam, Frank W. 1982. Traces of Eve's faces. *Psychology Today,* October: 88.

———. 1989. *Diagnosis and Treatment of Multiple Personality Disorder*. New York: Guilford Press.

———, J. J. Guroff, E. K. Silberman, L. Barban, and R. M. Post. 1986. The clinical phenomenology of multiple personality disorder: A review of 100 recent cases. *J. Clinical Psychiatry* 47:285–93.

Puységur, Armand Marie Jacques de Chastenet, marquis de. 1784. *Mémoires pour servir à l'histoire et à l'établissement du magnétisme animal*. Paris: Dentu.

———. 1785. *Suite des mémoires pour servir à l'histoire et à l'établissement du magnétisme animal*. Paris and London: N.p.

————. 1811. *Recherches, expériences et observations physiologiques sur l'homme dans l'état de somnambulisme naturel et dans le somnambulisme provoqué par l'acte magnétique.* Paris: J. G. Dentu.

————. 1812. *Les fous, les insensés, les maniaques et les frénétiques ne seraient-ils que des somnambules désordonnés?* Paris: J. G. Dentu.

————. 1813. *Appel aux savans observateurs du dix-neuvième siècle, de la décision portée par leurs prédécesseurs contre le magnétisme animal; continuation du journal du traitement du jeune Hébert; et fin du traitement du jeune Hébert.* Paris: J. G. Dentu.

————. 1814. [1st ed. 1807.] *Les verités cheminent, tot ou tard elles arrivent.* Paris: Dentu.

————. 1820. *Du magnétisme animal considéré dans ses rapports avec diverses branches de la physique générale.* 2d ed. Paris: J. G. Dentu.

Puységur, Jacques Maxime Paul de Chastenet, comte de. 1784. *Rapport des cures opérées à Bayonne par le magnétisme animal, adressé à M. l'abbé de Poulouzat, conseiller clerc au Parlement de Bordeaux, par le comte de Puységur, avec notes de M. Duval d'Eprémesnil, conseiller au Parlement de Paris.* Bayonne and Paris: Prault.

Pyne, Thomas. 1849. *Vital Magnetism; A Remedy.* 4th ed. London: Samuel Highley.

Quen, Jacques M. 1976. Mesmerism, medicine, and professional prejudice. *New York State Journal of Medicine* 76:2218–22.

————, ed. 1986. *Split Minds/Split Brains.* New York: New York University Press.

Rahn, Johann Heinrich. 1789. *Über Sympathie und Magnetismus. Aus den Lateinischen übersazt und mit anmerkungen begleitet von Heinrich Tabor.* Heidelberg: F. L. Pfahler.

Rausky, Franklin. 1977. *Mesmer ou la révolution therapeutique.* Paris: Payot.

Reichenbach, Karl von. 1849. *Physikalish-physiologische Untersuchungen über die Dynamide des Magnetismus, der Electrizität, der Wärme, des Lichtes, der Krystallisation, des Chemismus in ihren Beziehungen zur Lebenskraft.* 2 vols. 2d ed. Braunschweig: Friedrich Vieweg.

————. 1850. *Psycho-physiological Researches on the Dynamides or Imponderables, Magnetism, Electricity, Heat, Light, Crystallization, and Chemical Attraction, in Their Relation to the Vital Force.* Translated by William Gregory. London: Taylor, Walton and Maberly.

————. 1854. *Der sensitive Mench und sein Verhalten zum Ode. Eine Reihe experimenteller Untersuchungen über ihre gegenseitigen Kräfte und Eigenschaften,* 2 vols. Stuttgart and Tübingen: J. G. Cotta.

Report of the Committee Appointed by Government to Observe and Report upon Surgical Operations by Dr. J. Esdaile, upon Patients under the Influence of Alleged Mesmeric Agency. Calcutta: Military Orphan Press.

Reese, David Meredith. 1838. *Humbugs of New-York: Being a Remonstrance Against Popular Delusion; Whether in Science, Philosophy, or Religion.* New York: John S. Taylor.

Résie, Lambert-Elizabeth d'Aubert, comte de. 1853. *Lettre à M. l'abbé Croizet, curé de Neschers . . . sur le magnétisme et la danse des tables.* Clermont-Ferrand: Hubler, Bayle et Dubois.

———. 1857. *Histoire et traité des sciences occultes, ou examen des croyances populaires sur les êtres surnaturels, la magie, la sorcellerie, la divination, etc., depuis le commencement du monde jusqu'à nos jours.* 2 vols. Paris: Louis Vivès.

Retz, Noël de Rochefort. 1782. *Lettre sur le secret de M. Mesmer, ou Réponse d'un médecin à un autre, qui avait demandé des eclaircissements à ce sujet.* Paris: Méquignon.

Ribot, Théodule Armand. 1885. *Les maladies de la personnalité.* Paris: Alcan.

Ricard, Jean Joseph Adolphe. 1844. *Physiologie et hygiène du magnétiseur; régime dietétique du magnétisé; mémoires et aphorismes de Mesmer, avec des notes.* Paris: G. Baillière.

Rice, Nathan Lewis. 1849. *Phrenology Examined, and Shown to Be Inconsistent with the Principles of Physiology, Mental and Moral Science, and the Doctrines of Christianity: Also an Examination of the Claims of Mesmerism.* New York: R. Carter & Brothers. Cincinnati: J. D. Thorpe.

[Richemont, Panon des Bassyns, comte de.] 1853. *Le mystère de la danse des tables dévoilé par ses rapports avec les manifestations spirituelles d'Amérique. Par un Catholique.* Paris: Devarenne.

Richer, Paul. 1881. *Etudes cliniques sur l'hystéroépilepsie ou grand hystérie.* Paris: A. Delahaye & E. Lecrosnier.

Richet, Charles. 1875. Du somnambulisme provoqué. *Journal de l'anatomie et de la physiologie normales et pathologiques de l'homme et des animaux* 11:348–78.

———. 1880. Du somnambulisme provoqué. *Revue philosophique* 10:337–74, 462–84.

———. 1883. La personnalité et la mémoire dans le somnambulisme. *Revue philosophique* 15:225–42.

———. 1886. Un fait de somnambulisme à distance. *Revue philosophique* 21:199–200.

———. 1888. Expériences sur le sommeil à distance. *Revue philosophique* 25:434–52.

Rochas d'Aiglun, Eugène August Albert de. 1866. *Les effluves odiques: Conférences faites en 1866, par le Baron de Reichenbach à l'Académie des Sciences de Vienne; précédées d'une notice historique sur les effets mechaniques de l'Od.* Paris: Ernest Flammarion.

———. 1887. *Les Forces non définies: Recherches historiques et expérimentales.* Paris: G. Masson.

———. 1891. *Le fluide des magnétiseurs. Précis des expériences du Baron de Reichenbach sur les propriétés physiques et physiologiques, classées et annotées par le Lt.-Colonel A. de Rochas.* Paris: Georges Carré.

———. 1892. *Les états profonds de l'hypnose.* Paris: Chamuel.

———. 1893. *Les états superficiels de l'hypnose.* Paris: Chamuel.

———. 1895. *L'extériorisation de la sensibilité. Etude expérimentale et historique.* Paris: Bibliothèque Chacornac.

———. 1896. *L'extériorisation de la motricité. Recueil d'expériences et d'observations.* Paris: Chamuel.

Rogers, Edward Coit. 1852. *Philosophy of Mysterious Agents, Human & Mundane: or, the Dynamic Laws and Relations of Man. Embracing the Natural Philosophy of Phenomena, Styled "Spiritual Manifestations."* Boston: John P. Jewett.

———. 1853a. *Philosophy of Mysterious Agents, Human and Mundane: or, the Dynamic Laws and Relations of Man, Embracing the Natural Philosophy of Phenomena Styled "Spiritual Manifestations."* Boston: John P. Jewett.

———. 1853b. *A Discussion of the Automatic Powers of the Brain; Being a Defence against Rev. Charles Beecher's Attack upon the Philosophy of Mysterious Agents, in His Review of "Spiritual Manifestations."* Boston: John P. Jewett.

Rosen, George. 1946. Mesmerism and Surgery: A Strange Chapter in the History of Anesthesia. *J. Hist. Medicine* 1:527–50.

Ross, Colin A. 1989. *Multiple Personality Disorder: Diagnosis, Clinical Features, and Treatment.* New York: John Wiley.

———. 1991a. The dissociated executive self and the cultural dissociation barrier. *Dissociation* 4:55–61.

———. 1991b. Epidemiology of multiple personality disorder and dissociation. *Psychiatric Clinics of North America* 14:503–17.

Rostan, Louis. 1825. *Du magnétisme animal.* Paris: Rignoux.

Roubaud, Alexandre Félix. 1853. *La danse des tables, phénomènes physiologiques démontrés par le Dr. Félix Roubaud.* Paris: Librairie nouvelle.

Roullier, August. 1817. *Exposition physiologique des phénomènes du magnétisme animal et du somnambulisme; contenant des observations pratiques sur les avantages et l'emploi de l'un et de l'autre dans le traitement des maladies aiguës et chroniques.* Paris: J. G. Dentu.

Samson, George Whitefield. 1852. *"To Daimonion," or The Spiritual Medium. Its Nature Illustrated by the History of Its Uniform Mysterious Manifestations When Unduly Excited. In Twelve Familiar Letters to an Inquiring Friend. By Traverse Oldfield.* Boston: Gould and Lincoln.

Sandby, George. 1848. *Mesmerism and Its Opponents.* London: Longman, Brown, Green, and Longmans.

———. 1853a. The mesmerisation and movement of tables. *The Zoist* 11:175–85.

———. 1853b. Can Professor Farady never be wrong? *The Zoist* 11:320–24.

Satanic Agency and Table Turning. A Letter to the Rev. Francis Close in Reply to His Pamphlet, "Table-turning Not Diabolical." 1853. London: Bosworth.

Schneck, Jerome M. 1954. *Studies in Scientific Hypnosis.* New York: Nervous and Mental Disease Monographs.

Schneider, Emil. 1950. *Der animale magnetismus: Seine Geschichte und seine Beziehungen zur Heilkunst.* Zurich: Konrad Lampert.

Schott, Heinz. 1982. Die Mitteilung des Lebensfeuers: Zum therapeutischen Konzept von Franz Anton Mesmer (1734–1815). *Medizin-Historisches Journal* 17.

Schott, Heinz, ed. 1985. *Franz Anton Mesmer und die Geschichte des Mesmerismus.* Stuttgart: Franz Steiner.

Schreiber, Flora Rheta. 1973. *Sybil.* Chicago: Henry Regnery.

Schroeder, H. R. Paul. 1890. *Die Heilmethode des Lebensmagnetismus. Theorie und Praxis besprochen und mit eine Nachweise über den wesentlichen Unterschied zwischen Hypnotismus und Heilmagnetismus versehen.* Leipzig: N.p.

————. 1899. *Geschichte des Lebensmagnetismus und des Hypnotismus. Vom Uranfang bis auf den heutigen Tag.* Leipzig: Arwed Strauch.

Schürer von Waldheim, Fritz. 1930. *Anton Mesmer, ein Naturforschen ersten Ranges.* Vienna: The Author.

Scoresby, W. 1849. *Zoistic magnetism: Being the Substance of Two Lectures, Descriptive of Original Views and Investigations Respecting This Mysterious Agency; Delivered, by Request, at Torquay, on the 24th of April and 1st of May, 1849.* London: Longman, Brown, Green, and Longmans.

[Servan, Joseph Michel Antoine.] 1784. *Doutes d' un provincial, proposés à messiers le médecins-commissaires chargés par le Roi de l' examen du magnétisme animal.* Lyon and Paris: Prault.

Sidis, Boris. 1911. [1st ed. 1898.] *The Psychology of Suggestion: A Research into the Subconscious Nature of Man and Society.* New York: D. Appleton.

Silas, Ferdinand. 1853. *Instructions explicatives des tables tournantes, d' après les publications allemandes, américaines, et les extraits des journaux allemands, français, et américains. Précédées d' une introduction sur l' action motrice du fluide magnétique, par Henri Delaage.* Paris: Houssiaux.

Sizemore, Chris Costner. 1989. *A Mind of My Own.* New York: William Morrow.

————. 1992. Personal communication. January 30, 1992.

Sizemore, Chris Costner, and Elen Sain Pittillo. 1977. *I'm Eve.* Garden City, N.Y.: Doubleday.

Sizer, Nelson. 1884. *Forty Years in Phrenology; Embracing Recollections of History, Anecdote, and Experience.* New York: Fowler and Wells.

Skae, D. 1845. Case of intermittent mental disorder of the tertian type, with double consciousness. *Northern Journal of Medicine* 4:10–13.

Smith, Gison. 1845. *Lectures on Clairmativeness: or, Human Magnetism.* New York: Searing & Prall.

Snow, Herman. 1853. *Spirit Intercourse: Containing Incidents of Personal Experience While Investigating the New Phenomena of Spirit Thought and Action; with Various Spirit Communications through Himself as Medium.* Boston: Crosby, Nichols.

Société Exégétique & Philantropique, Stockholm. 1788. *Lettre sur la seule explication satisfaisante des phénomènes du magnétisme animal et du somnambulisme déduite des vrais principes fondés dans la connaissance du créateur, de l' homme, et de la nature, et confirmée par l' expérience.* Stockholm: L'Imprimerie royale.

Société harmonique des amis réunis de Strasbourg. 1786. *Exposé de différentes cures opérées depuis le 25. d'Août 1785.* [Strasbourg:] Lorenz & Schouler.

―――. 1787a. *Exposé de différentes cures opérées depuis le 25. d'août 1785.* 2d ed. revised, corrected, and augmented. Strasbourg: Librairie Académique.

―――. 1787b. *Suite des cures faites par différents magnétiseurs, membres de la Société harmonique des amis-réunis de Strasbourg,* Vol. 2. Strasbourg: Lorenz and Schouler.

―――. 1789. *Annales de la Société harmonique des amis réunis de Strasbourg, ou Cures que des membres de cette société ont opérées par le magnétisme animal.* Vol. 3. Strasbourg: N.p.

Spicer, Henry. 1853a. *Sights and Sounds: The Mystery of the Day: Comprising an Entire History of the American "Spirit" Manifestations.* London: Thomas Bosworth.

―――. 1853b. *Facts and Fantasies: A Sequel to Sights and Sounds; the Mystery of the Day.* London: Thomas Bosworth.

Stearns, Samuel. 1791. *The Mystery of Animal Magnetism Revealed to the World, Containing Philosophical Reflections on the Publication of a Pamphlet Entitled, A True and Genuine Discovery of Animal Electricity and Magnetism: Also, an Exhibition of the Advantages and Disadvantages That May Arise in Consequence of Said Publication.* London: M. R. Parsons.

Stemman, Roy. 1972. *One Hundred Years of Spiritualism.* London: Spiritualist Association of Great Britain.

Stewart, Dugald. 1827. *Elements of the Philosophy of the Human Mind.* Vol. 3. Philadelphia: Carey, Lea, & Carey.

Stieglitz, Johann. 1814. *Über den thierischen Magnetismus.* Hannover: Hahn.

Stone, Michael H. 1974. Mesmer and his followers: The beginnings of sympathetic treatment of childhood emotional disorders. *History of Childhood Quarterly* 1.

Stone, William Leete. 1837. *Letter to Doctor A. Brigham, on Animal Magnetism: Being an Account of a Remarkable Interview between the Author and Miss Loraina Brackett While in a State of Somnambulism.* New York: George Dearborn.

Strombeck, Friedrich Karl von. 1814. *Histoire de la guérison d'une jeune personne, par le magnétisme animal, produit par la nature elle-meme. Par un témoin oculaire de ce phénomène extraordinaire. Traduit de l'Allemand.* Paris: Librairie Grecque, Latine, Allemande.

Sunderland, LaRoy. 1843. *Pathetism; with Practical Instructions: Demonstrating the Falsity of the Hitherto Prevalent Assumptions in Regard to What Has Been Called "Mesmerism" and "Neurology," and Illustrating Those Laws Which Induce Somnambulism, Second Sight, Sleep, Dreaming, Trance, and Clairvoyance, with Numerous Facts Tending to Show the Pathology of Monomania, Insanity, Witchcraft, and Various Other Mental or Nervous Phenomena.* New York: P. P. Good.

―――. 1845. *"Confessions of a Magnetizer" Exposed.* Boston: Redding.

Supplément aux deux rapports de MM. les commissaires de l'Académie & de la Faculté de Médecine, & de la Société Royale de Médecine. 1784. Amsterdam: Gueffier.

Szapary, Ferencz Grof. 1840. *Ein Wort über animalischen Magnetismus, Seelenkorper und Lebensessenz; nebst Beschreibung des ideo-somnambulen Zustandes des Fräuleins Therese von B—y zu Vasarhely im Jahre 1838, und einem Anhang.* Leipzig: F. A. Brockhaus.

———. 1854a. *Das Tischrücken. (Fortsetzung.) Geistige Agapen. Psychographische Mittheilungen der Pariser Deutsch-Magnetischen Schule des Grafen F. von Szapary.* Paris: The Author.

———. 1854b. *Table-Moving. Somnambulish-Magnetische Traumbedeutung.* Paris: Bonaventure and Ducessois.

———. 1854c. *Magnétisme et magnéto-thérapie.* 2d ed. Paris: The Author.

Table Moving by Animal Magnetism Demonstrated; with Directions How to Perform the Experiment. Also, A Full and Detailed Account of the Experiments Already Performed. [1853.] London: N.p.

Table Moving, Its Causes and Phenomena; with Directions How to Experiment. Profusely Illustrated, on Plate Paper, by Nicholls. 1853. [London:] J. Wesley.

Table Turning and Table talking, Containing Detailed Reports of an Infinite Variety of Experiments Performed in England, France, and Germany with Most Marvellous Results; Also Minute Directions to Enable Every One to Practise Them, and the Various Explanations Given of the Phenomena by the Most Distinguished Scientific Men of Europe. Second Edition with Professor Faraday's Experiments and Explanation. [1853.] London: Henry Vizetelly.

[Tanchou.] 1846. *Enquête sur l'authenticité des phénomènes électriques d'Angélique Cottin.* Paris: Baillière.

Tardieu, A. 1878. *Étude médico-légale sur les attentats aux moeurs.* 7th ed. Paris: J. B. Baillière.

[Tardy de Montravel, A. A.] 1785. *Essai sur la théorie du somnambulisme magnétique.* London: N.p.

[———.] 1786a. *Journal du traitement magnétique de la demoiselle N. Lequel a servi de base à l'Essai sur la théorie du somnambulisme magnétique.* London: N.p.

[———.] 1786b. *Suite du traitement magnétique de la demoiselle N., lequel a servi de base à l'Essai sur la théorie du somnambulisme magnétique.* London: N.p.

[Tascher, Paul.] 1854. *Grosjean à son évêque au sujet des tables parlantes.* Paris: Ledoyen.

[———.] 1855. *Second lettre de Grosjean à son évêque au sujet des tables parlantes, des possessions et autres diableries.* Paris: Ledoyen.

Tatar, Maria M. 1978. *Spellbound: Studies on Mesmerism and Literature.* Princeton, N.J.: Princeton University Press.

Taylor, Eugene. 1983. *William James on Exceptional Mental States: The 1896 Lowell Lectures.* New York: Charles Scribner's Sons.

———. 1987. Prospectus for a person-centered science: The unrealized potential of psychology and psychical research. *J. Amer. Society for Psychical Research* 81:313–31.

Taylor, W. S., and Mabel Martin. 1944. Multiple personality. *J. Abnormal and Social Psychology* 39:281–300.

Teste, Alphonse. 1840. *Manuel pratique de magnétisme animal. Exposition méthodique des procédés employés pour produire les phénomènes magnétiques et leur application à l'étude et au traitement des maladies.* Paris: J. B. Baillière.

––––––. 1843. *A Practical Manual of Animal Magnetism; Containing an Exposition of the Methods Employed in Producing the Magnetic Phenomena; with its Application to the Treatment and Cure of Disease.* Translated by D. Spillan. London: H. Baillière.

––––––. 1845. *Le magnétisme animal expliqué, ou leçons analytiques sur la nature essentielle du magnétisme, sur ses effets, son histoire, ses applications, les diverses manières de la pratiquer, etc.* Paris: J. B. Baillière.

Thalbourne, Michael A. 1982. *A Glossary of Terms Used in Parapsychology.* London: Heinemann.

Thatcher, Virginia S. 1953. *History of Anesthesia, with Emphasis on the Nurse Specialist.* Philadelphia: Lippincott.

Thigpen, Corbett, and Hervey Cleckley. 1954. A case of multiple personality. *J. Abnormal and Social Psychology* 49:135–51.

––––––. 1957. *The Three Faces of Eve.* New York: McGraw-Hill.

Thomas d'Onglée, François Louis. 1785. *Rapport au public de quelques abus en médecine; avec des réflexions & notes historiques, critiques & médicales.* Paris: Herissant.

Thouret, Michel. 1784. *Recherches et doutes sur le magnétisme animal.* Paris: Prault.

Thouret, Michel, and Charles Andry. 1782. *Observations et recherches sur l'usage de l'aimant en médecine; ou Mémoire sur le magnétisme médicinal.* Paris: L'Imprimerie de monsieur.

Thuillier, Jean. 1988. *Franz Anton Mesmer, ou l'extase magnétique.* Paris: Editions Robert Laffont.

Tischner, Rudolf. 1928. *Franz Anton Mesmer: Leben, Werk und Wirkungen.* Munich: Müncher Drucke.

Tischner, Rudolf, and Karl Bittel. 1941. *Mesmer und sein Problem: Magnetismus—Suggestion—Hypnose.* Stuttgart: Hippokrates-Verlag, Marquardt & Cie.

Tissot, Honoré. 1841. *L'antimagnétisme animal . . . Ouvrage utile et nécessaire spécialement aux ecclésiastiques et aux médecins.* Bagnols: Alban Broche.

Topham, William, and W. Squire Ward. 1842. *Account of a Case of Successful Amputation of the Thigh, during the Mesmeric State, without the Knowledge of the Patient; Read to the Royal Medical and Chirurgical Society of London, on Tuesday, the 22nd of November 1842.* London: H. Baillière.

Townshend, Chauncy Hare. 1840. *Facts in Mesmerism, with Reasons for a Dispassionate Inquiry into It.* London: Longman, Orme, Brown, Green, and Longmans.

––––––. 1844. *Facts in Mesmerism, with Reasons for a Dispassionate Inquiry into It.* New York: Harper & Brothers.

––––––. 1853. On table moving. *The Zoist* 11:185–89.

———. 1854. *Mesmerism Proved True, and the Quarterly Reviewer Reviewed.* London: Thomas Bosworth.

Treichler, Hans Peter. 1988. *Die magnetische Zeit: Alltag und Lebensgefühl im frühen 19. Jahrhundert.* Zürich: SV international Schweizer Verlagshaus.

Trismegiste, Johannes [pseud.]. 1855. *Les merveilles du magnétisme et les mystères des tables, tournantes et parlantes.* Paris: Passard.

Underhill, Samuel. 1868. *Underhill on Mesmerism, with Criticisms on Its Opposers, and a Review of Humbugs and Humbuggers, with Practical Instructions for Experiments in the Science, Full Directions for Using it as a Remedy in Disease . . . the Philosophy of Its Curative Powers; How to Develop a Good Clairvoyant; the Philosophy of Seeing Without Eyes. The Proofs of Immortality Derived from the Unfoldings of Mesmerism. . . .* Chicago: The Author.

Van der Hart, Onno, and Barbara Friedman. 1989. A reader's guide to Pierre Janet on dissociation: A neglected intellectual heritage. *Dissociation* 2:3–16.

Viatte, Auguste. 1928. *Les sources occultes du romantisme. Illuminisme—théosophie. 1770–1820.* Vol. 1. Paris: Honoré Champion.

[Villers, Charles de]. 1787. *Le magnétiseur amoureux, par un membre de la société harmonique du régiment de Metz.* Geneva (Besançon): N.p.

[———.]. 1824. *Le magnétiseur amoureux.* 2 vols. 2d ed. Paris: Dentu.

Vinchon, Jean. 1936. *Mesmer et son secret.* Paris: A. Lagrand.

Virey, Julien Joseph. 1818. *Examen impartial de la médecine magnétique, de sa doctrine, de ses procédes et de ses cures.* Paris: C. L. F. Panckoucke.

Wallace, Alfred Russell. 1866. *The Scientific Aspect of the Supernatural: Indicating the Desirableness of an Experimental Enquiry by Men of Science into the Alleged Powers of Clairvoyants and Mediums.* London: N.p.

———. 1874. A defence of modern spiritualism. *Fortnightly Review* 15:630–57.

———. 1875. *On Miracles and Modern Spiritualism. Three Essays.* London: J. Burns.

Walmsley, D. M. 1967. *Anton Mesmer.* London: Robert Hale.

Watkins, John G. 1976. Ego states and the problem of responsibility: A psychological analysis of the Patty Hearst case. *J. Psychiatry and the Law* 4:471–89.

———. 1977. The psychodynamic manipulation of ego states in hypnotherapy. In *Therapy in Psychosomatic Medicine*, vol. 2, edited by F. Antonelli. Rome: International College of Psychosomatic Medicine.

———. 1978. *The Therapeutic Self.* New York: Human Sciences Press.

Watkins, John G., and Helen H. Watkins. 1979. Theory and practice of ego-state therapy: A short-term therapeutic approach. In *Short-term Approaches to Psychotherapy*, edited by H. Grayson. New York: Human Sciences Press.

[Wendel-Wurtz, abbé]. 1817. *Superstitions et prestiges des philosophes, ou les démonolatres du siècle des lumières par l'Auteur des Précurseurs de l'Ante-Christ.* Lyon: Rusand.

Wesermann, H. M. 1822. *Der Magnetismus und die allgemeine Weltsprache.* Creveld: Johann Heinrich Runcke. Cologne: Johann Peter Bachem.

Wilson, James Victor. 1879. [1st ed. 1847.] *How to Magnetize, or Magnetism and Clairvoyance. A Practical Treatise on the Choice, Management and Capabilities of Subjects, with Instructions on the Method of Procedure.* Rev. ed. New York: S. R. Wells.

Wilson, John. 1839. *Trials of Animal Magnetism on the Brute Creation.* London: Sherwood, Gilbert, & Piper.

Wirth, Johann Urich. 1836. *Theorie des Somnambulismus oder des thierischen Magnetismus. Ein Versuch die Mysterien des magnetischen Lebens, den Rapport der Somnambulen mit dem Magnetiseur, ihre Ferngesichte und Ahnungen, und ihren Verkehr mit der Geisterwelt vom Standpunkte vorurtheilsfreier Kritik aus zu erhellen und erkälren für Gebildete überhaupt, und für Mediciner und Theologe insbesondere.* Leipzig and Stuttgart: J. Scheible.

Wolfart, Karl Christian. 1815. *Erläuterungen zum Mesmerismus.* Berlin: Nikola.

Würtz, Georg Christophe. 1787. *Prospectus d' un nouveau cours théorique et pratique de magnétisme animal, réduit à des principes simple de physique, de chymie, et de médecine.* Strasbourg: Treuttel.

Wyckoff, James. 1975. *Franz Anton Mesmer: Between God and Devil.* Englewood Cliffs, N.J.: Prentice-Hall.

Wydenbruck, Nora. 1947. *Doctor Mesmer: An Historical Study.* London: John Westhouse.

Yap, P. M. 1960. The possession syndrome. *J. Mental Science* 106:114–37.

Zöllner, Johann Carl Friedrich. 1878–1881. *Wissenschaftlichen Abhandlungen.* Vols. 1–3. Leipzig: L Staackmann.

———. 1880. *Transcendental Physics: An Account of Experimental Investigations. From the Scientific Treatises of Johann Carl Friedrich Zöllner. Translated from the German, with a Preface and Appendices, by Charles Carleton Massey.* London: N.p.

Index

65
73
84

338.39